T0134726

Springer Tracts in Mechanical Engineering

Springer Tracts in Mechanical Engineering (STME) publishes the latest developments in Mechanical Engineering - quickly, informally and with high quality. The intent is to cover all the main branches of mechanical engineering, both theoretical and applied, including:

- Engineering Design
- Machinery and Machine Elements
- Mechanical Structures and Stress Analysis
- Automotive Engineering
- Engine Technology
- Aerospace Technology and Astronautics
- Nanotechnology and Microengineering
- Control, Robotics, Mechatronics
- MEMS
- Theoretical and Applied Mechanics
- Dynamical Systems, Control
- Fluids Mechanics
- Engineering Thermodynamics, Heat and Mass Transfer
- Manufacturing
- Precision Engineering, Instrumentation, Measurement
- Materials Engineering
- Tribology and Surface Technology

Within the scope of the series are monographs, professional books or graduate textbooks, edited volumes as well as outstanding PhD theses and books purposely devoted to support education in mechanical engineering at graduate and post-graduate levels.

Indexed by SCOPUS, zbMATH, SCImago.

Please check our Lecture Notes in Mechanical Engineering at http://www.springer.com/series/11236 if you are interested in conference proceedings.

To submit a proposal or for further inquiries, please contact the Springer Editor **in your country**:

Dr. Mengchu Huang (China)
Email: mengchu.Huang@springer.com
Priya Vyas (India)
Email: priya.vyas@springer.com
Dr. Leontina Di Cecco (All other countries)
Email: leontina.dicecco@springer.com
All books published in the series are submitted for consideration in Web of Science.

More information about this series at http://www.springer.com/series/11693

Iulian Popescu · Xenia Calbureanu ·
Alina Duta

Problems of Locus Solved by Mechanisms Theory

Iulian Popescu
Faculty of Mechanics
University of Craiova
Craiova, Romania

Xenia Calbureanu
Faculty of Mechanics
University of Craiova
Craiova, Romania

Alina Duta
Faculty of Mechanics
University of Craiova
Craiova, Romania

ISSN 2195-9862 ISSN 2195-9870 (electronic)
Springer Tracts in Mechanical Engineering
ISBN 978-3-030-63081-2 ISBN 978-3-030-63079-9 (eBook)
https://doi.org/10.1007/978-3-030-63079-9

This Springer imprint is published by the registered company Springer Nature Switzerland AG
The registered company address is: Gewerbestrasse 11, 6330 Cham, Switzerland

Preface

This book is based on our research in the field of planar and spatial mechanisms, which has resulted in the generation of a wide range of rod curves. During this research, we also addressed the locus problems and found out that such difficult problems in geometry can be solved easily by applying some of the methods of the Theory of Mechanisms. For researchers that are not very familiar with the Theory of Mechanisms, this book presents the necessary basics. Despite giving a special emphasis to different loci that are specific to the triangle and the quadrilateral and generating some aesthetic ones, this book also describes some other cases. The original mechanisms that are obtained are described in detail, showing numerous curves resulting as loci.

Craiova, Romania

Iulian Popescu
Xenia Calbureanu
Alina Duta

Contents

Chapter 1
Introduction

Abstract The locus is defined and correlated with the notions of generated trajectories and curves. It is considered a geometric Figure in which one side becomes fixed, and the others are moving, so that certain specified points describe the trajectories, i.e. curves assimilated to geometric places, such as conical curves: ellipse, parabola, hyperbola. Drawing these curves starting from geometric problems is easy by using the Theory of Mechanisms. The equivalent mechanisms are built and analyzed. For those who are not familiar with the Theory of Mechanisms, the strictly necessary elements are given in order to be able to construct the equivalent mechanisms and then to determine the trajectories of some points. The diagrams of the kinematic elements with rotational motion R, and translational motion (sliding), P are indicated, showing how the elements can be connected, also the cases when the lengths of some elements are equal to zero are presented too. The method of contours developed by Tchebichev is shown and examples of mechanisms and generated curves are given [1, 2, 3] (Cebâşev in Izobrannâe trud, Izd, Nauka, Moskva, 1953; Popescu in Mecanisme, vol. I, II. Tipografia Universităţii din Craiova, 1995; Popescu in Proiectarea mecanismelor plane, Craiova, Editura Scrisul Românesc, 1977).

1.1 The Locus Problem in Geometry

In Geometry, there are many problems about looking for the loci described by points that meet certain geometric conditions. The *locus* is defined as the total of the points in space that fulfill a certain geometric property. The solving methods are purely geometric, based on theorems or other properties known in geometry. However, many of such problems cannot be solved easily due to the limited possibilities of geometry, and most of the problems are limited to simple loci: straight, circle arcs, ellipses, paraboles, hyperboles.

The ellipse is the locus of the points in the plane for which the sum of the distances to two fixed points is constant, the ellipse being a curve or a trajectory. In other words, a curve, or a trajectory can be assimilated as a locus. If we consider a quadrilateral with a fixed side, the rest of the sides being in motion, then the trajectories of some

I. Popescu et al., *Problems of Locus Solved by Mechanisms Theory*,
Springer Tracts in Mechanical Engineering,
https://doi.org/10.1007/978-3-030-63079-9_1

points, for example the mobile point of intersection of the mediators, is a locus, i.e. a curve or a trajectory.

In our research based on Mechanisms Theory we have found that if the methods from this discipline are applied to locus problems, the results can be easily obtained without resorting to classical geometric methods [4]. In the present book, we want to show how to obtain the desired loci using some considerations of the Theory of Mechanisms.

The work was intended for a wider range of scholars so we considered that it is necessary to treat very briefly the methods in the kinematics of the mechanisms, providing a chapter with strictly necessary simply described data. These mechanisms do not have to be built, today there are machines with numerical controls that can generate complex surfaces. Such mechanisms are used here only as a theoretical possibility to obtain the desired geometric locations. They can also be used to build certain toys.

1.2 Some Basics from the Mechanisms Theory

The mechanisms are subassemblies of some cars, which move according to certain laws, such as the crank mechanism from the engines of cars. Their pieces are called *elements*, being symbolized in the theory of mechanisms by straight segments, without widths and thicknesses. Each mechanism receives movement from a *leading element* (motor, piston, spring) and transmits it with other kinematic parameters to the last driven translational element. There are also mechanisms with several leading elements.

In Fig. 1.1 there are two leading elements. The element AB in Fig. 1.1a rotates around a fixed point on the bat (the one struck), its position being given by the angle φ, known, the direction of rotation being the trigonometric one (as in the figure) or clockwise, contrary to those mentioned in the figure. Figure 1.1b shows a leading element with translational motion (*sliding*), whose position is given by the known stroke S.

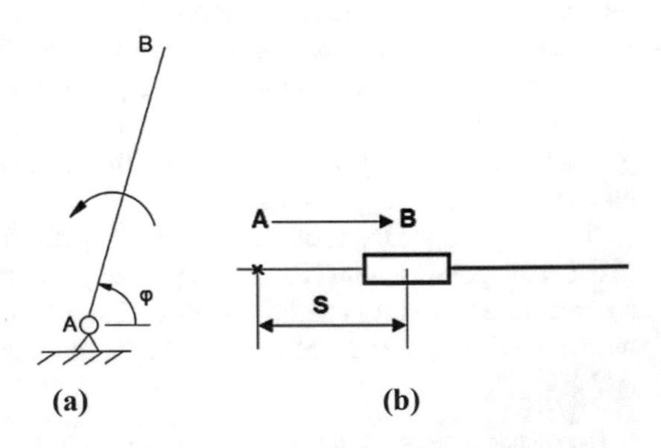

Fig. 1.1 **a**, **b** Leading kinematic elements

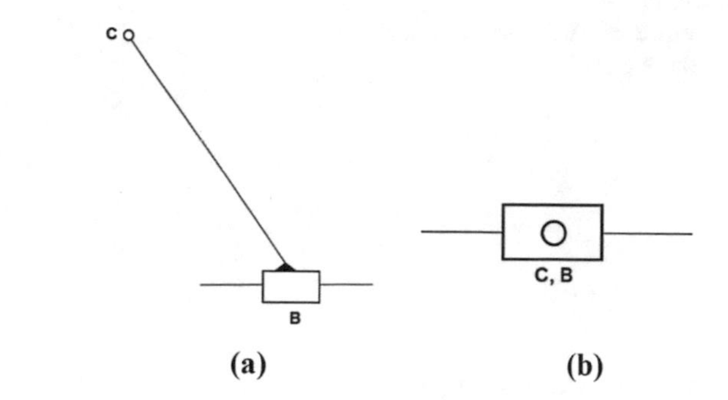

Fig. 1.2 **a**, **b** Kinematic link coupled with two pairs

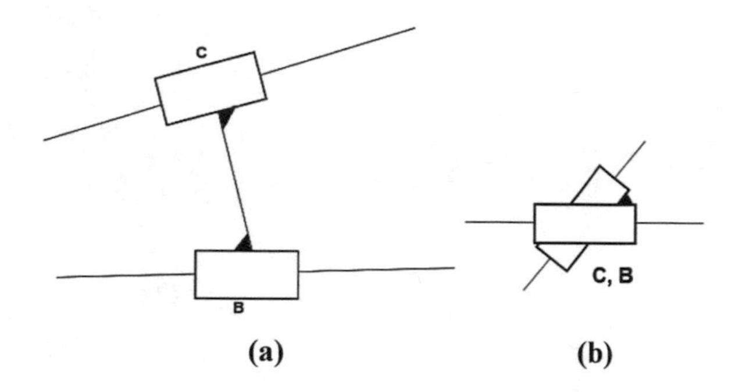

Fig. 1.3 **a**, **b** Kinematic link coupled with two pairs with non-parallel axes

Revolute pair:

A certain element of a mechanism links with other elements as in Fig. 1.2. The element in Fig. 1.2a is connected in C by a *rotation couple* (a bolt) with another element, and in B by a slide through which another element passes. The length of CB can be greater or smaller than that of Fig. 1.2a, so that it can also reach zero, that is, the connection point in C (the bolt), reaches the slide as in Fig. 1.2b. This is a *particular case, with the zero length of an element.*

In the case of the Fig. 1.3 the BC link (Fig. 1.3a) is connected with two pairs with non-parallel axes. Also, in this case, the length of the element can be equal to zero, symbolized in the Fig. 1.3b, with the blackened part showing that there is a weld that stiffens the sliders.

In the Fig. 1.4 shows how two elements linked to each other through a in C, and in B and D it is connected with other elements also through the revolute pairs.

Two elements can be connected to each other by a pair (slide)—Fig. 1.5a, and the length of the DC link can be zero (Fig. 1.5b). The BC element cannot be zero because in this case, the slider in C would have no guidance.

If two elements are linked to each other by means of a revolute pair and in the exterior with two pairs (Fig. 1.6a), the particular cases can be obtained when the CD element has a zero length (Fig. 1.6b), respectively which length BC is equal to zero (Fig. 1.6c). It looks like to be a paradox; it is possible that both lengths are equal to

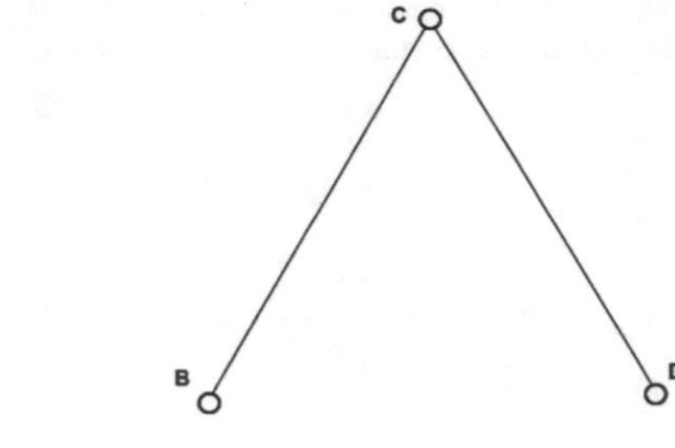

Fig. 1.4 Two elements with three pairs

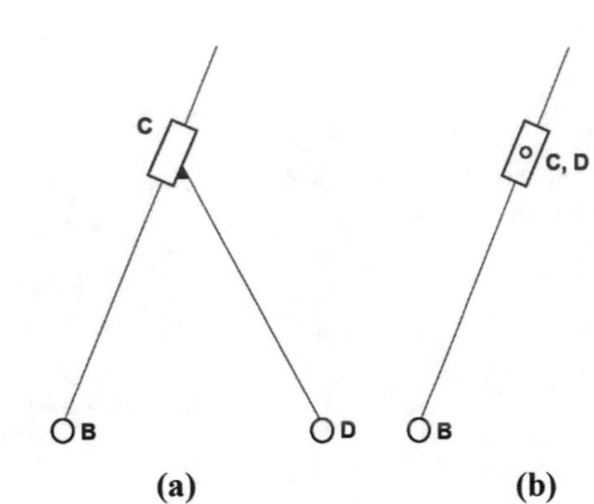

Fig. 1.5 a, **b** General case and particular case

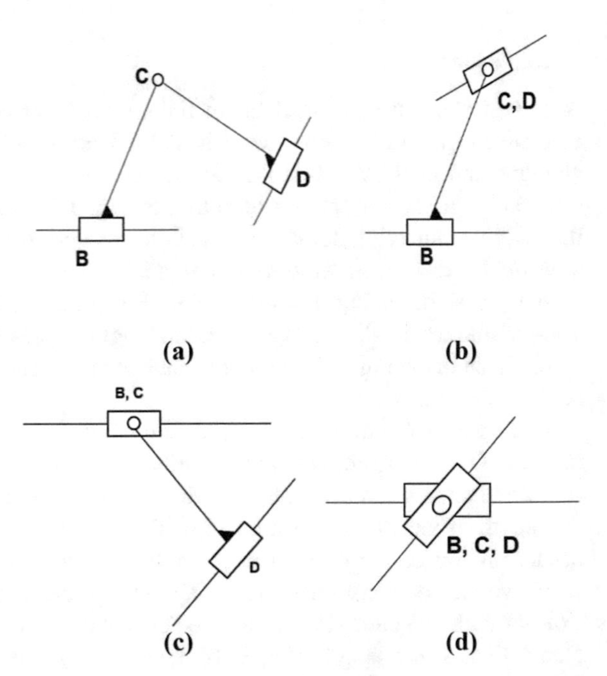

Fig. 1.6 a, **b**, **c**, **d** Particular cases

zero (Fig. 1.6d). The prismatic pairs are rotated to each other and two other elements are moved through their guidelines [2, 3].

The sequence of elements and pairs form a kinematic chain.

The geometry is used for the calculations required to find geometric places. This method was created by the Pafnutii Lvovici Cebâşev. Cebâşev refers to computational problems in a paper titled “About Parallelograms, Containing Three Elements". In this work, he used a previous study from December 5th in 1878 (the work with parallelograms is written between 1879 and 1880). This is a fragment of great importance in this work (Fig. 1.7) [2, pp. 873, 3]:

Here’s the translation of the text: “*By projecting the broken CAA1 line along the coordinate axes, we obtain the necessary relationships to determine the coordinates of A1 for different sizes of β, α:*”.

The text specifies the α angles and β angles.

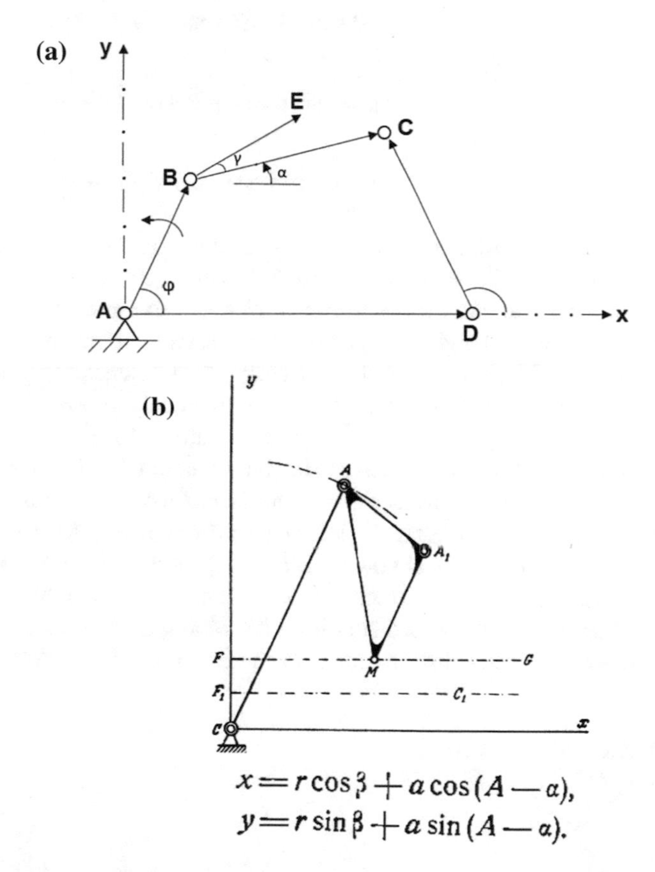

Проектируя ломаную линию CAA_1 на оси координат, мы находим такие равенства для определения координат точки A_1 при различных

Fig. 1.7 **a** Four-bar mechanism, **b** the Cebâşev contour method [4]

Here's how this method applies to the mechanism given in Fig. 1.7, (called the articulated quadrilateral mechanisms). We consider the kinematic elements of the mechanism as vectors and the vector contours are projected on the system axes. There are two vector outlines reaching the point C: ABC and ADC, which, on axes, lead to the nonlinear algebraic system given by Eqs. 1.1 and 1.2, from which the angles are calculated. In order to find the trajectory of an E-point belonging to the ECB element, the EA outline is projected onto the axes to obtain relations Eqs. 1.3 and 1.4, that is, the coordinates of the traversing point. This leads to a first locus problem: to find the geometric location of the point E of an ECB plate element that moves with points B and C on two circles.

$$AB\ \cos\varphi + BC\ \cos\alpha = AD + DC\ \cos\beta \tag{1.1}$$

$$AB\ \sin\varphi + BC\ \sin\alpha = DC\ \sin\beta \tag{1.2}$$

$$\mathrm{x_E} = \mathrm{AB}\ \cos\varphi + \mathrm{BE}\ \cos(\alpha + \gamma) \tag{1.3}$$

$$y_E = AB\ \sin\varphi + BE\ \sin(\alpha + \gamma) \tag{1.4}$$

In the Fig. 1.8 shows the locus determined for the following initial data: AB = 22; BC = 35; CD = 37; AD = 50; BE = 40; $\gamma = 270$.

The mechanism for position $\varphi = 80$ was also represented.

In this work lenghts are given in millimeters ant the angles in degrees.

It should be noted that the trajectory (locus) depends on the initial data, i.e. the dimensions of the figure, which must be previously known. In general, geometrically, from Eqs. 1.1–1.4 results a 6th-grade equation (for this case) called the curve rod equation, which is a tricyclic sextic, but without knowing what the curve looks like. For example, if the dimensions of the mechanism are maintained, changing only to 45 degrees, a completely different curve is obtained (Fig. 1.9).

For γ =0 there is obtained the locus from the Fig. 1.10, where the BE element are superposed over BC element, but that is longer than this.

The method presented here for determining different loci is more thorough than those in geometry that provide only the equation of the curve without showing it.

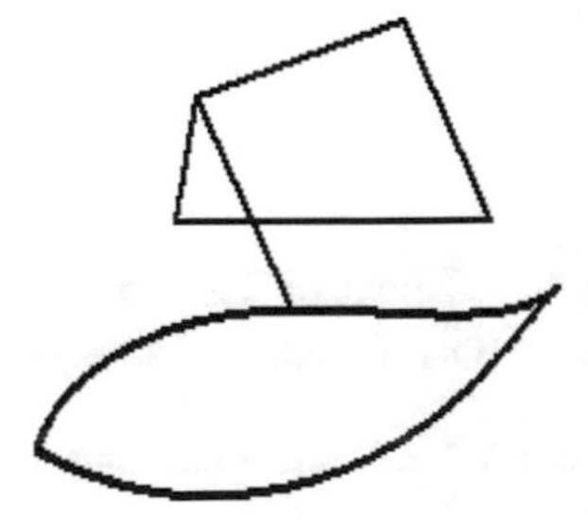

Fig. 1.8 The locus for $\gamma = 270$

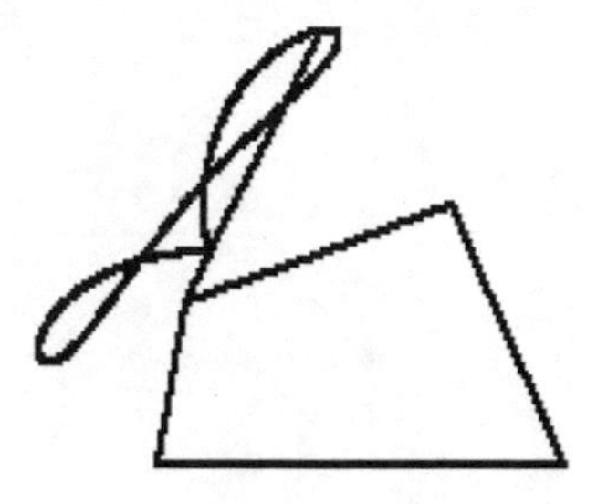

Fig. 1.9 The locus for $\gamma = 45$

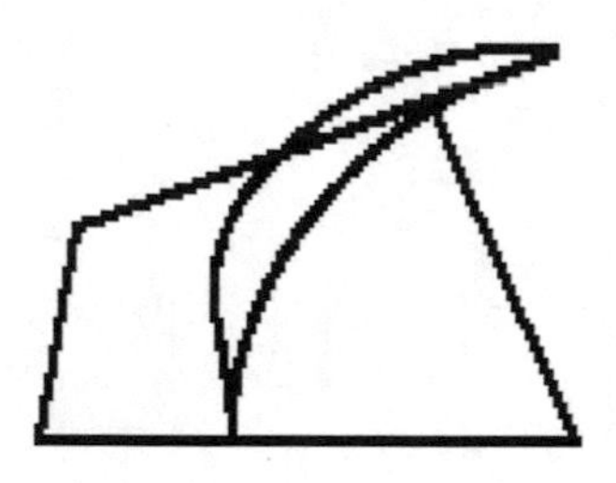

Fig. 1.10 The locus for $\gamma = 0$

References

1. Cebâşev PL (1953) Izobrannâe trud. Izd, Nauka, Moskva
2. Popescu I (1995) Mecanisme, vol I, II. Tipografia Universităţii din Craiova
3. Popescu I (1977) Proiectarea mecanismelor plane. Craiova, Editura Scrisul Românesc
4. Popescu I (2016) Locuri geometrice şi imagini estetice generate cu mecanisme. Editura Sitech, Craiova

Chapter 2
Loci Generated by the Point of a Line Which Moves One End on a Circle and the Other on a Line

Abstract We start from a simple geometric problem, when a straight line slides with one end on a circle and the other on a straight line (both fixed) and we look for the trajectory of a point on this straight line. We construct the equivalent mechanism which is the rod-crank mechanism, we write the relationships for the positions, we draw the mechanism in several positions and then we obtaine a succession of the curves generated by the points on this line. The case is generalized by replacing the fixed line with a mobile one, resulting a lot of curves, variable depending on the correlation of the angles of the two leading elements. The resulting curves are different from one case to another, being open curves and some having several branches [1, 2] (Luca et al. in Studies regarding of aesthetics surfaces with mechnanisms. In: Mathematical Methods for Information Science and Economics. Proceedings of the 17th WSEAS International Conference on Applied Mathematics (AMATH '12), Montreux, Switzerland, pp. 249–254, 2012; Sass et al. in J Ind Design Eng Graphics. Papers of the International Conference on Engineering Graphics and Design, Sect. 3: Engineering Computer Graphics, Nr. 12, (1):41–46, 2017).

- **Case 1**

The line BC in Fig. 2.1 has end B moving on a circle, and end C on a fixed line, EC. The resulting mechanism, given in Fig. 2.2 is of the type R-RRP, i.e. the connecting rod-crank mechanism extensively studied in the works on mechanisms [3] [*Note: R—rotation coupling that allows only one element to rotate in relation to the other, P—prismatic torque, which allows only one translation between two elements*].

The following relations are written:

$$x_B = x_A + AB\cos\varphi \tag{2.1}$$

$$y_B = y_A + AB\sin\varphi \tag{2.2}$$

I. Popescu et al., *Problems of Locus Solved by Mechanisms Theory*,
Springer Tracts in Mechanical Engineering,
https://doi.org/10.1007/978-3-030-63079-9_2

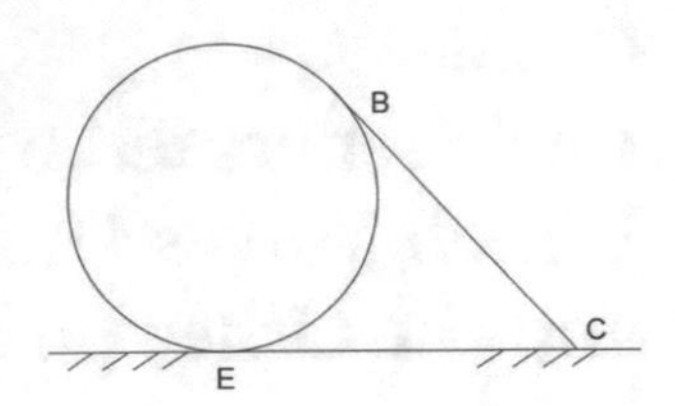

Fig. 2.1 The geometric construction

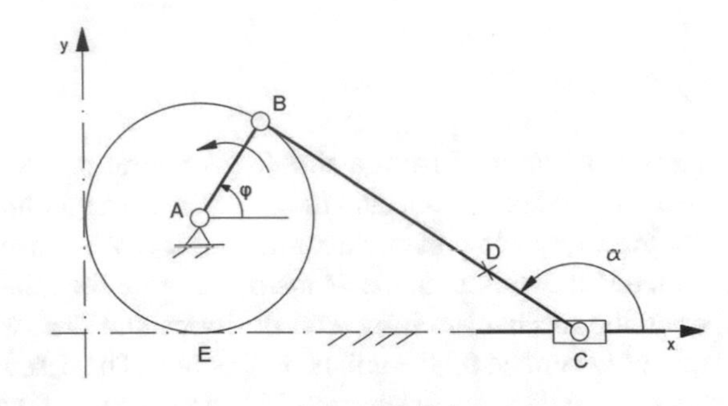

Fig. 2.2 The equivalent mechanism

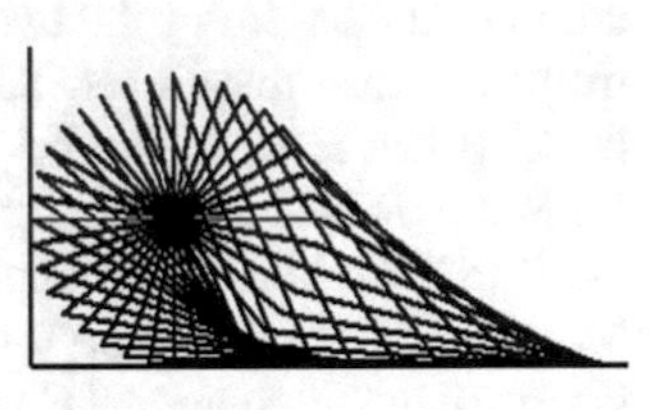

Fig. 2.3 Successive positions

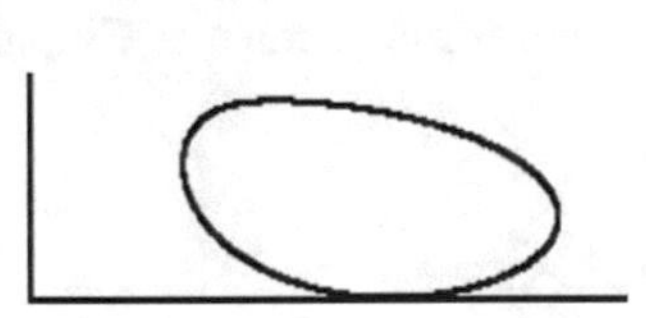

Fig. 2.4 Locus of D point

$$x_C + BC\cos\alpha = x_B \tag{2.3}$$

$$BC\sin\alpha = y_B \tag{2.4}$$

$$x_D = x_C + DC\cos\alpha \tag{2.5}$$

$$y_D = DC\sin\alpha \tag{2.6}$$

The geometric place of D is requested. Figure 2.3 shows the successive positions of the mechanism, and Fig. 2.4 gives the geometric place described by point D, for: $X_A = Y_A = AB = 35$ mm; BC = 70 mm; DC = 40 mm (sign—in front of the radical when calculating α). The locus is a 4th grade curve.

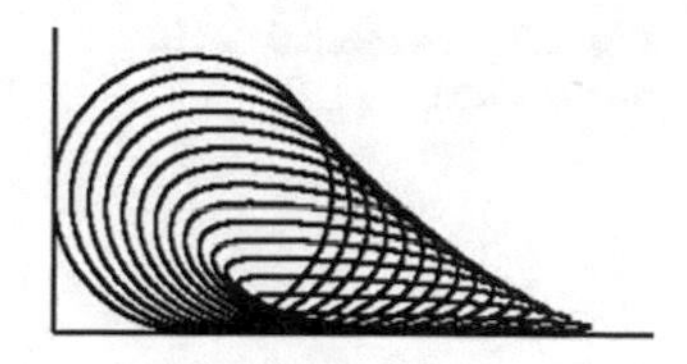

Fig. 2.5 Locus of some points on BC

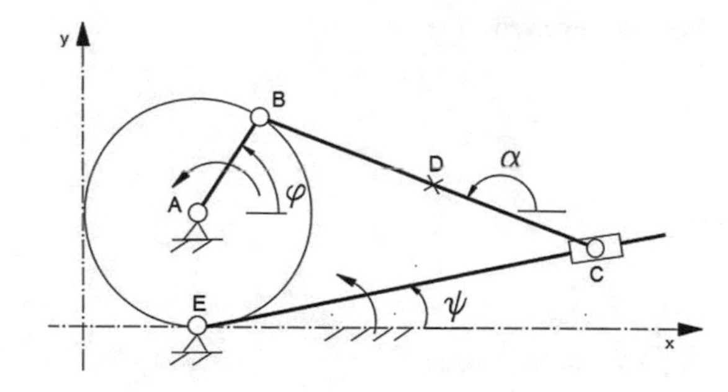

Fig. 2.6 The RR-RRP mechanism

Moving point D on BC with a 5 mm pace, resulted in the range of geometric loci in Fig. 2.5.

- **Case 2**

Consider the case in Fig. 2.6, where the line EC is mobile, rotating around E. The mechanism is of type RR-RRP.

The following relations are written:

$$x_B = x_E + EC\cos\psi + BC\cos\alpha \tag{2.7}$$

$$y_B = EC\sin\psi + BC\sin\alpha \tag{2.8}$$

The coordinates of B from case 1 are known, so we calculate α and the EC stroke. By dividing the relations, we obtain the trigonometric equation:

$$(x_B - x_E - BC\cos\alpha)\, tg\,\psi = y_B - BC\sin\alpha \tag{2.9}$$

which results in α, then EC is calculated. XE = AB was adopted.

It is also added the known relation of linear correlation between the angles of the crank AB and EC: $\psi = q.\varphi$

The dimensions $AB = X_A = Y_A = X_E = 35$ mm; BC = 70 mm; CD = 40 mm were adopted.

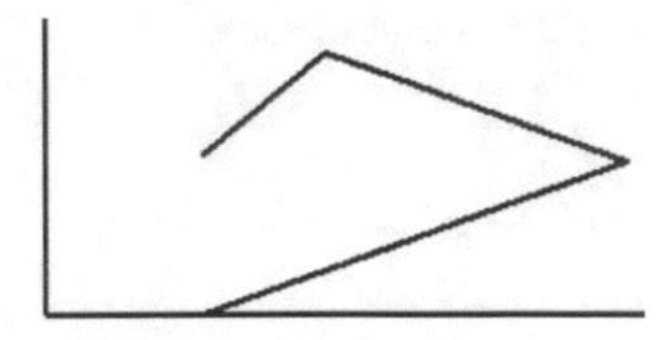

Fig. 2.7 The mechanism in one position

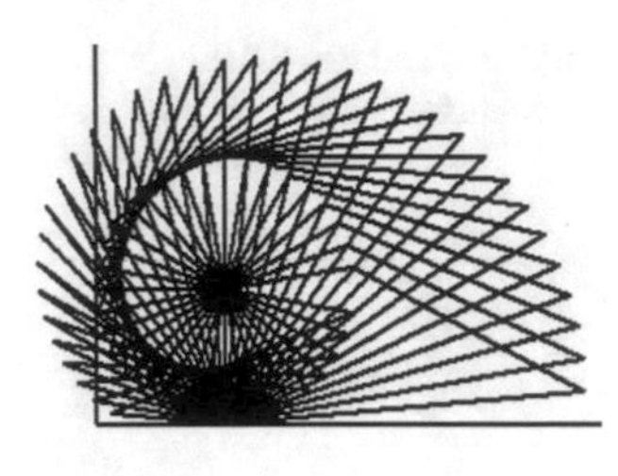

Fig. 2.8 Successive positions

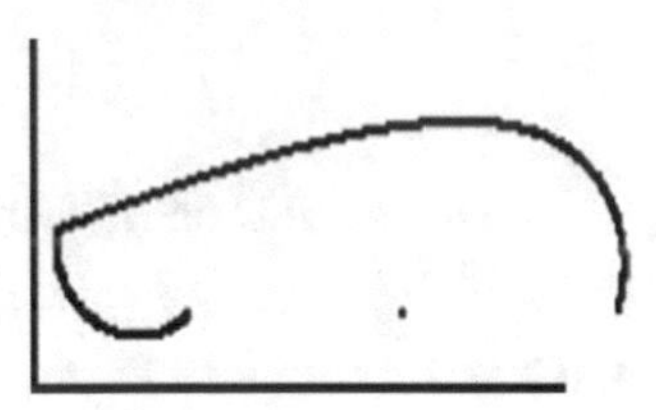

Fig. 2.9 Loci of D point

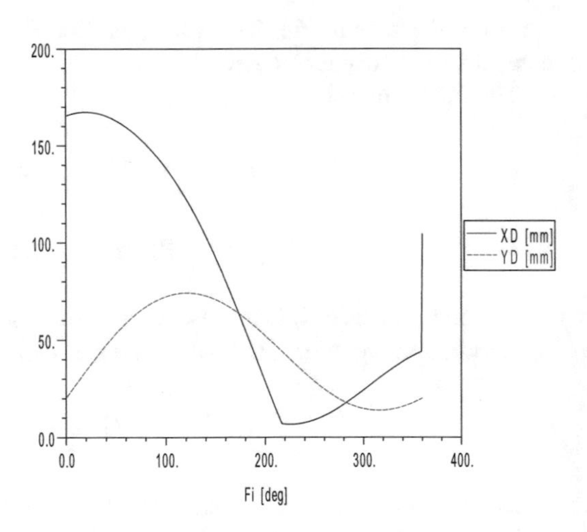

Fig. 2.10 The variation of XD and YD

When calculating α the sign (−) to the sine was taken in front of the radical, and the sign for calculating the cosine (+).

Figure 2.7 shows the mechanism in one position, and in Fig. 2.8 in successive positions.

For q = 0.5, the locus in Fig. 2.9 and the curves for the coordinates of D in Fig. 2.10 resulted.

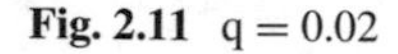

Fig. 2.11 q = 0.02

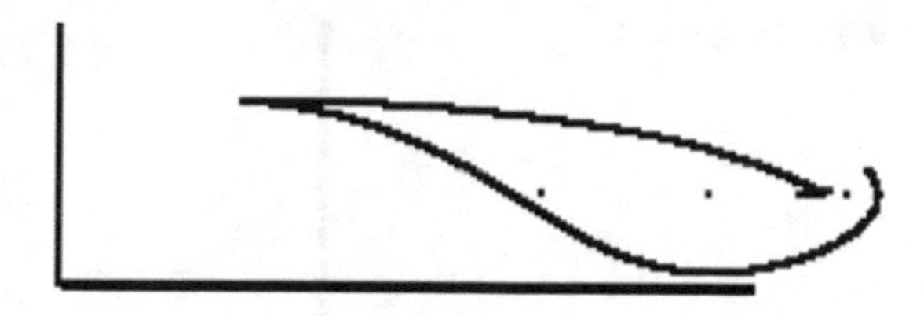

Fig. 2.12 q = −0.02

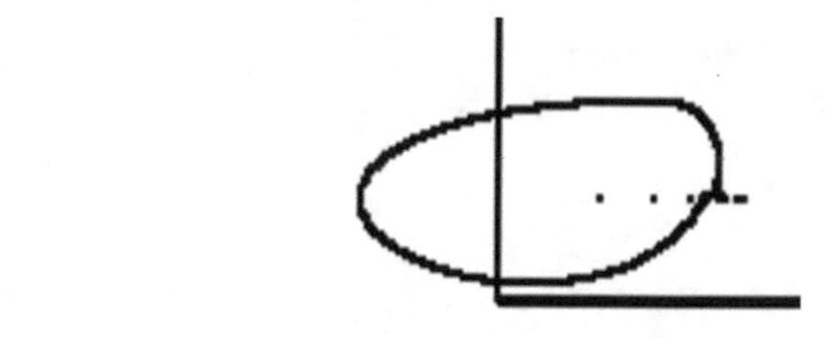

Fig. 2.13 q = 0.05

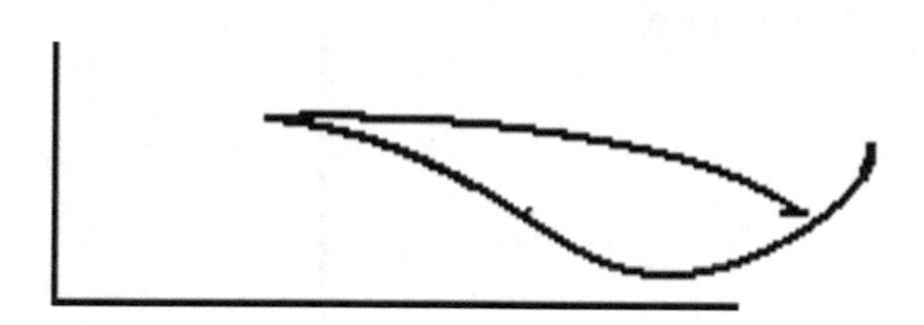

Fig. 2.14 q = −0.05

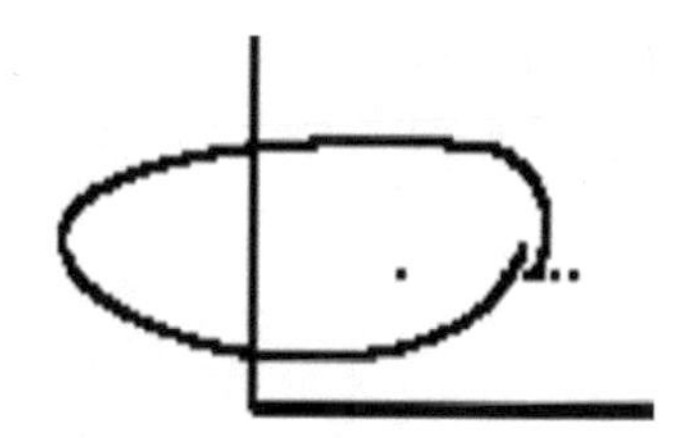

Next, the resulting loci for point D are given at different values of the correlation coefficient q both positive (AB and EC rotate in the same direction) and negative (AB and EC rotate in the opposite direction) (Figs. 2.11, 2.12, 2.13, 2.14, 2.15, 2.16, 2.17, 2.18, 2.19, 2.20, 2.21, 2.22, 2.23, 2.24, 2.25, 2.26, 2.27, 2.28, 2.29, and 2.30).

It is found that they have loci with one or more branches, all being open curves.

Fig. 2.15 q = 0.09

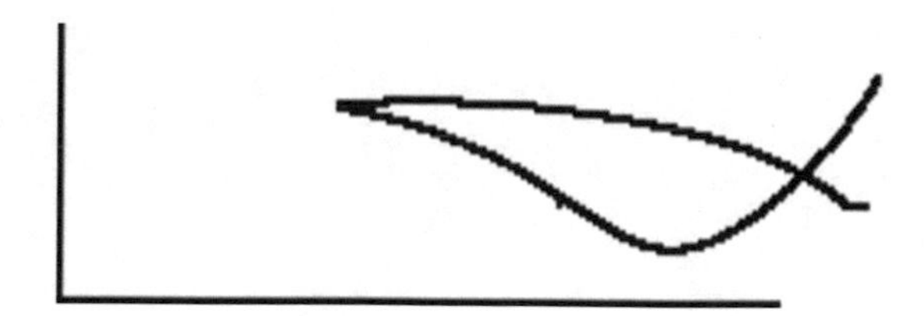

Fig. 2.16 q = 0.15

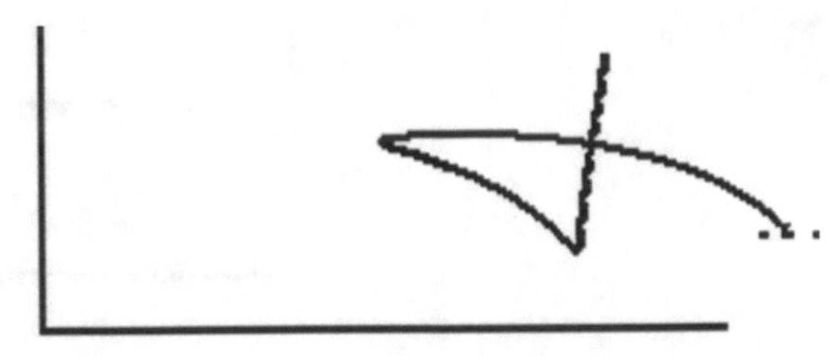

Fig. 2.17 q = −0.15

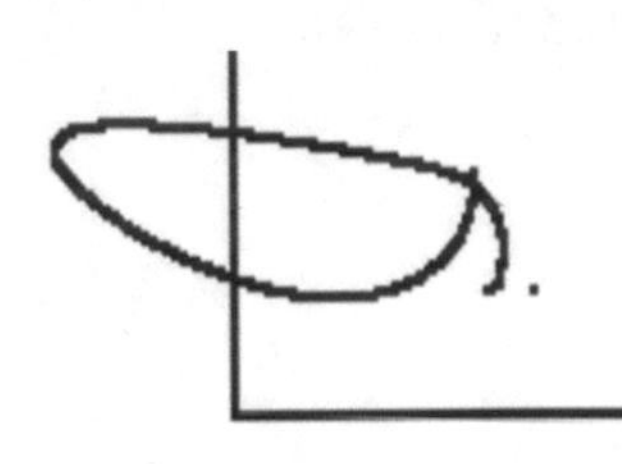

Fig. 2.18 q = 0.2

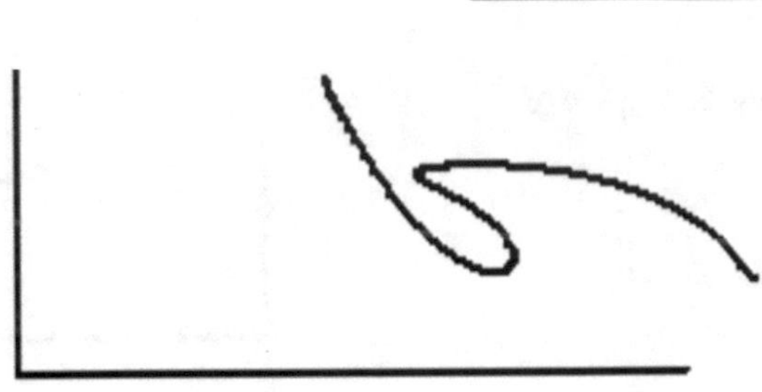

Fig. 2.19 q = 0.3

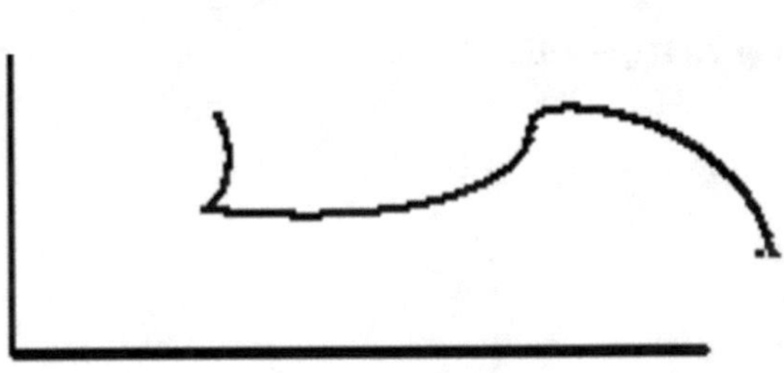

Fig. 2.20 q = −0.3

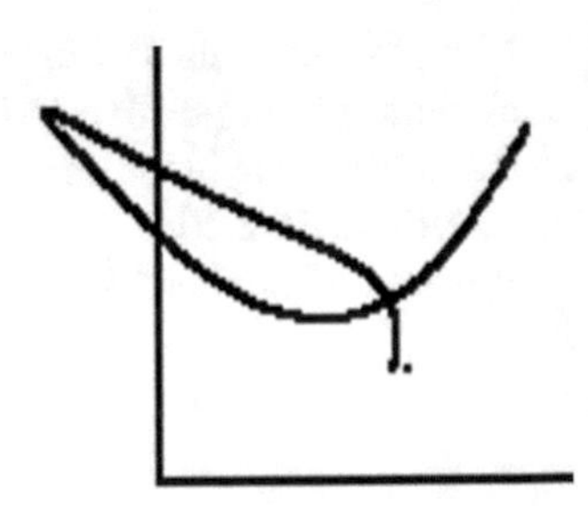

Fig. 2.21 q = 0.4

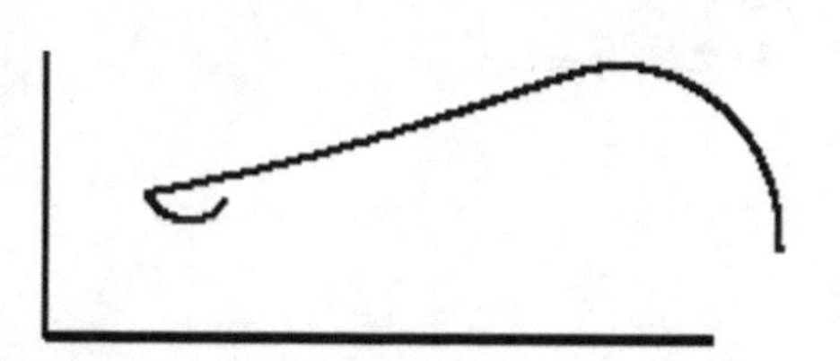

Fig. 2.22 q = −0.4

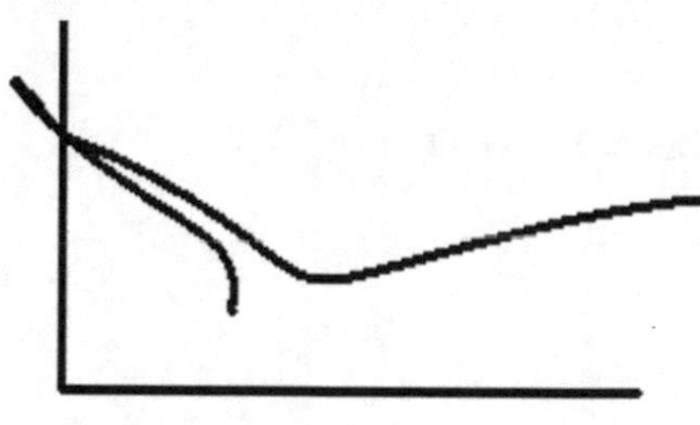

Fig. 2.23 q = −0.6

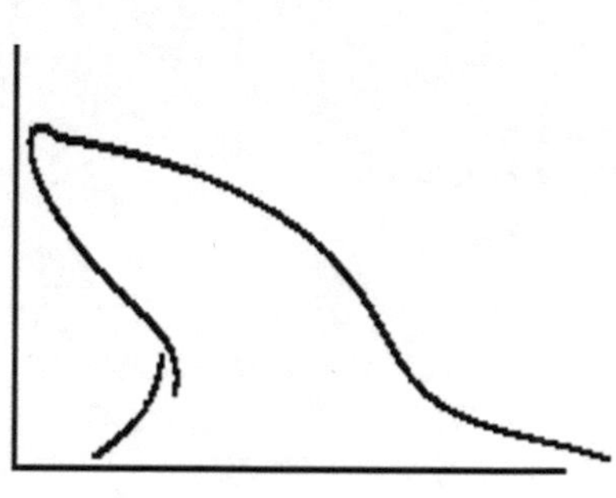

Fig. 2.24 q = 0.7

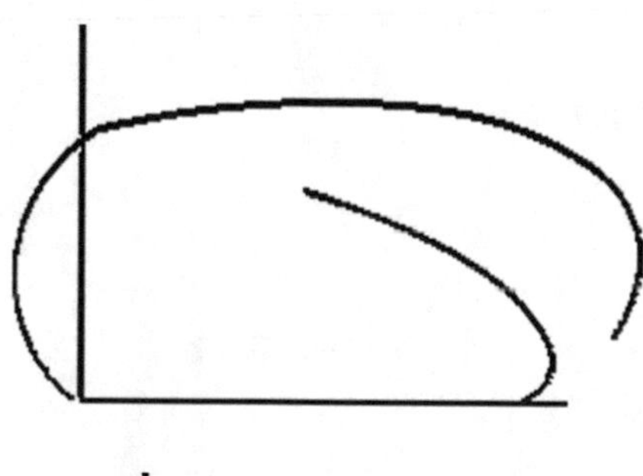

Fig. 2.25 q = −0.8

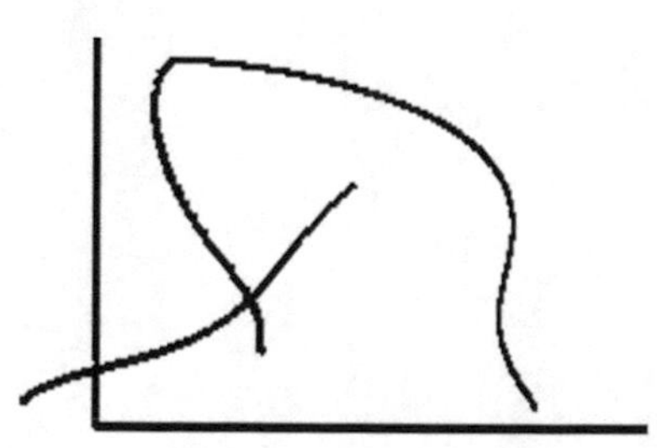

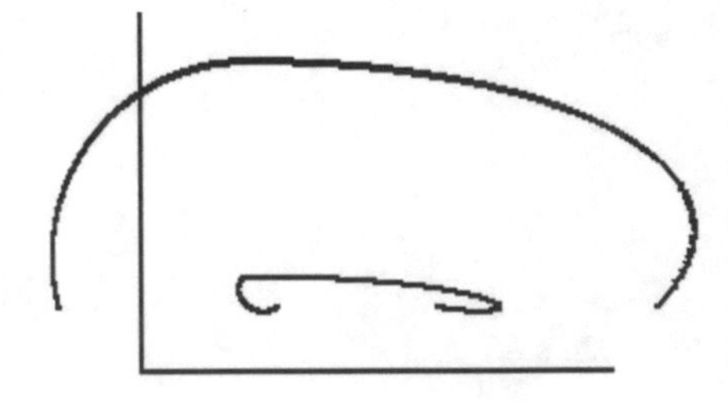

Fig. 2.26 q = 1

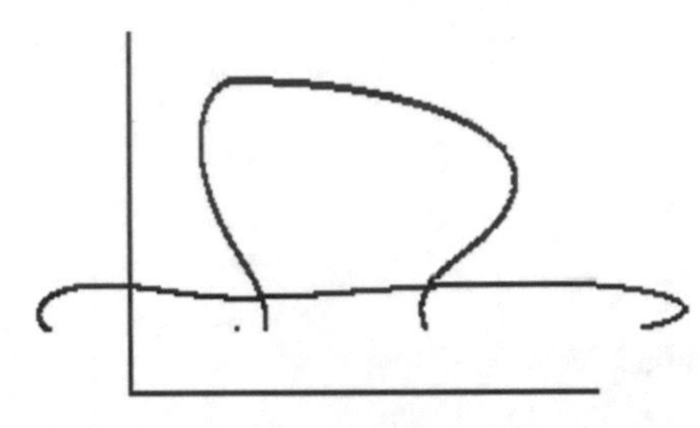

Fig. 2.27 q = −1

Fig. 2.28 q = 1.5

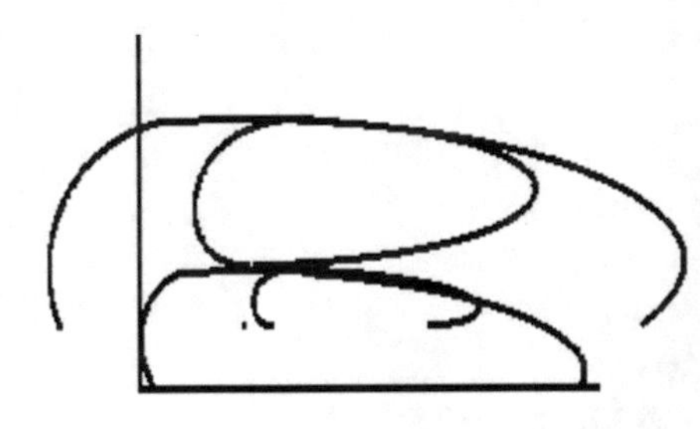

Fig. 2.29 q = 2

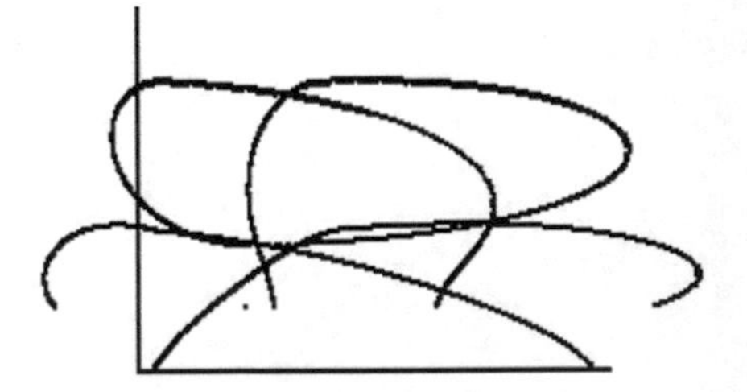

Fig. 2.30 q = −2

References

1. Calbureanu M, Malciu R, Lungu M, Calbureanu D (May, 2008) The influence of the lubricant from a rectilinear pair above the work accuracy of the elastic elements from the high precision mechanisms. Appl Theoret Mech Manuscript 3(5):176–185, ISSN: 1991–8747 (received Oct. 16, 2007) (revised Apr. 18, 2008)
2. Luca L, Popescu I, Ghimiş Ş (December, 2012) Studies regarding of aesthetics surfaces with mechnanisms. In : Mathematical methods for information science and economics. Proceedings of the 17th WSEAS international conference on applied mathematics (AMATH '12), pp 29–31, pp 249–254. Montreux, Switzerland
3. Sass L, Duţă A, Popescu I (2017) Nature gives us beauty, insoiring the technical creativity. J Ind Design Eng Graphics (1):41–46. Papers of the international conference on engineering graphics and design, section 3: engineering computer graphics, Nr. 12

Chapter 3
Loci Generated by the Point of Intersection of Two Lines

Abstract We consider two straight lines (cranks) with one end fixed at the base (frame) and we are looking for the trajectory of their intersection point when both straight lines rotate. The mechanism is constructed for the case when a line has a finite length and shows the successive positions and the trajectory of their intersection point [1], which is an incomplete circle. Both straight lines with infinite lengths are considered, resulting another equivalent mechanism with two leading elements. Depending on the correlation between the angles of rotation of these lines, many trajectories result, some being known curves, others unknown [3], with several incomplete branches [2]. The successive positions of the equivalent mechanism are also given. In another case, one of the lines is no longer connected to the base by a R coupling, but by a P coupling, so it slides on a fixed line. Numerous successive curves and positions are also obtained. Another case has in B only two couplings, the trajectory being a straight line.

- **Case 1**

Consider two lines AB and BC, which rotate around the fixed points A and C and the locus of their intersection point B is required (Fig. 3.1). The length of BC is constant.

The equivalent mechanism is given in Fig. 3.2 and is of type R—PRR.

Based on Fig. 3.2 the following relations are written:

$$x_B = AB\ \cos\varphi = x_C + BC\ cos\alpha \tag{3.1}$$

$$y_B = AB\ \sin\varphi = BC\ sin\alpha \tag{3.2}$$

We get to the equation:

$$(x_C + BC\ \cos\alpha)\text{tg}\ \varphi = BC\ sin\alpha \tag{3.3}$$

I. Popescu et al., *Problems of Locus Solved by Mechanisms Theory*,
Springer Tracts in Mechanical Engineering,
https://doi.org/10.1007/978-3-030-63079-9_3

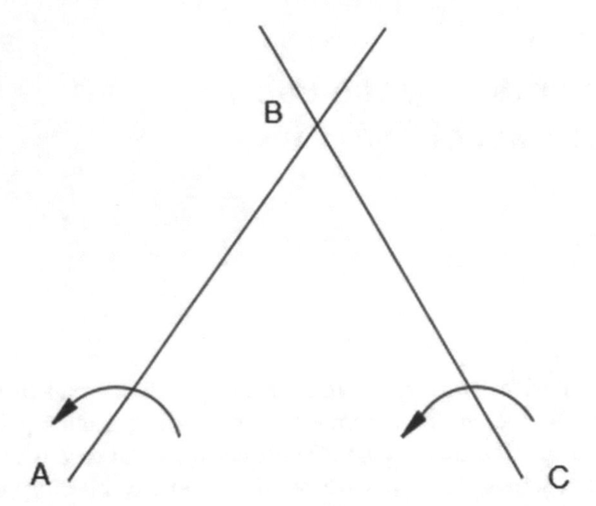

Fig. 3.1 The geometric construction

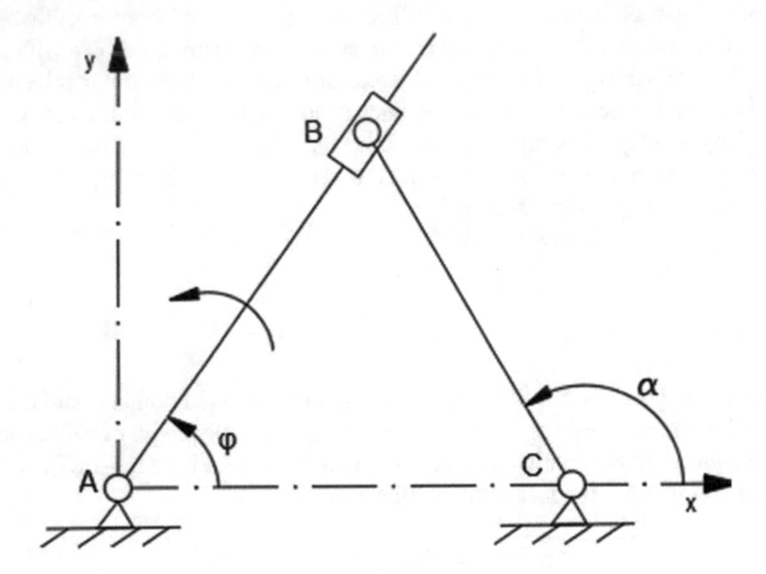

Fig. 3.2 The equivalent mechanism

from which the angle α results.

From Fig. 3.2 it is noted that point B describes a circle around C, because BC is constant. For BC = 50 mm and XC = 20 mm we have the circle in Fig. 3.3 and the successive positions of the mechanism of Fig. 3.4. It is observed that the drawn circle is incomplete, that is the mechanism does not work in a subinterval of the cycle.

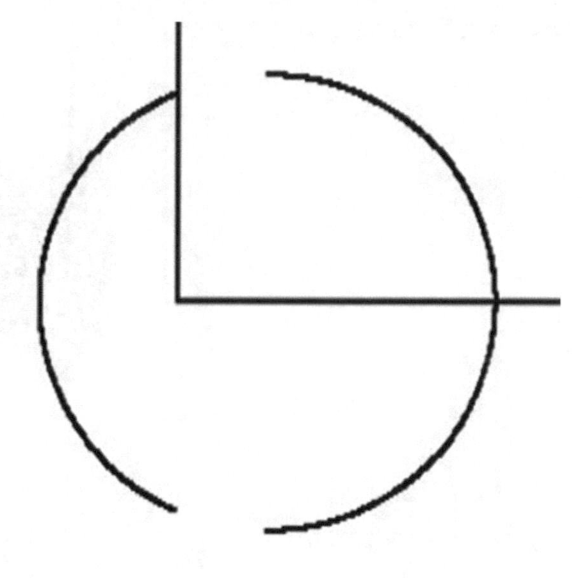

Fig. 3.3 Locus for point B

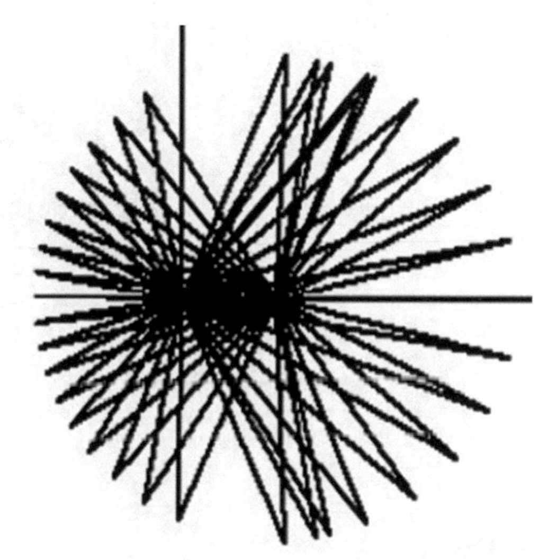

Fig. 3.4 Successive positions

From Fig. 3.2 it is found that the distance XC influences the operating range of the mechanism. This can also be seen in Figs. 3.5 and 3.6, where the successive positions show the mechanism locking in certain subintervals of φ.

The length of AB is variable as in Fig. 3.7.

In the subintervals where the inrushes in the diagram appear, the mechanism seizes, because the length AB tends to infinity, having $\varphi = 90$ and 270°, as shown in Fig. 3.4. In fact, the mechanism works based on inertia.

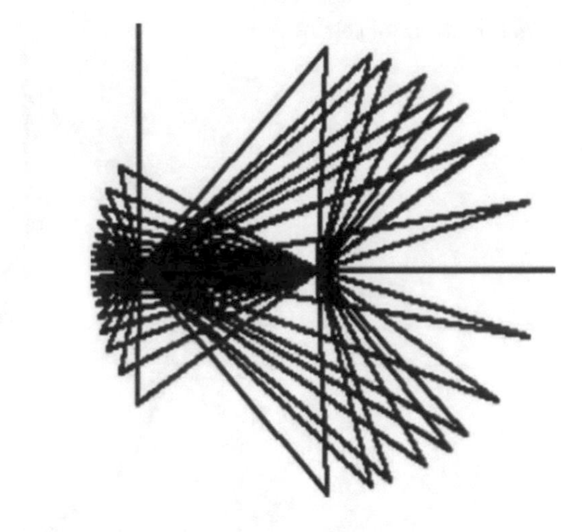

Fig. 3.5 BC = 50; XC = 40

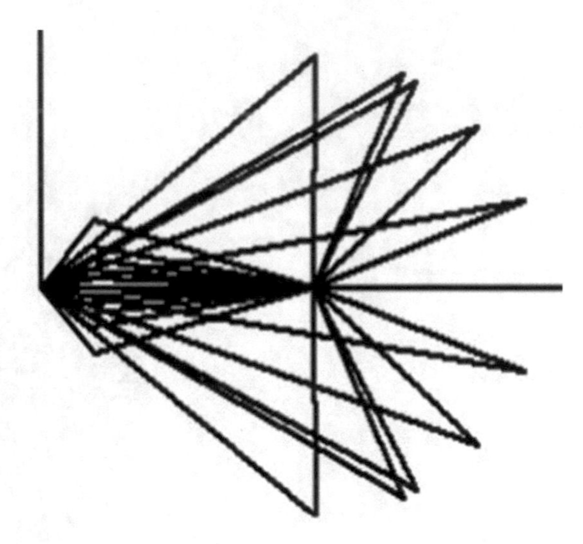

Fig. 3.6 BC = 50; XC = 60

- **Case 2**

Starting from Fig. 3.1, the mechanism of Fig. 3.8 can also be found.

In this case, the mechanism has two conducting elements AB and CD, being of the RR-PRP type. Here the length of BC is variable. In order to calculate the positions of point B, the above relations are used, to which the linear correlation relation between the angles of the two cranks is added: $\alpha = q.\varphi$.

The following are examples of loci resulting at different values of the correlation coefficient q, both for the case when the cranks rotate in the same direction and for the case when they rotate in the opposite direction. Again there are inrushes in the operation of the mechanism, that is on certain subintervals of the crank rotation cycle

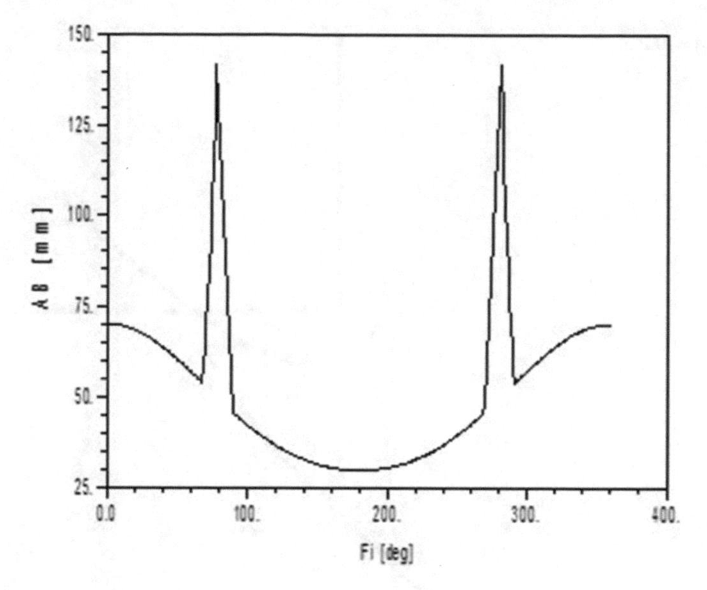

Fig. 3.7 BC = 50; XC = 20

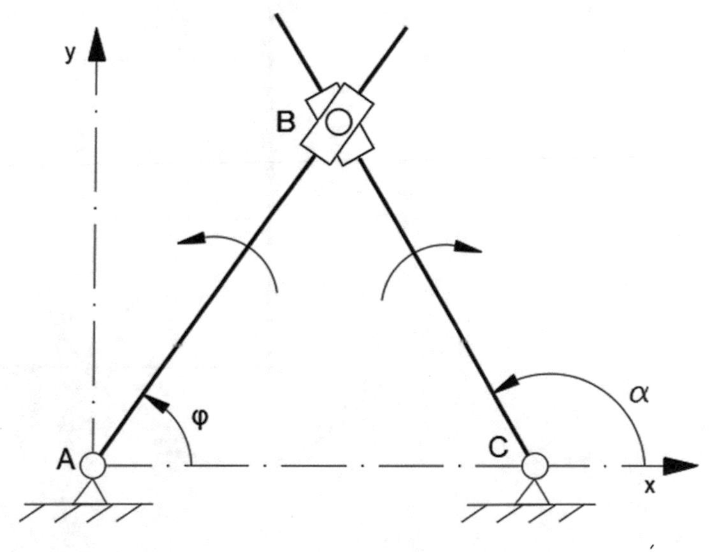

Fig. 3.8 Another mechanism

the mechanism does not work (Figs. 3.9, 3.10,3.11, 3.12, 3.13, 14, 3.15, 3.16, 3.17, 3.18, 3.19, 3.20, 3.21, 3.22, 3.23, 3.24, and 3.25).

From these images many conclusions can be drawn. Thus, at high values of q, i.e. the crank BC rotates more times than the crank AB, the geometric place has many branches, because the seizure sub-intervals of the movement of the mechanism are

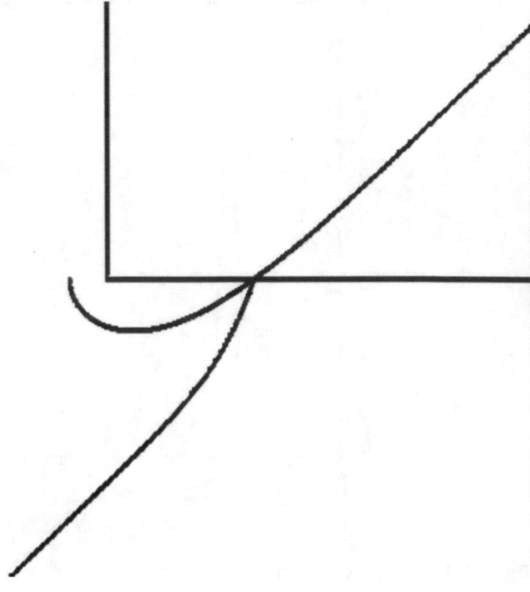

Fig. 3.9 $q = 0.2$

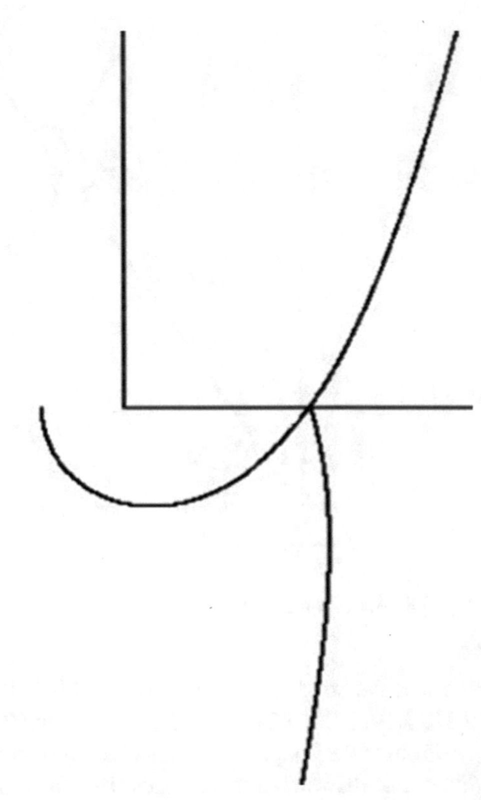

Fig. 3.10 $q = 0.3$

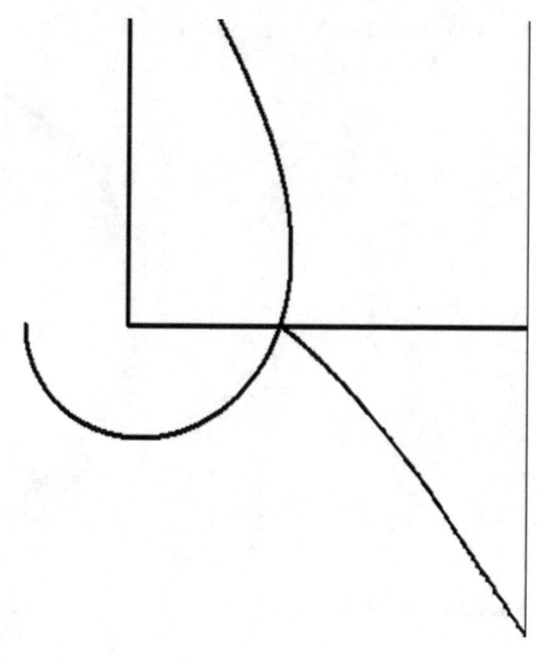

Fig. 3.11 $q = 0.4$

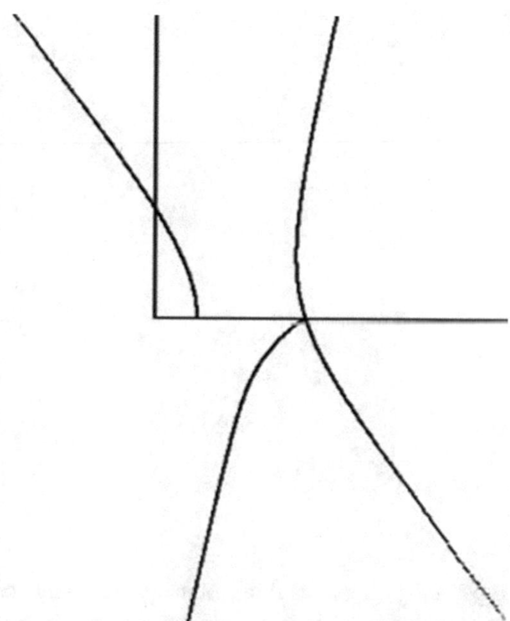

Fig. 3.12 $q = -0.4$

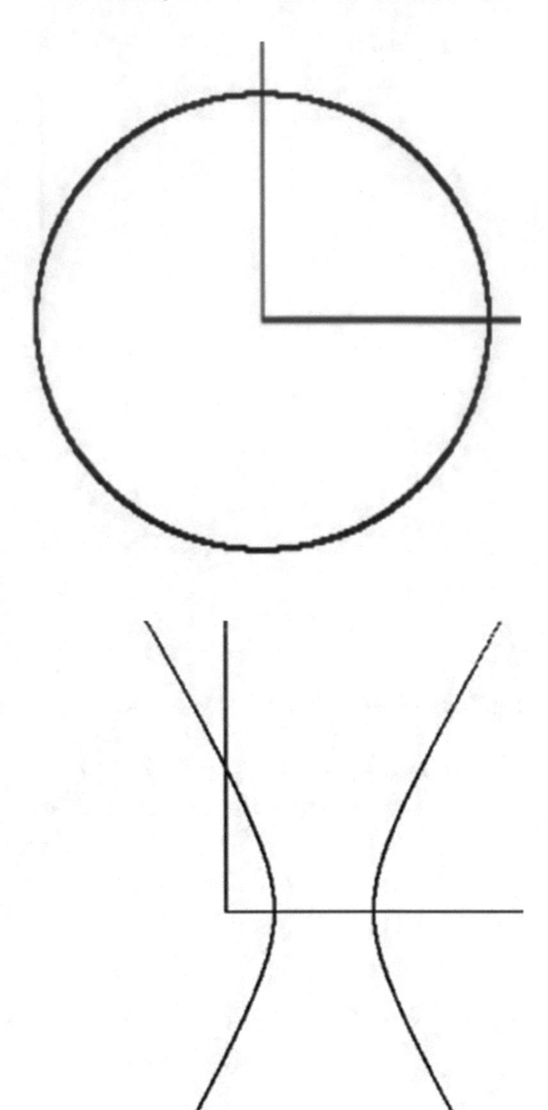

Fig. 3.13 q = 0.5 circle

Fig. 3.14 q = −0.5 hyperbola

repeated q times, resulting in many fragments of curves. It is also noted that circles, hyperboles and other known curves appear, which are particular cases that can be explained geometrically. It is also observed that there are no similarities between the curves generated at the same q, but with different signs.

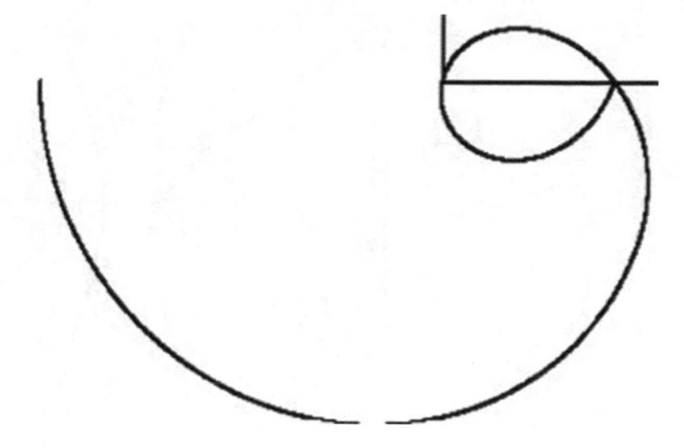

Fig. 3.15 q = 0.7

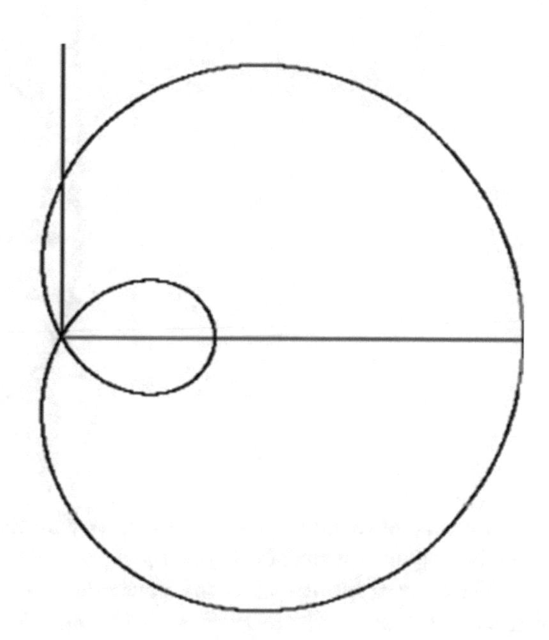

Fig. 3.16 q = 1.5 Pa's snail

Hereinafter there are some pictures with the successive positions of the sides AB and BC so that one can observe the influence of q on the areas where the mechanism does not work (Figs. 3.26, 3.27, 3.28, 3.29, 3.30, 3.31, 3.32, 3.33, and 3.34).

- **Case 3**

In this case, the bar BC no longer rotates around C, but moves on the fixed line AC (Fig. 3.35), the mechanism being of the type RP-PRP.

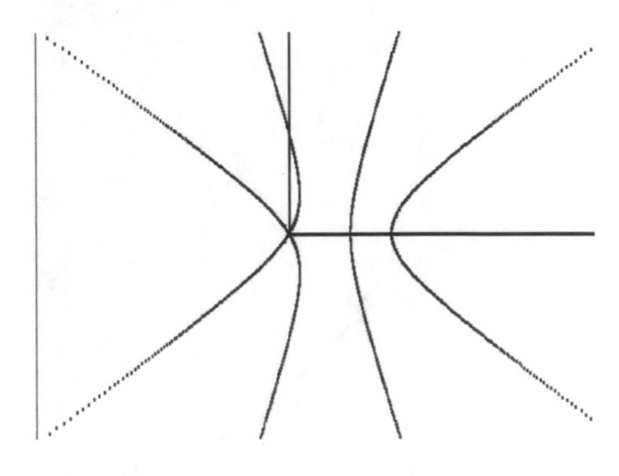

Fig. 3.17 q = −1.5

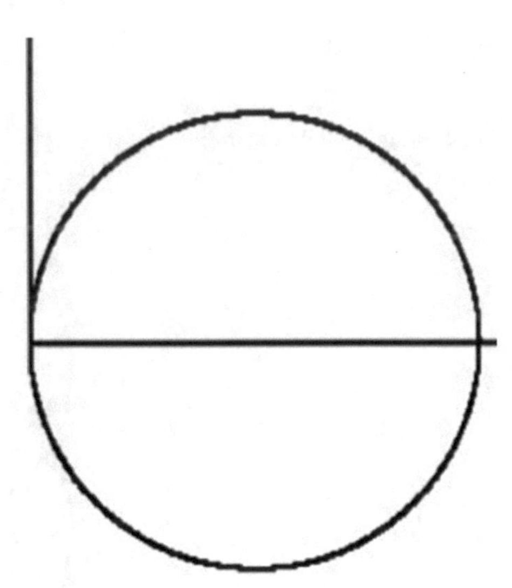

Fig. 3.18 q = 2 circle

The above relations are used, together with the following correlation relation of the two outer movements: $X_C = q.\varphi$.

Next, the resulting loci for different values of the correlation coefficient are given. It was taken $\alpha = 110°$ (Figs. 3.36, 3.37, and 3.38).

The loci are discontinuous but similar to different values of q, having three branches due to inrushes near the critical values. Next there are some cases with the successive positions of the mechanism at different values of q (Figs. 3.39, 3.40, 3.41, and 3.42).

We can notice the symmetries of these positions for the different directions of movement of the slide in C. In Fig. 3.43 the variations of the AB and BC races are given, so that the areas with inrushes can be traced to the critical values of φ

.

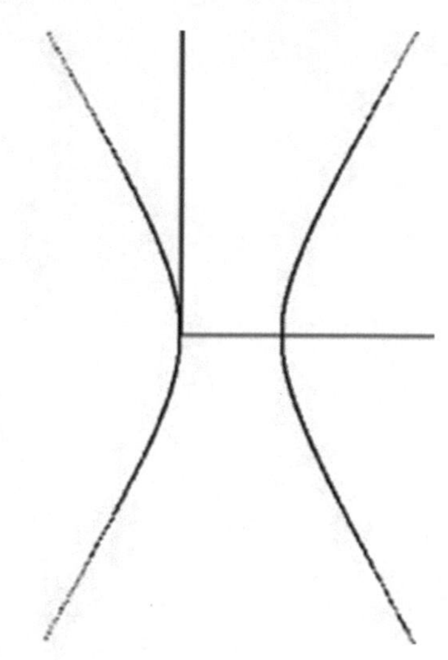

Fig. 3.19 q = −2 Hyperbola

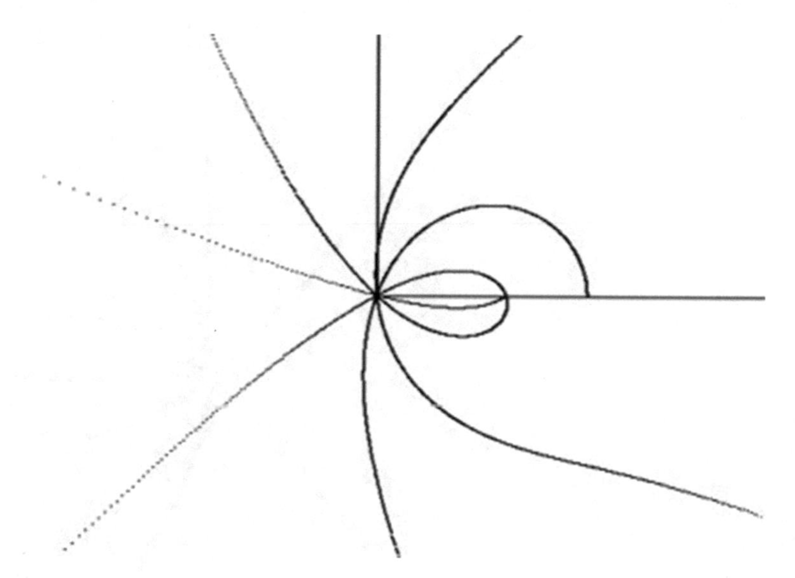

Fig. 3.20 q = 2.6

- **Case 4**

In this case, point C slides on the abscissa, $\alpha = 110^\circ$ = constant, with only one steering element, AB. Length BC = 50 mm = constant. The mechanism is of the type R-PRP (Fig. 3.44).

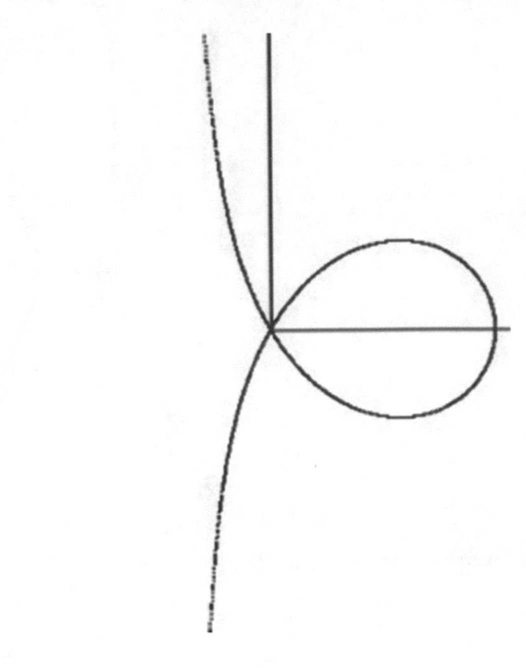

Fig. 3.21 q = 3 strophoid

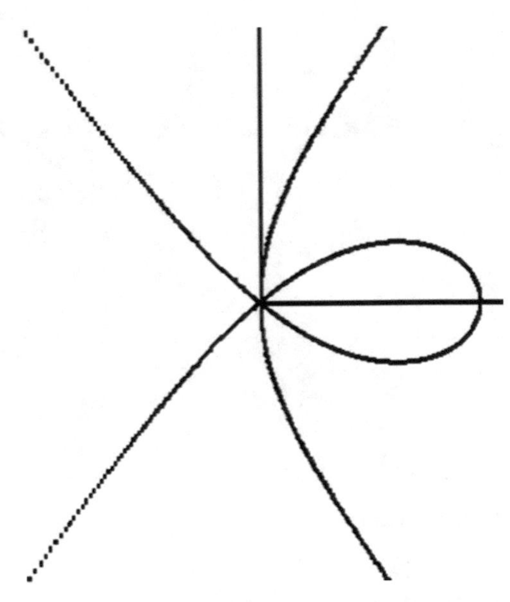

Fig. 3.22 q = 4

The above relations from which AB and XC are calculated are used. It is clear that the locus of B is a straight line, parallel to AC (Fig. 3.45).

This line is also observed in the successive positions given in Fig. 3.46.

The variations of the AB and XC strokes are given in Fig. 3.47, which results in a pause in the operation of the mechanism for φ around 180°.

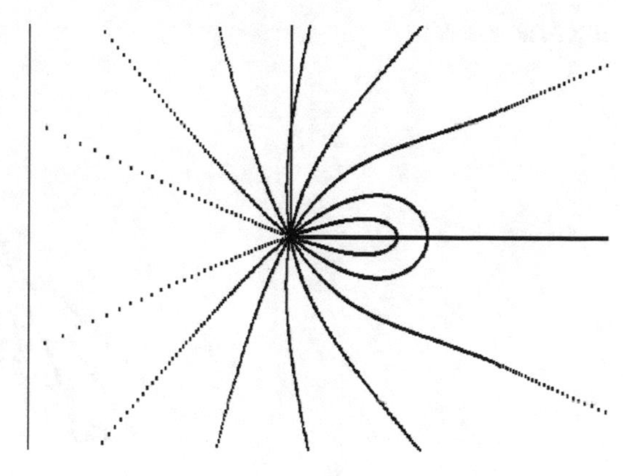

Fig. 3.23 $q = 4.5$

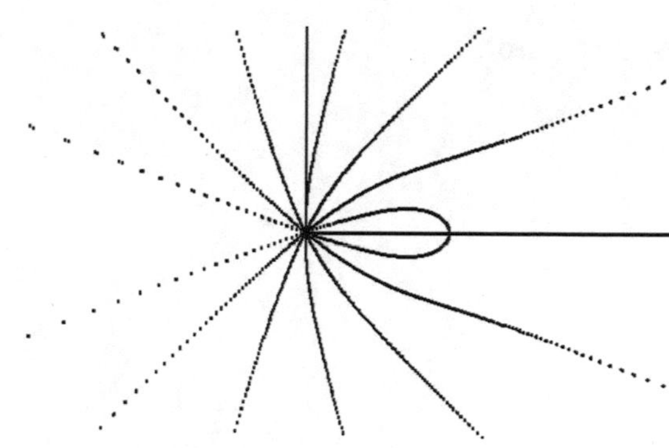

Fig. 3.24 $q = 8$

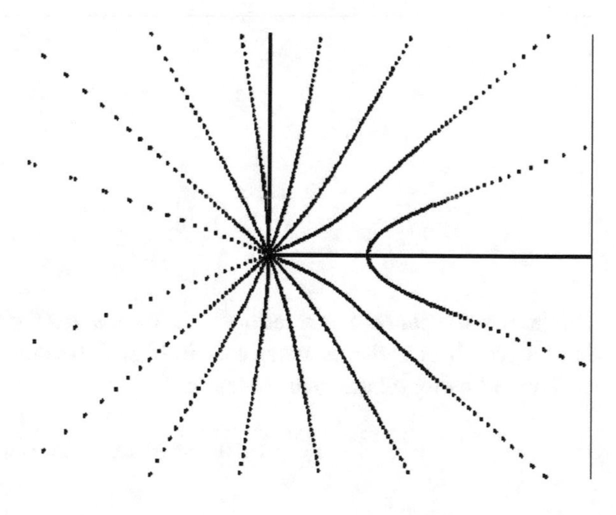

Fig. 3.25 $q = -8$

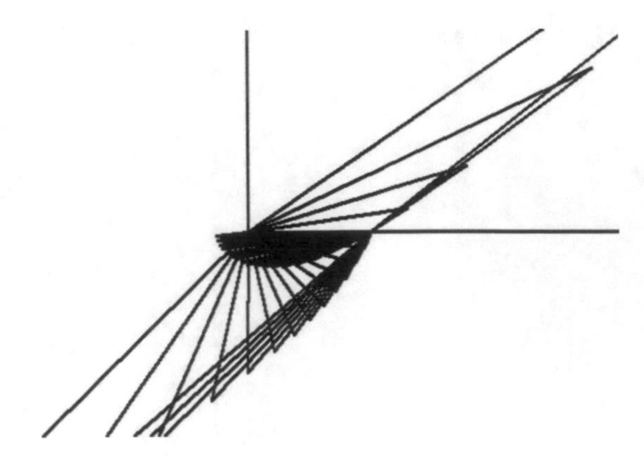

Fig. 3.26 q = 0.2

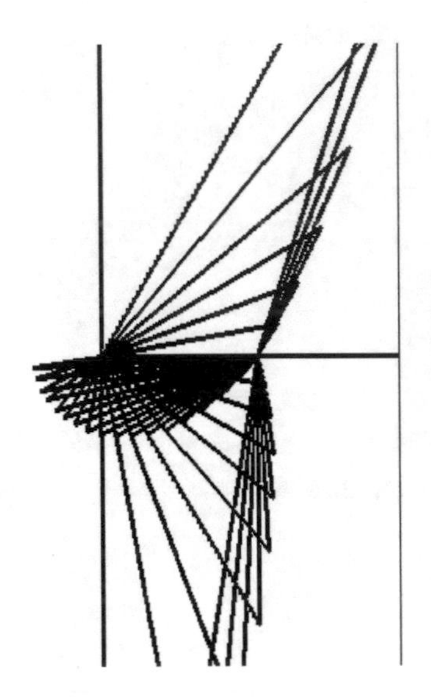

Fig. 3.27 q = 0.3

- **Case 5**

In this case, the bar BC has rotational motion towards the movable point C (Fig. 3.47), not towards B as in the previous case in Figs. 3.44 and 3.48.

The following relations are written:

$$x_B = AB\cos\varphi = x_C + BC\cos\alpha \tag{3.4}$$

$$y_B = AB\sin\varphi = BC\sin\alpha \tag{3.5}$$

$$\alpha = \gamma + \varphi \tag{3.6}$$

Fig. 3.28 q = 0.4

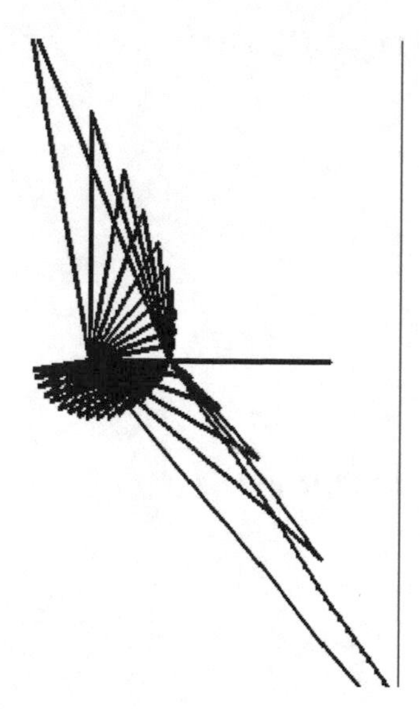

Fig. 3.29 q = 0.5

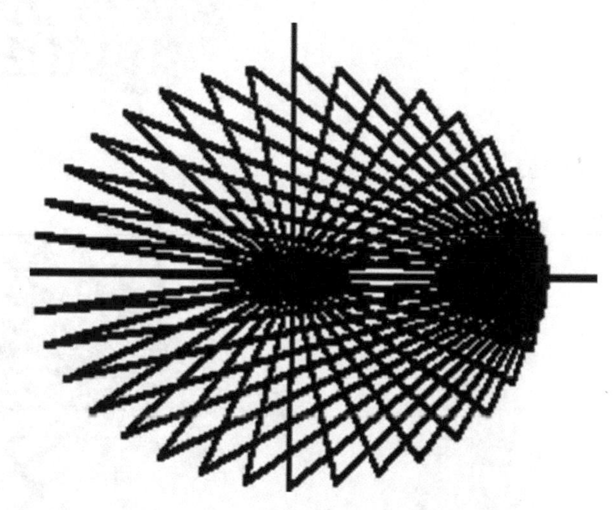

Fig. 3.30 q = 0.6

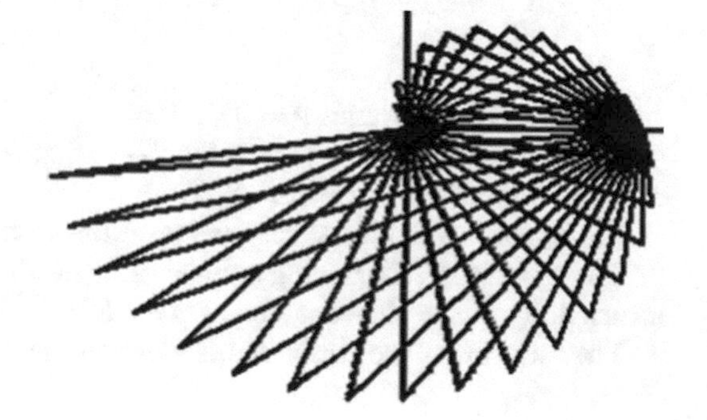

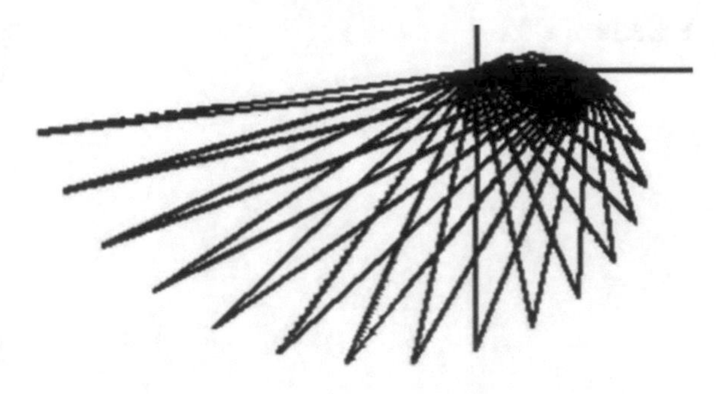

Fig. 3.31 q = 0.8

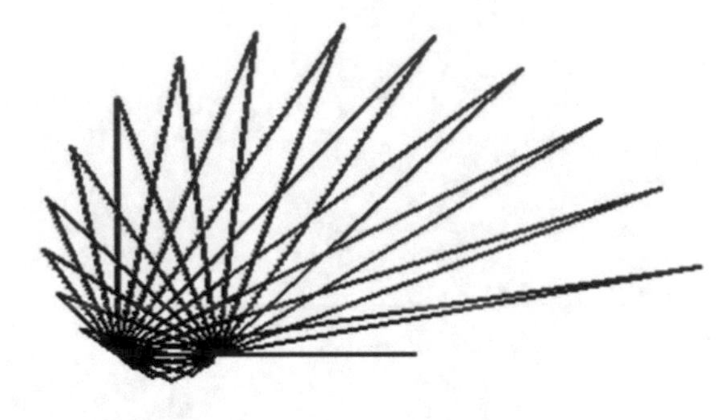

Fig. 3.32 q = 1.2

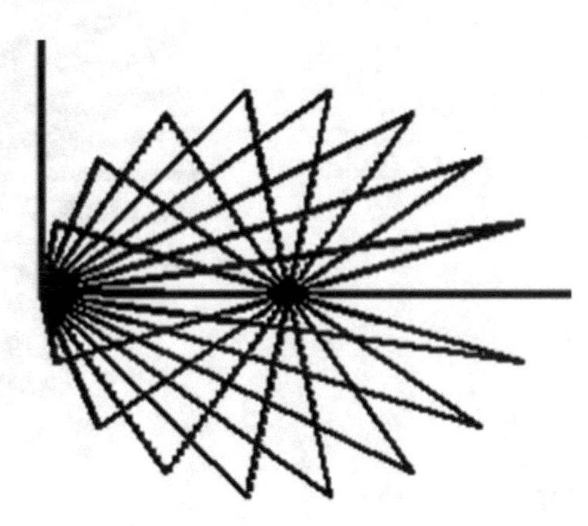

Fig. 3.33 q = 2

If given: BC = 50 mm, $\gamma = 35°$.

The searched locus is given in Fig. 3.49, that is, a curve with two branches. The curve is"Kappa" type (see Sect. 8.1).

Figure 3.50 shows the successive positions of the mechanism.

The variation of the tracing point coordinates is shown in Fig. 3.51, observing an inrush of Y_B in the adjacent area of $\varphi = 180°$.

The variation of the stroke AB is given in Fig. 3.52, finding the same inrush.

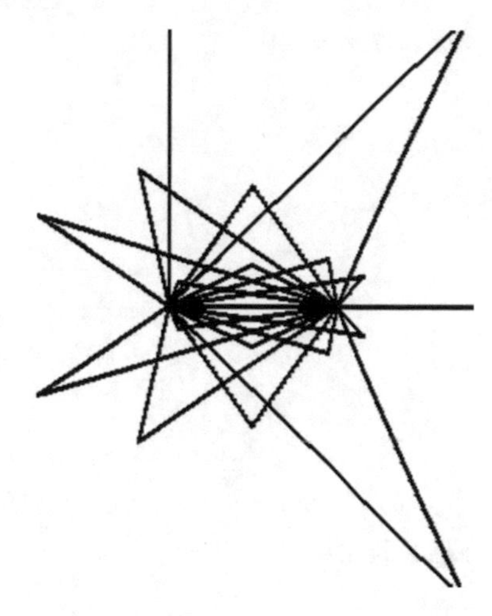

Fig. 3.34 q = 5

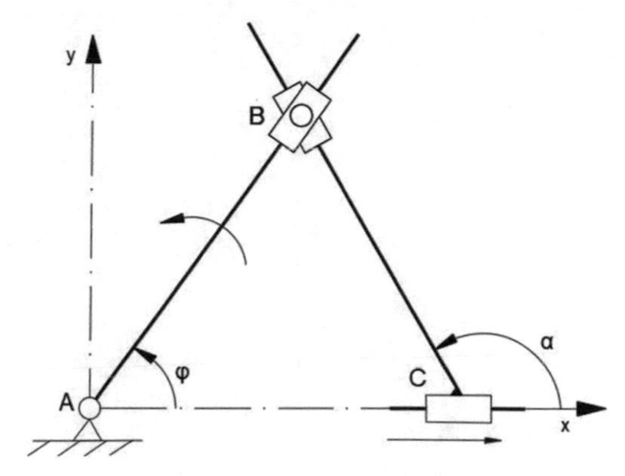

Fig. 3.35 Another mechanism

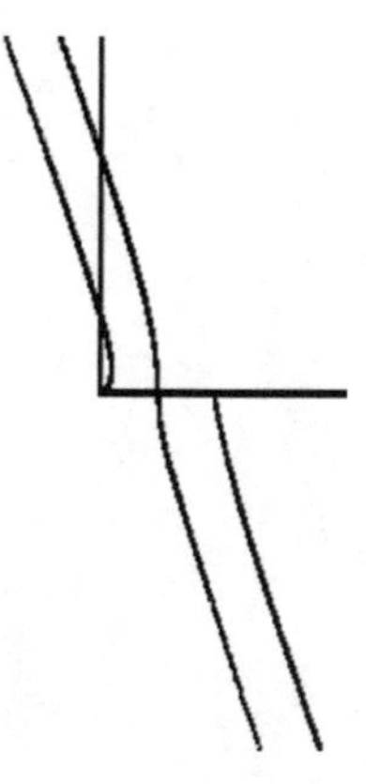

Fig. 3.36 q = 4

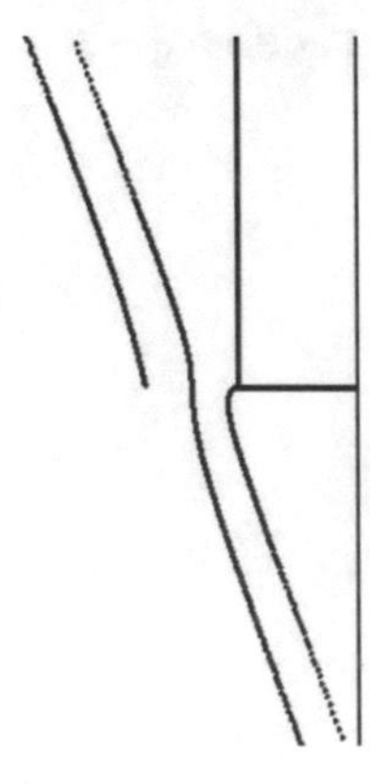

Fig. 3.37 $q = -4$

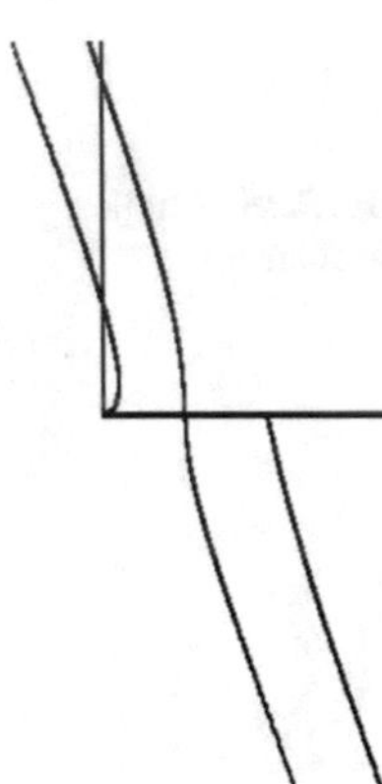

Fig. 3.38 $q = 7$

Fig. 3.39 $q = 4$

Fig. 3.40 q = −4

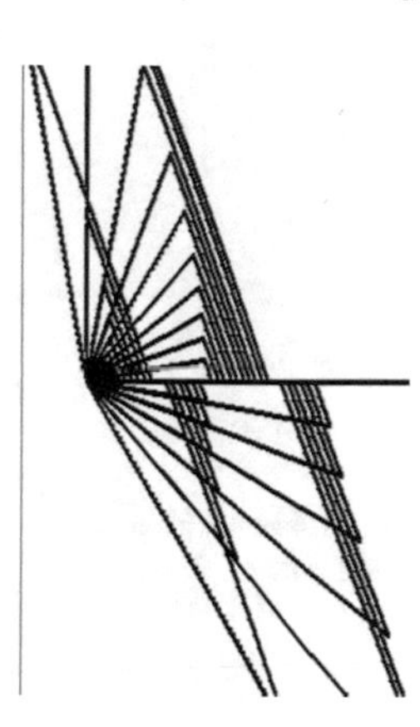

Fig. 3.41 q = 12

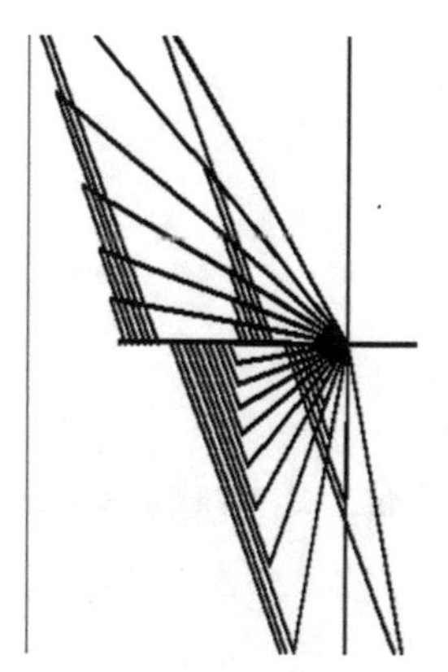

Fig. 3.42 q = −12

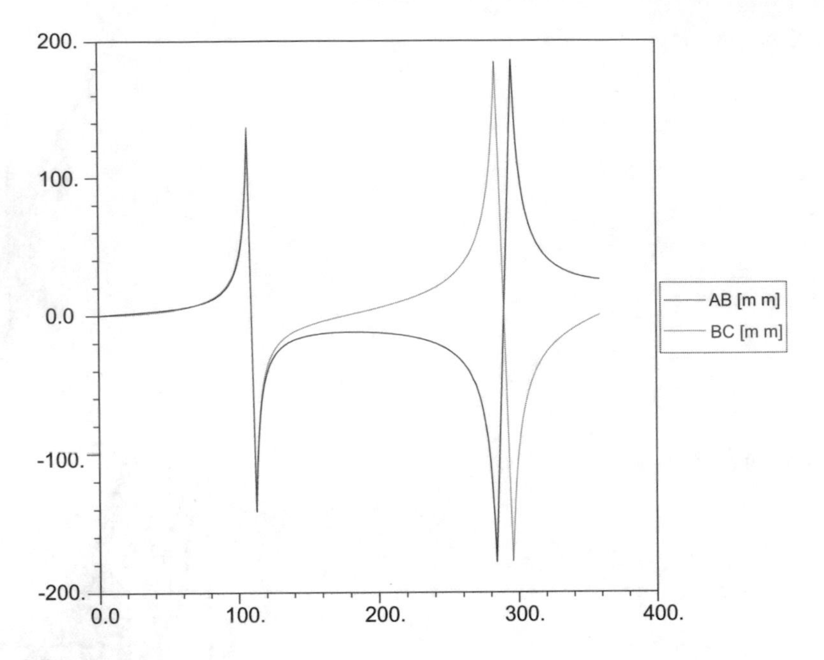

Fig. 3.43 q = 4

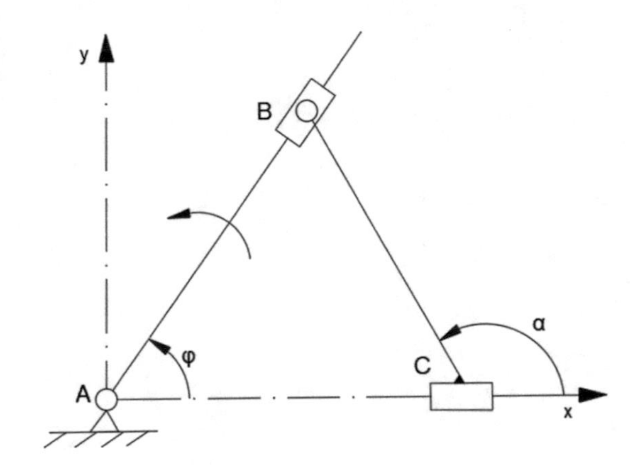

Fig. 3.44 Another mechanism

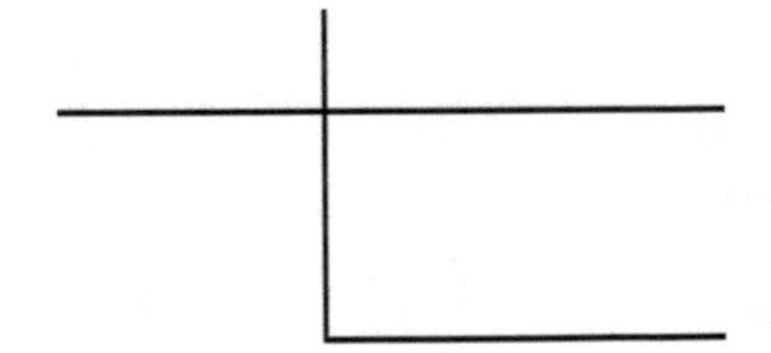

Fig. 3.45 Locus of B point

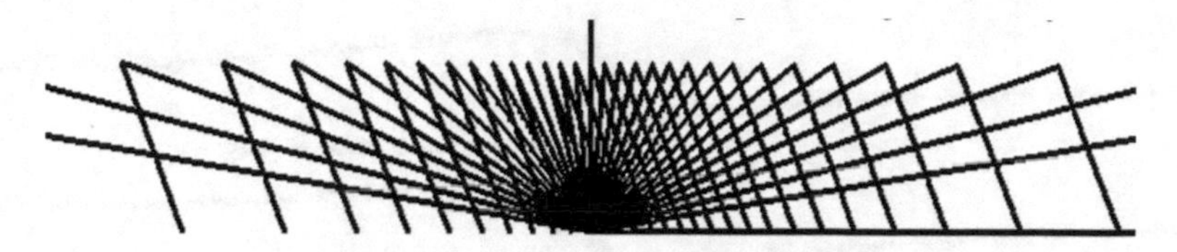

Fig. 3.46 Successive positions

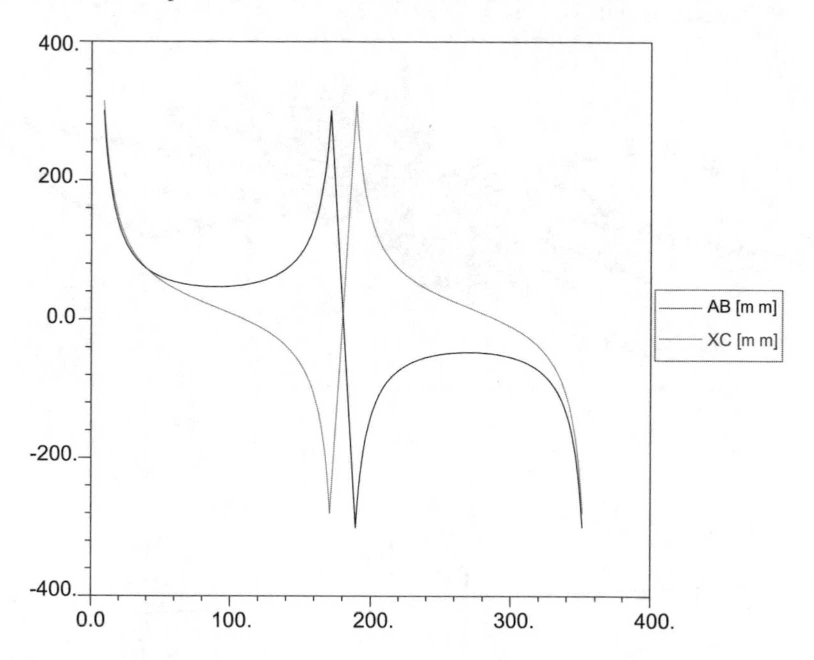

Fig. 3.47 AB and X_C strokes

Fig. 3.48 Another mechanism

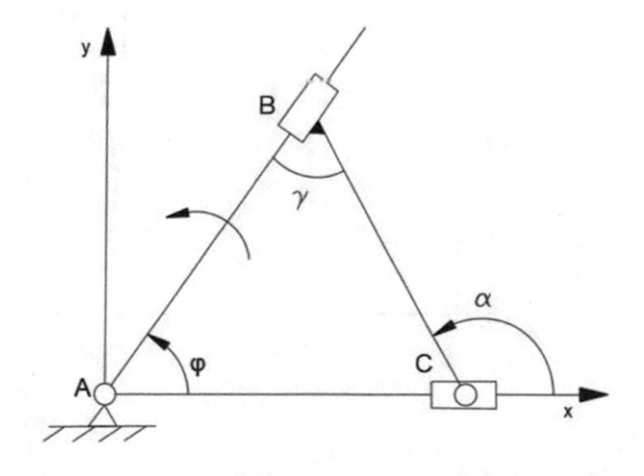

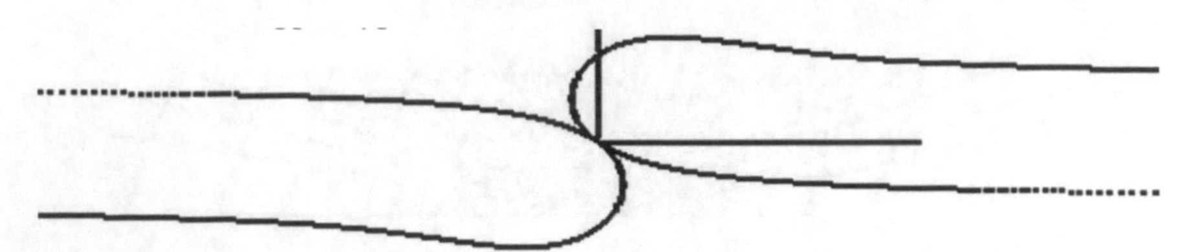

Fig. 3.49 Locus of point B

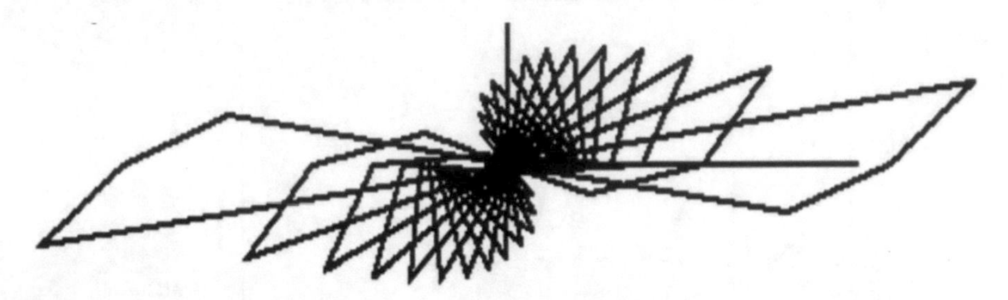

Fig. 3.50 Successive positions

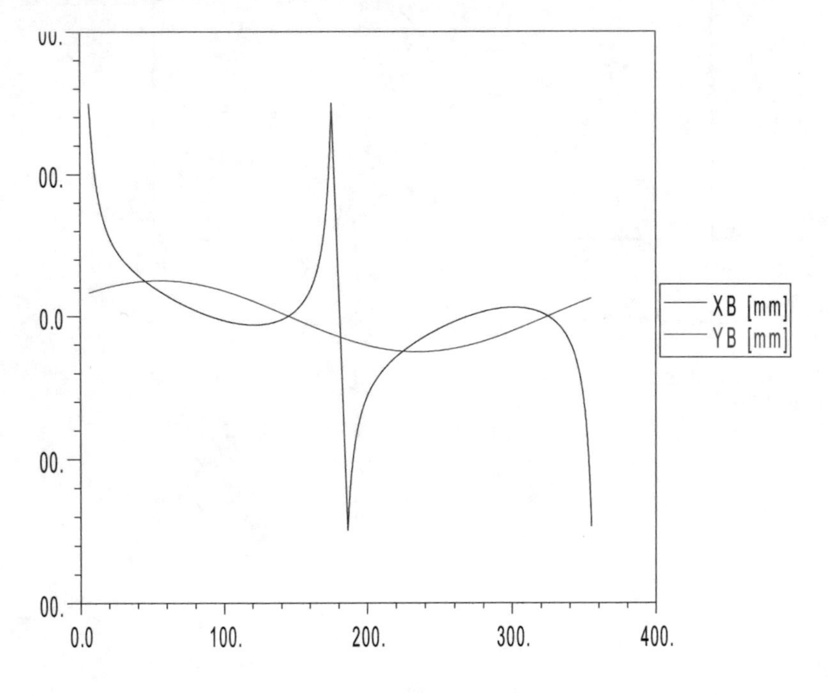

Fig. 3.51 Coordinates of point B

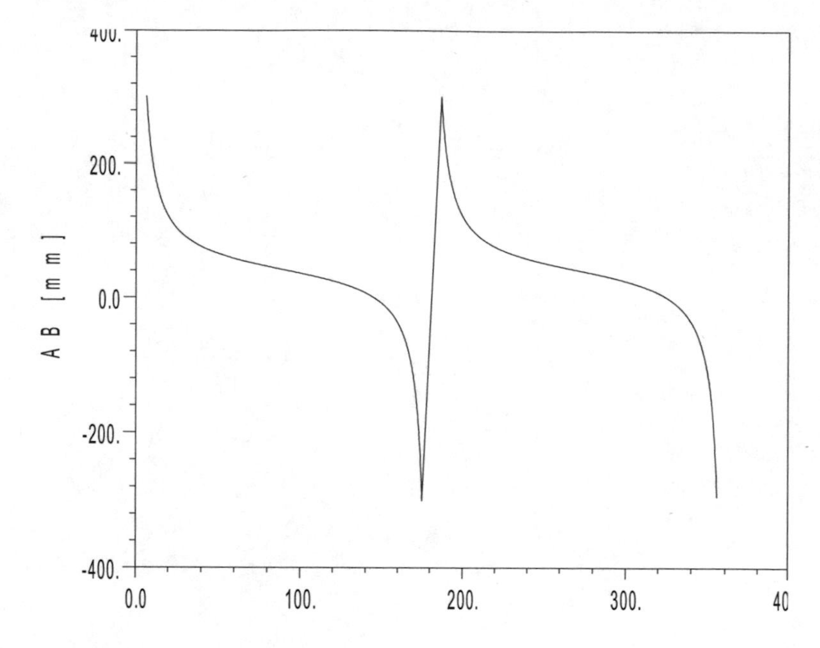

Fig. 3.52 Variation of stroke AB

References

1. Popescu I, Romanescu AE, Sass L (23–25, September, 2016) A locus problem solved through methods from the theory of mechanisms. În: International conference: innovative manufacturing engineering and energy, IManE, p 7. Kallithea Chalkidiki Greece
2. Popescu I, Luca L (2015) Geometric locus generated by a mechanism with three dyads. În: Analele Universităţii «Constantin Brâncuşi» din Târgu-Jiu, Seria Inginerie, nr. 3, pp 29–37
3. Sass L, Duţă A, Popescu I (2017) Nature gives us beauty, insoiring the technical creativity. J Ind Design Eng Graphics (1):41–46. Papers of the international conference on engineering graphics and design, section 3: engineering computer graphics, Nr. 12

Chapter 4
Loci Generated by the Points on a Line Which Move on Two Concurrent Lines

Abstract We consider a straight line that can have a rotation motion around a fixed point and another one which slides with one end on the first one and the other end slides on a fixed line, we try to obtain the trajectories of several points situated on the mobile line. Numerous trajectories are obtained depending to the angle of the fixed line with the abscissa [1]. Successive positions are also given. The curves have two branches. When the fixed line have rotation motion too, there are two leading elements and the trajectories depend on the correlation between their movements [3].

4.1 Case 1

Consider two lines on which the line BC slides: line AB rotates around A, and the line passing through C is fixed. The loci of some points on BC must be found (Fig. 4.1).

The obtained mechanism is given in Fig. 4.2, being of type R-PRP, the steering element being AB.

The following relations are written:

$$\mathrm{AB}\cos\varphi + BC\cos\alpha = \mathrm{x}_D + CD\cos\gamma \tag{4.1}$$

$$\mathrm{AB}\sin\varphi + BC\sin\alpha = CD\sin\gamma \tag{4.2}$$

We get:

$$(\mathrm{x_D} + \mathrm{CD}\cos\gamma - BC\cos\alpha)\mathrm{tg}\,\varphi = \mathrm{CD}\sin\gamma - BC\sin\alpha \tag{4.3}$$

$$\alpha = \varphi - \delta \tag{4.4}$$

$$\mathrm{x_E} = \mathrm{AB}\cos\varphi + BE\cos\alpha \tag{4.5}$$

I. Popescu et al., *Problems of Locus Solved by Mechanisms Theory*,
Springer Tracts in Mechanical Engineering,
https://doi.org/10.1007/978-3-030-63079-9_4

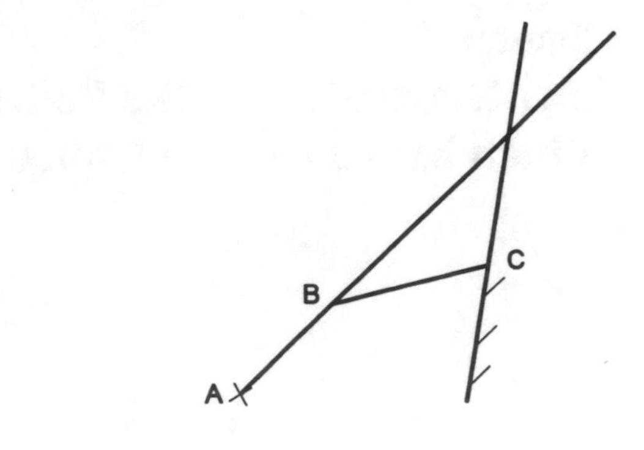

Fig. 4.1 The geometric construction

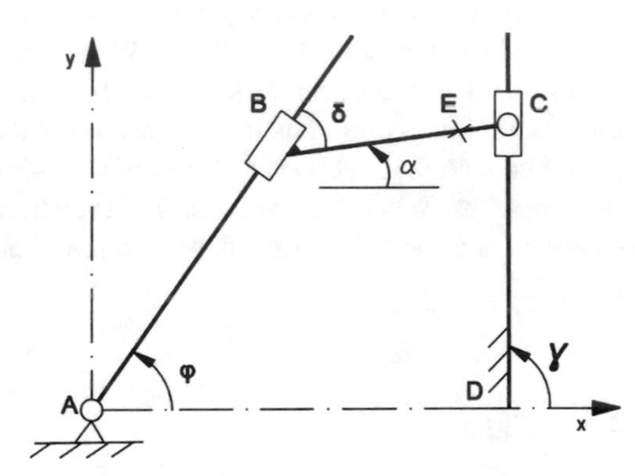

Fig. 4.2 The equivalent

$$y_E = \mathrm{AB} \sin \varphi + BE \sin \alpha \tag{4.6}$$

$$\mathrm{CD} = [(x_D - BC \cos \alpha) tg\ \varphi + BC \sin \alpha] / [\sin \gamma - \cos \gamma\ tg\ \varphi] \tag{4.7}$$

Initial data: XD = 50 mm; BC = 30 mm; BE = 20 mm; $\alpha = 20$; $\gamma = 80$; $\delta = 25$ degrees.

The mechanism of Fig. 4.3 for one position and Fig. 4.4 for successive positions are obtained.

From Fig. 4.4 it is seen that the crank AB rotates only in a subinterval of the cycle. The locus of point C is shown in Fig. 4.5 (the branch in the middle), being a line.

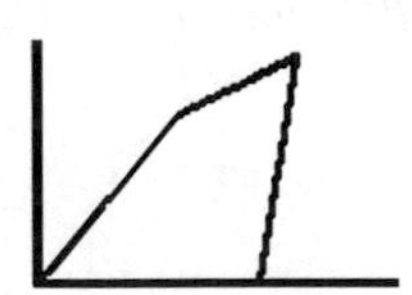

Fig. 4.3 The mechanism in one position

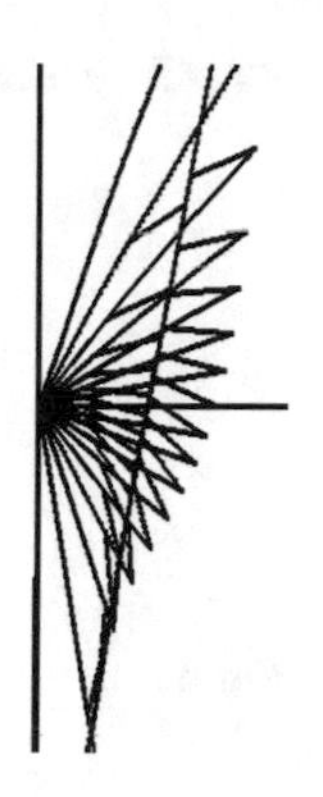

Fig. 4.4 Successive positions

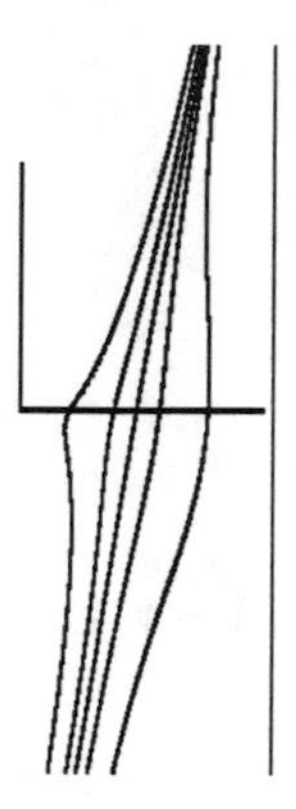

Fig. 4.5 Loci of points B, E and C

The locus of point B is a curve with two branches (the outer ones of Fig. 4.5), and E's locus is also a curve with two branches—inside the figure.

The following shows only the locus drawn by point E for its different values γ, as well as the successive positions of the mechanism (Figs. 4.6 and 4.7).

The locus is the same curve, but inclined at the angle γ. The following shows only the successive positions of the mechanism, in order to follow the change in relation to the trigonometric quadrant (Figs. 4.8, 4.9 and 4.10).

Then γ and length BC changed, obtaining other forms of loci.

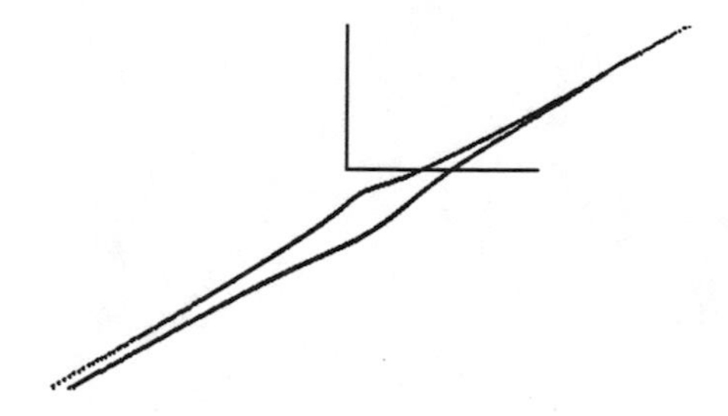

Fig. 4.6 $\gamma = 30$

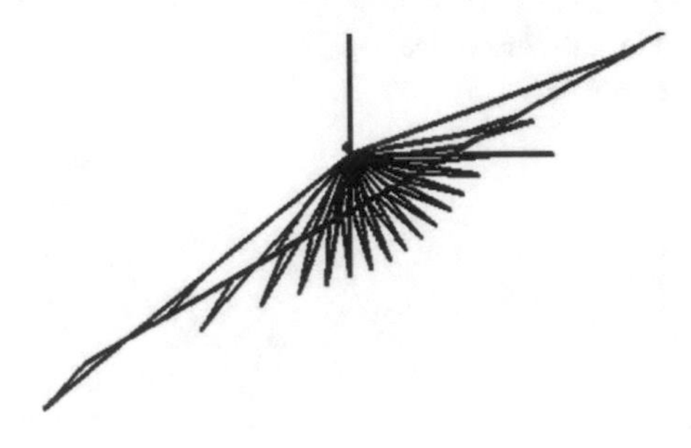

Fig. 4.7 $\gamma = 30$

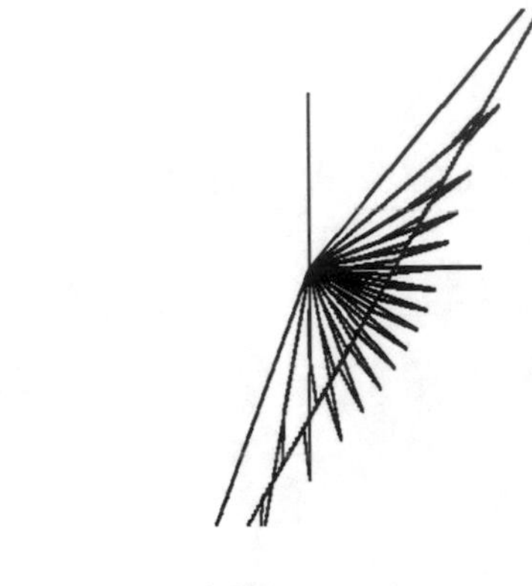

Fig. 4.8 $\gamma = 60$

Fig. 4.9 $\gamma = 200$

Fig. 4.10 $\gamma = 320$

Fig. 4.11 $\gamma = 75$; BC = 60

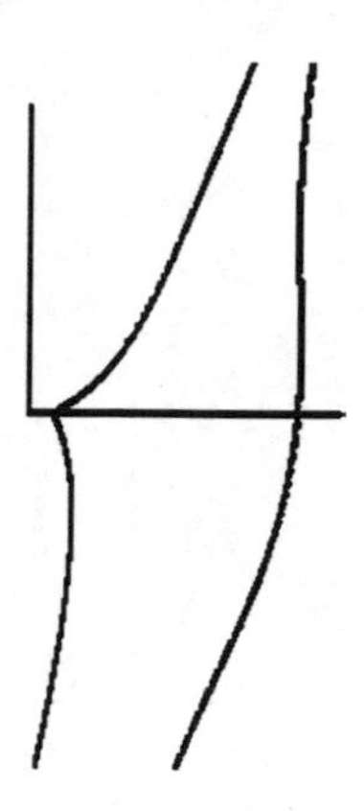

Similar curves are obtained also in Chap. 8.1 (Figs. 4.11, 4.12, 4.13, 4.14 and 4.15).

Fig. 4.12 $\gamma = 120$; BC = 75

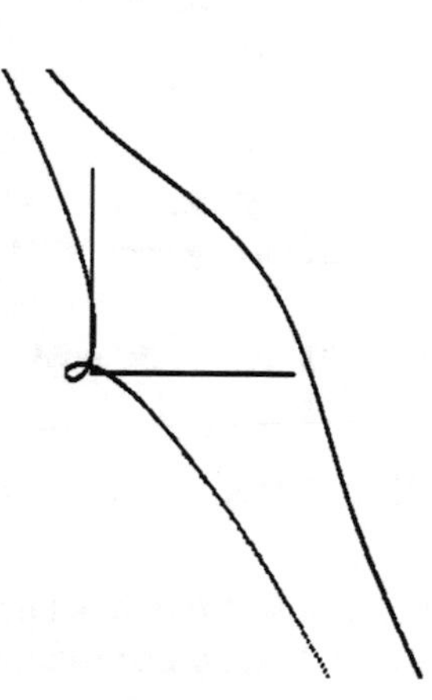

Fig. 4.13 $\gamma = 200$; BC = 80

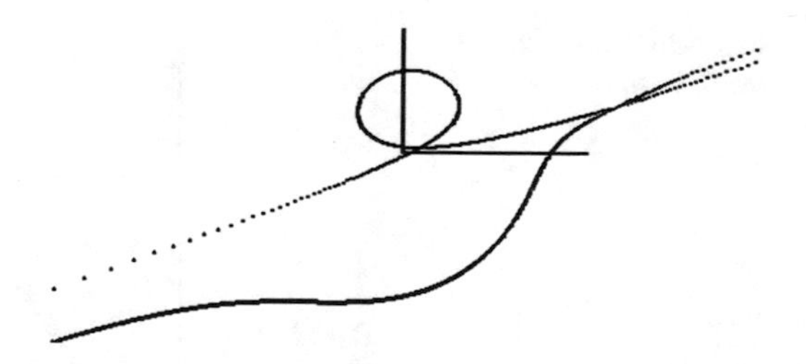

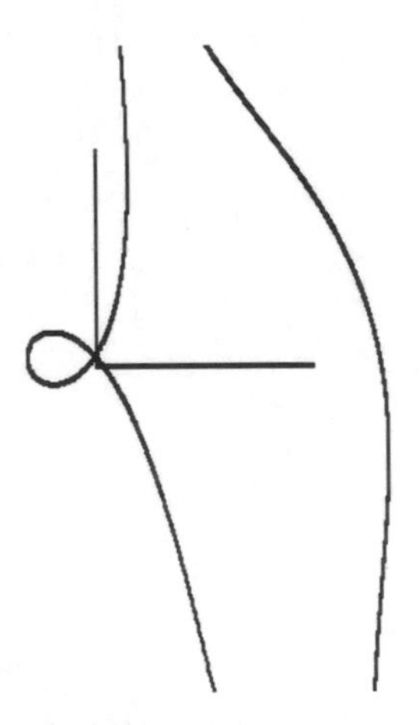

Fig. 4.14 $\gamma = 100$; BC = 100

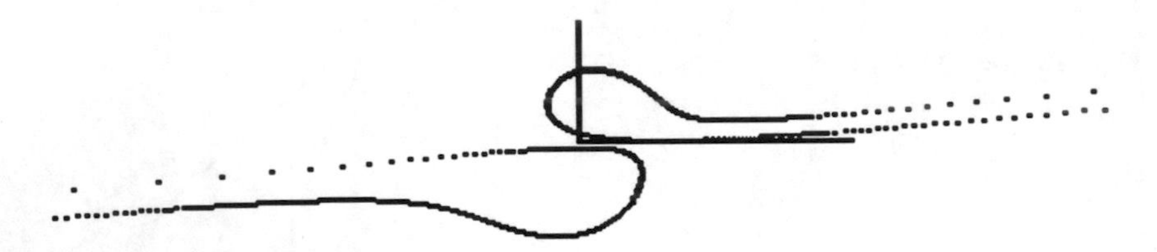

Fig. 4.15 $\gamma. = 5$; BC = 50

4.2 Case 2

We return to the initial problem in Fig. 4.1, but the line DC is no longer fixed, only rotates around D, resulting in the mechanism of Fig. 4.16 of the type RR-PRP.

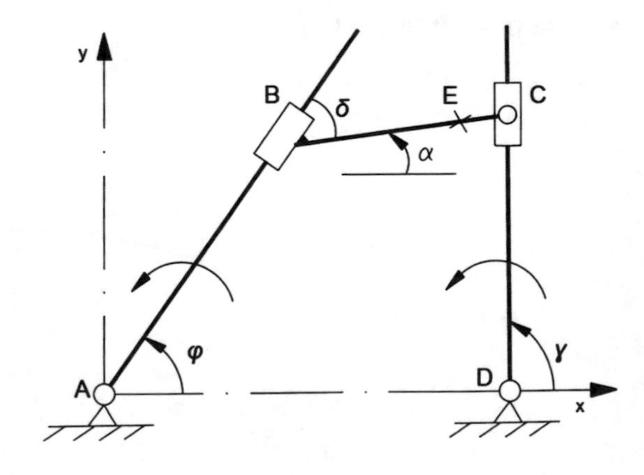

Fig. 4.16 The R-R-PRP mechanism

The above relations are used, to which it is added the correlation relation between φ and γ: $\gamma = q.\varphi$.

The initial data are the ones above. Then follow the loci and successive positions for certain values of q (Figs. 4.17, 4.18, 4.19, 4.20, 4.21, 4.22, 4.23, 4.24, 4.25, 4.26, 4.27, 4.28, 4.29, 4.30, 4.31, 4.32, 4.33, 4.34, 4.35, 4.36, 4.37).

The loci are very varied as shapes, being different according to the sign of q, that is to say the directions of movement of the steering elements, differing also the successive positions from one case to another [2].

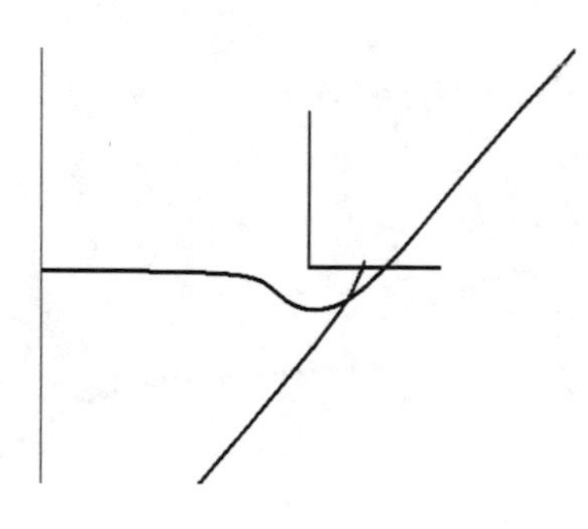

Fig. 4.17 q = 0,2

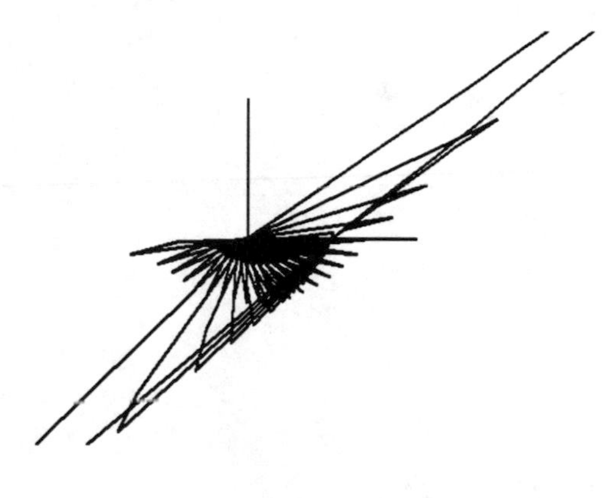

Fig. 4.18 q = 0,2

Fig. 4.19 q = −0,2

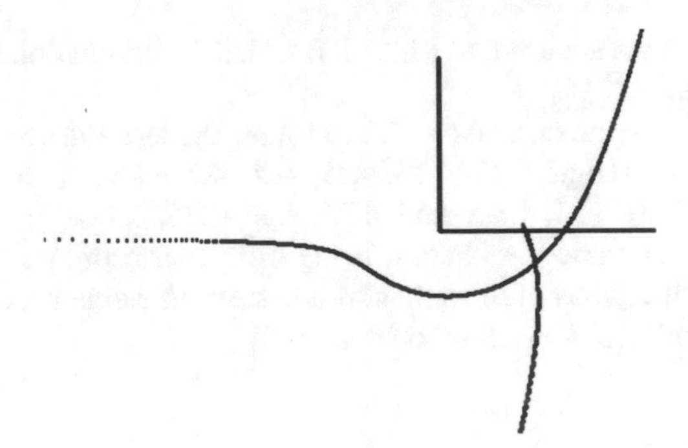

Fig. 4.20 $q = 0{,}3$

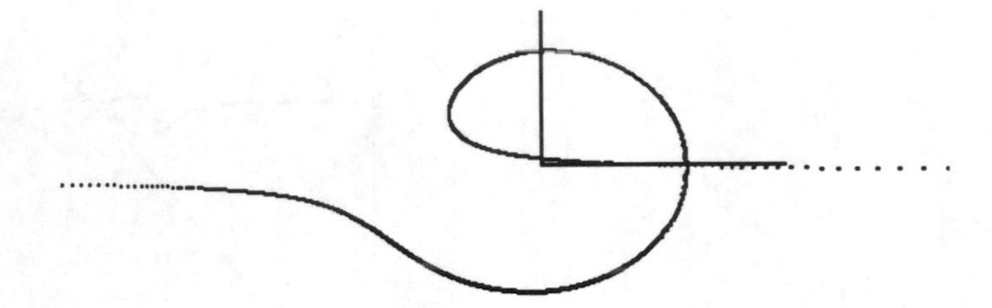

Fig. 4.21 $q = 0{,}5$

Fig. 4.22 $q = -0{,}5$

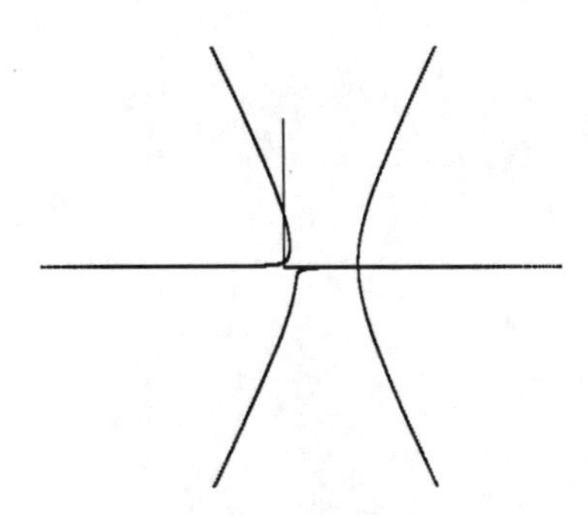

Fig. 4.23 $q = -0{,}5$

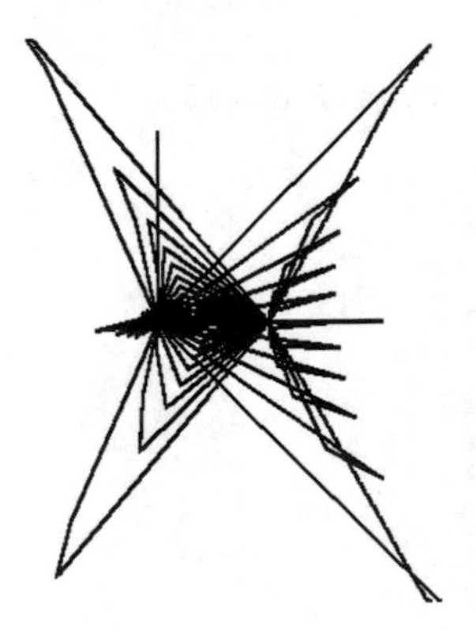

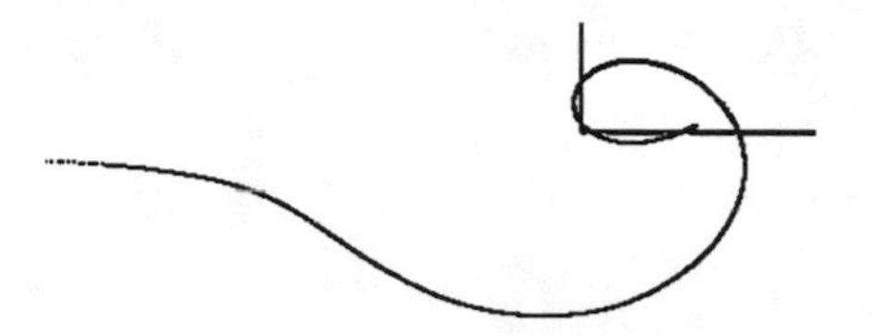

Fig. 4.24 q = 0,6

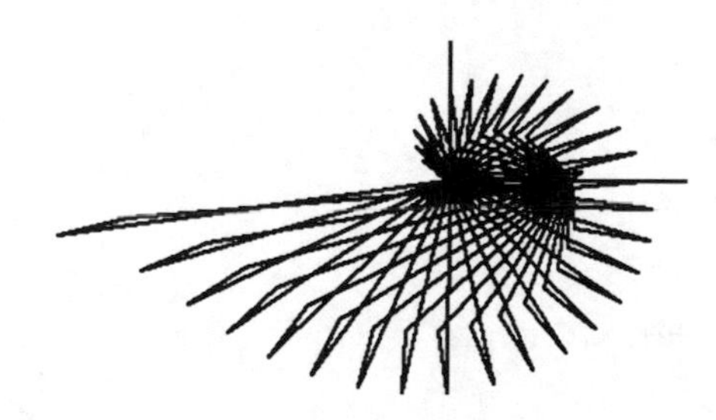

Fig. 4.25 q = 0,6

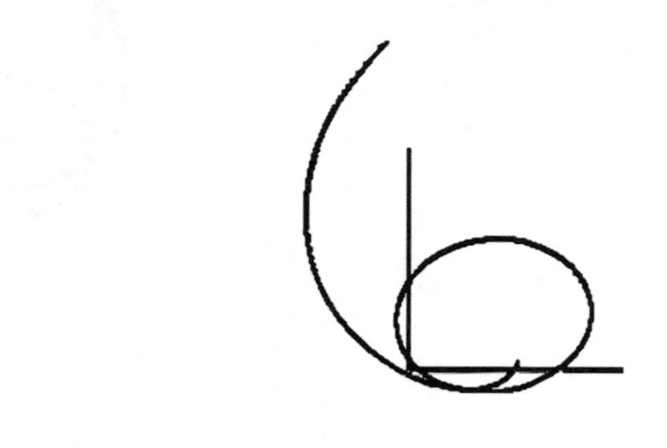

Fig. 4.26 q = 1,2

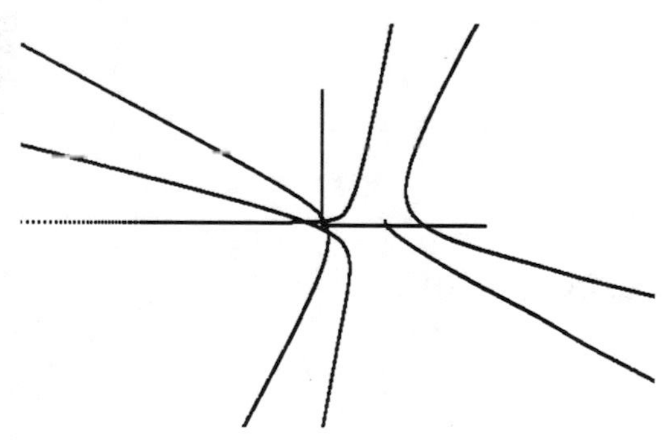

Fig. 4.27 q = −1,2

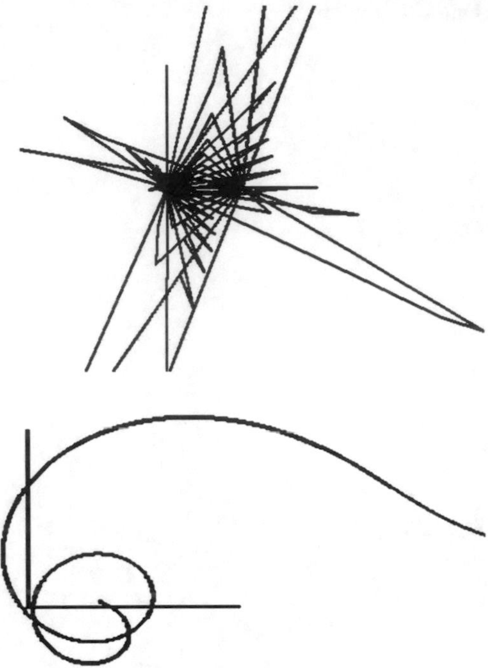

Fig. 4.28 q = −1,2

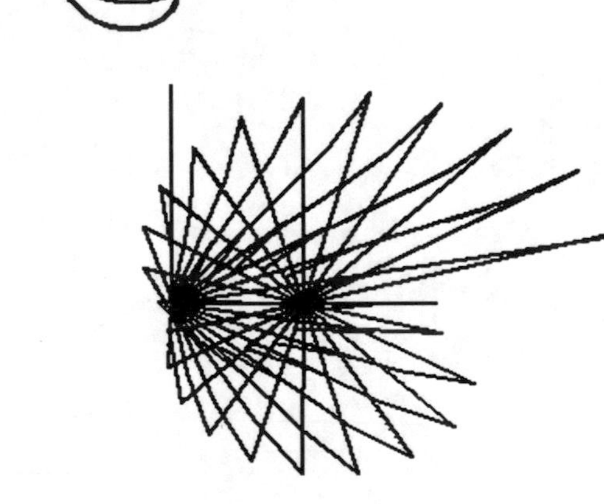

Fig. 4.29 q = 1,4

Fig. 4.30 q = 1,5

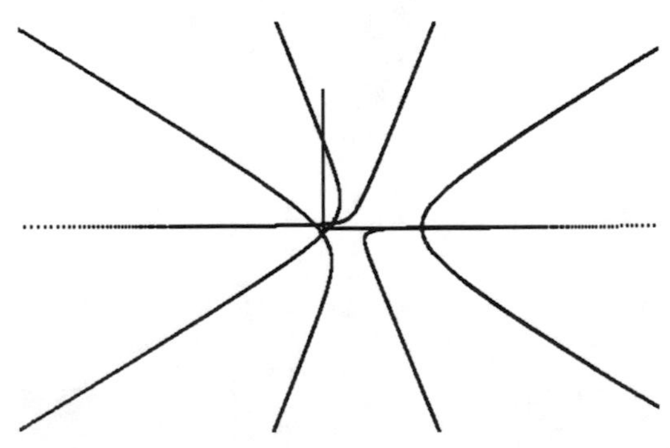

Fig. 4.31 q = −1,5

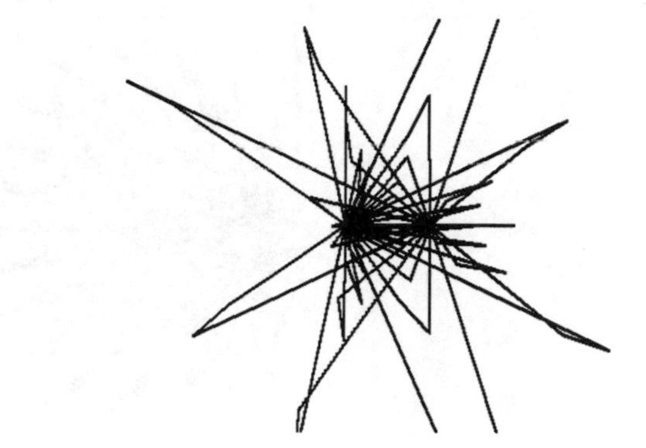

Fig. 4.32 q = −1,5

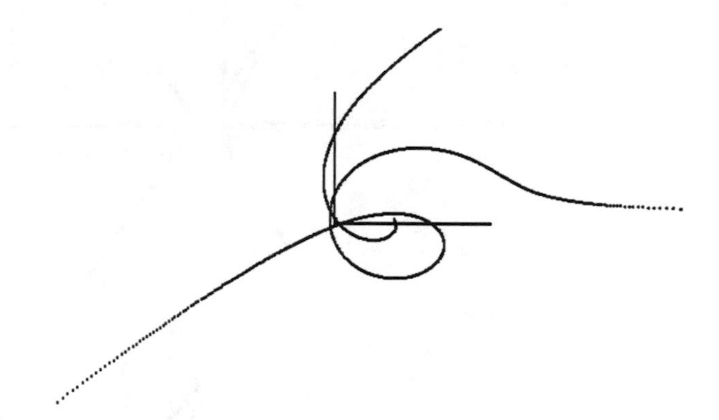

Fig. 4.33 q = 1,8

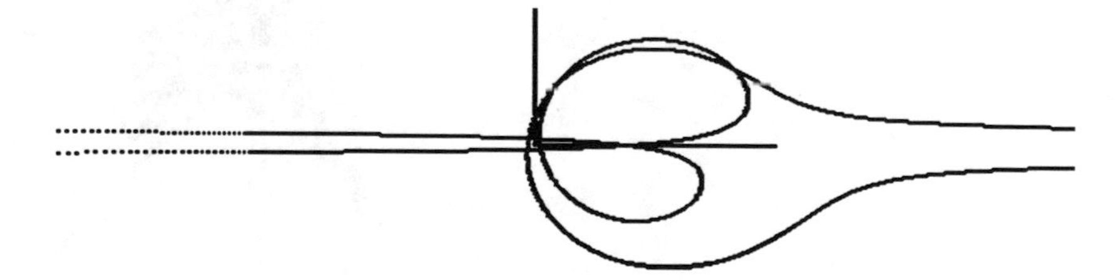

Fig. 4.34 q = 2

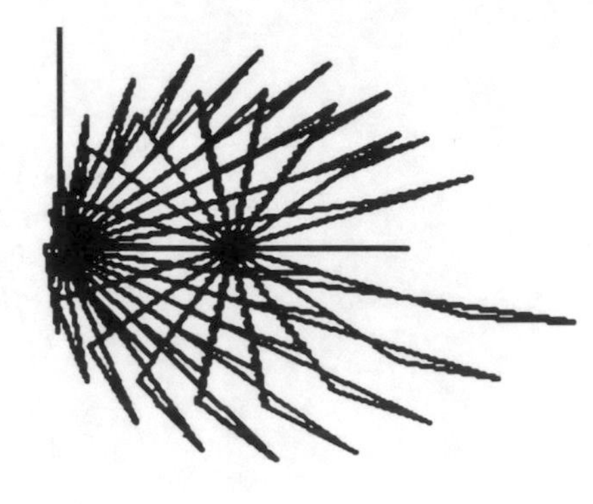

Fig. 4.35 q = 2

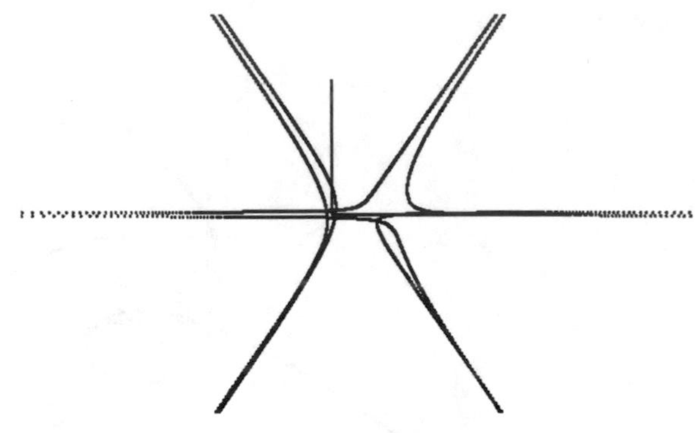

Fig. 4.36 q = −2

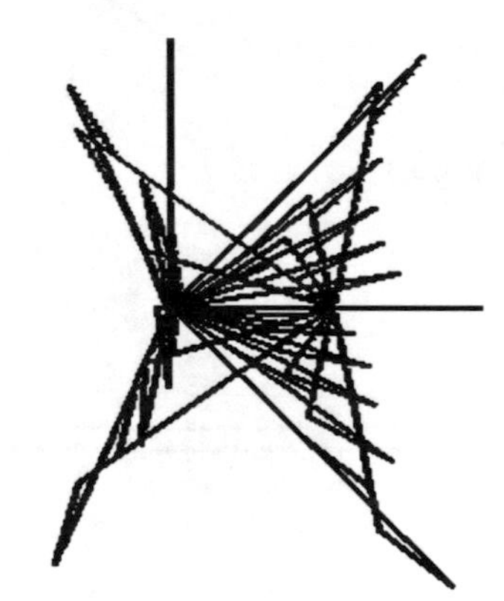

Fig. 4.37 q = −2

References

1. Popescu Iulian—Locuri geometrice şi imagini estetice generate cu mecanisme. Editura Sitech, Craiova, 2016
2. Popescu Iulian—Curbe de bielă în diferite plane şi locuri geometrice generate de mecanisme. Editura Sitech, Craiova, 2017
3. Popescu I, Luca L (2015) Geometric locus generated by a mechanism with three dyads. În: Analele Universităţii « Constantin Brâncuşi » din Târgu-Jiu, Seria Inginerie, nr. 3/2015, pp 29–37

Chapter 5
Loci Generated by the Points on a Bar Which Slides with the Heads on Two Fixed Lines

Abstract A straight-line slide with its heads on the axes of the system. The points on this line generate ellipses (example point C), composed of two halves, one for the + sign and the other for the—sign in front of a radical. Successive positions generate asteroids [1, 2]. Diagrams are given with the variations of some coordinates of the tracer point. From the ends of the sliding line we draw parallel lines to the axes of the reference system, establishing the trajectory of their intersection, the point D. The result is a circle containing the ellipse generated by the first point. In another case, the length of the line AB that has the ends on the two fixed lines was considered to be variable depending to the length of x_B, i.e. the stroke of the leading element, so $AB = q * x_B$, where q is conveniently adopted. It was found that in this case the trajectory of C point is a line parallel to the abscissa, and that of D point is an inclined line. The case where the fixed lines are inclined to the axes of the system was also studied, resulting in many curves dependent on the inclinations of these lines.

5.1 The Straight Lines are Considered to be the Axes of the Fixed Xoy System

- **Case 1. The trajectory of C point**

The bar AB slides with points A and B on the axes of the fixed xOy system. The locus of point C is required, when B slides on xx. The resulting mechanism is given in Fig. 5.1, being of type P-RRP.

The following relations are written:

$$x_A = x_B + AB\cos\alpha = 0 \tag{5.1}$$

$$y_A = AB\sin\alpha \tag{5.2}$$

I. Popescu et al., *Problems of Locus Solved by Mechanisms Theory*,
Springer Tracts in Mechanical Engineering,
https://doi.org/10.1007/978-3-030-63079-9_5

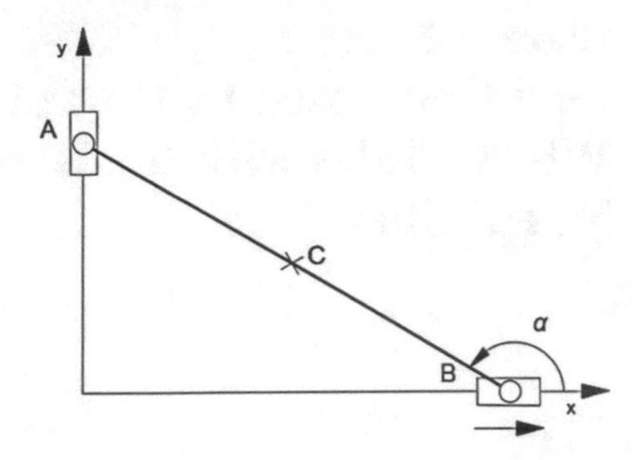

Fig. 5.1 The resulting mechanism

AB = 60 mm and BC = 20 mm are chosen, and the stroke of B is adopted between −100… +100 mm, but by imposing the length AB, the mechanism will only work in a certain subinterval of the proposed one, visible in the diagrams below (XB max = AB).

From the first relation it is determined cos α, then sin α is calculated, intervening the + and—signs. From the second relation Y_A results.

For BC = 20 mm, we have the locus in Fig. 5.2, that is a half ellipse, resulting in the + sign. For the—sign the half ellipse in Fig. 5.3 resulted, and for both signs we obtained the ellipse of Fig. 5.4. In the case of a physically achieved mechanism, it is positioned first above the x axis and then below the x axis.

If the point C is in the middle of the bar AB, then its locus is a circle (Fig. 5.5).

In the case of BC = 20 mm, the successive positions of the mechanism are those in Fig. 5.6. The coordinates of point C are in this case variable as in Fig. 5.7.

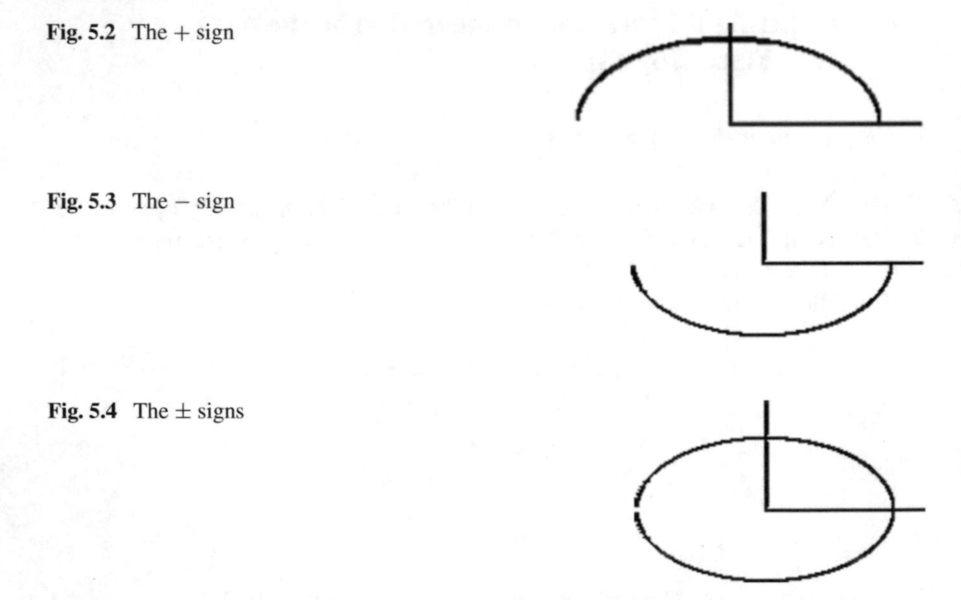

Fig. 5.2 The + sign

Fig. 5.3 The − sign

Fig. 5.4 The ± signs

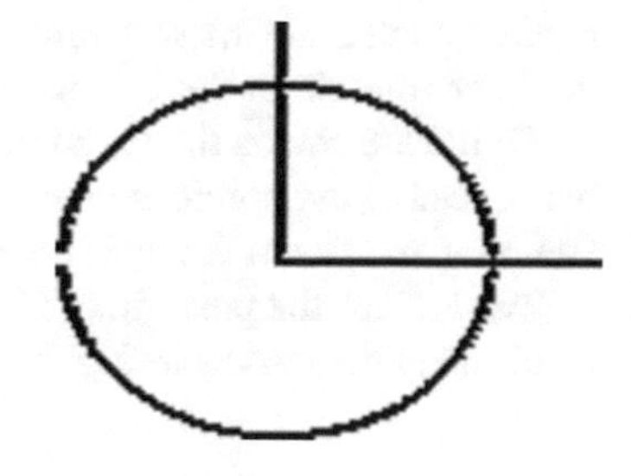

Fig. 5.5 BC = 0, 5.AB

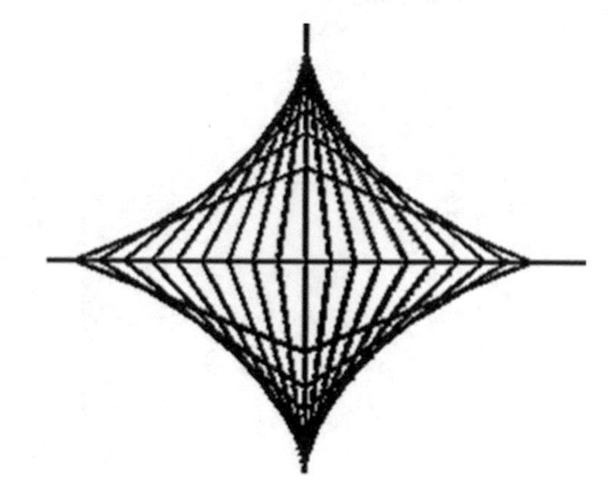

Fig. 5.6 Successive positions

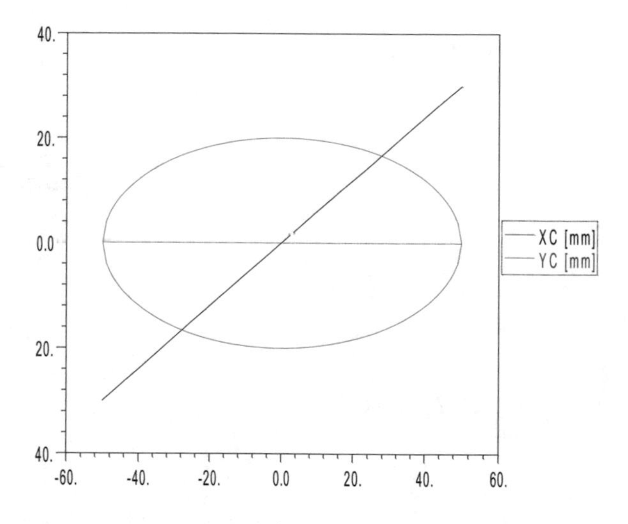

Fig. 5.7 X_C, Y_C according to X_B

It is observed that X_C varies linearly, hanging in a loop, and Y_C varies non-linearly, the curve being similar to the ellipse. For the sign + the curve of Y_C is the upper branch, then it passes to the sign—that is, from the right end of the drawn curve we reach the left end (the software unites these extreme values by a line) and we get to the lower branch of the YC curve.

Figure 5.8 shows the variation of the stroke of point A on the ordinate, the curve being read as the previous one: from the left end we reach the top branch to the right end, then returns to the left end and the lower branch is obtained.

The loci for the position of C from 10 to 10 mm along the length AB were drawn, resulting in the curves in Fig. 5.9 and then from 5 to 5 mm thus obtaining Fig. 5.10.

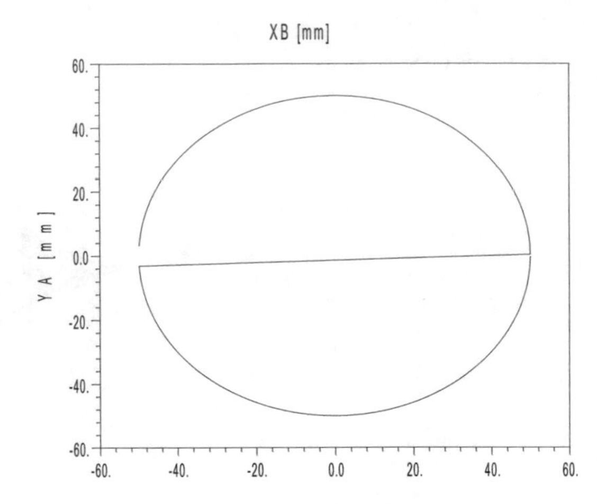

Fig. 5.8 YA variation

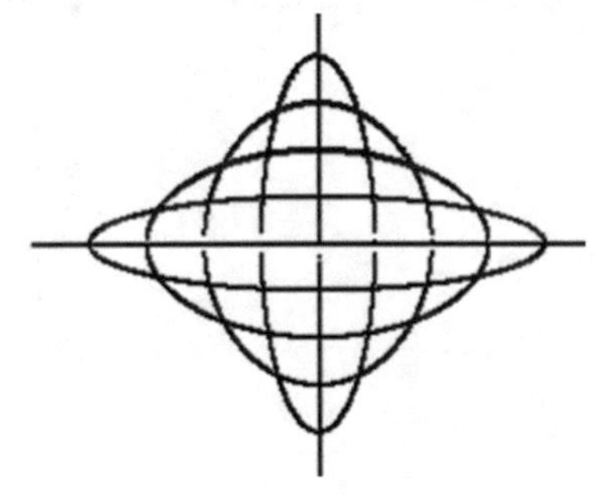

Fig. 5.9 Loci of points C (9 mm pace)

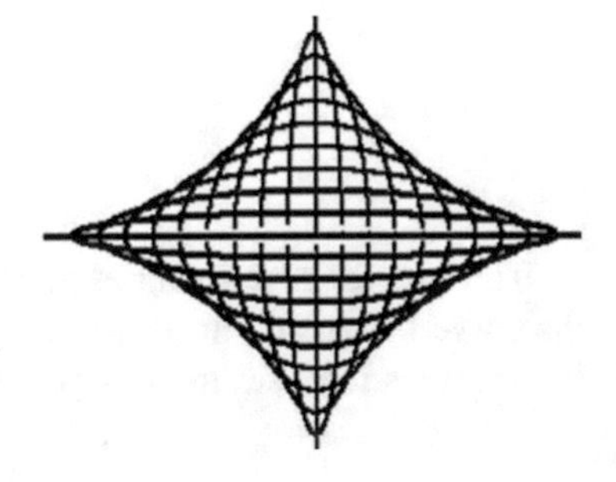

Fig. 5.10 Loci of points C (5 mm pace)

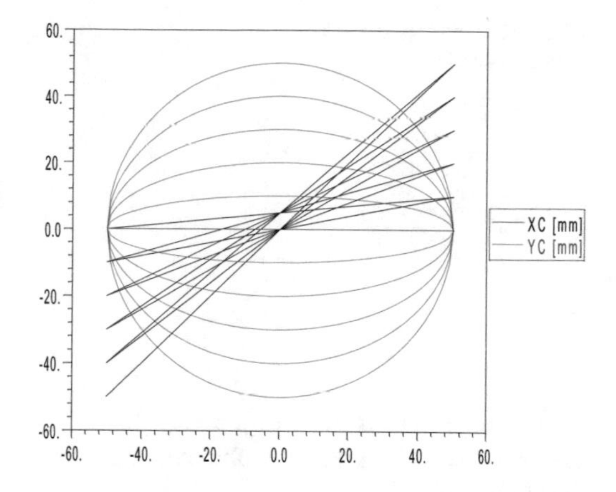

Fig. 5.11 Coordonates for point C (9 mm pace)

Figure 5.11 gives the variation of the coordinates of points C spaced by 10 mm, the diagram being read as above.

- **Case 2. The trajectory of D point**

If from A a parallel goes to the abscissa and from B a parallel to the ordinate, they intersect at point D. The mechanism becomes the one in Fig. 5.12, where the rotational torque in D is structurally parasitic.

The relations for point D are written:

$$x_D = x_B \tag{5.3}$$

$$y_D = y_A \tag{5.4}$$

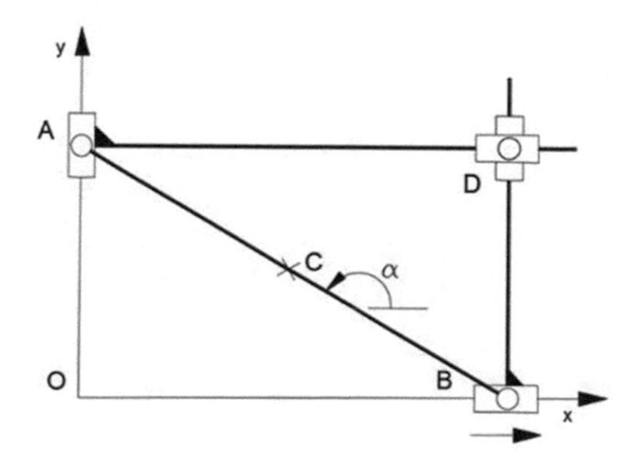

Fig. 5.12 The new mechanism

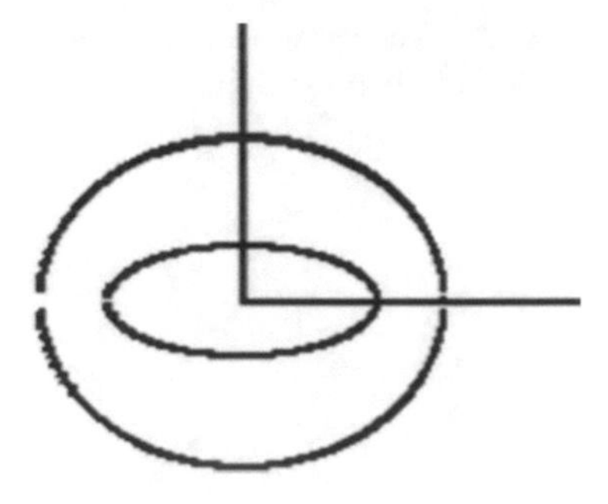

Fig. 5.13 The trajectories of points C and D

The trajectory of D in Fig. 5.13, which also shows the trajectory of C (the inner one) has been obtained. The obtained curve is a circle that includes inside the ellipse drawn by point C.

- **Case 3. The length AB is variable**

The length of AB is considered variable, i.e. AB $= q^*x_B$, where q is a coefficient theoretically adopted and concretely achieved by electromagnetic actuation. It is also used Fig. 5.12. From Eq. 5.1 it results the value of $\cos\alpha = \pm\, x_B/AB$ and having AB $= q^*x_B$ you get a $\cos\alpha = \pm\, 1/q$, therefore $q > 1$ to result in values for α.

The trajectories in this case are straight lines. Figure 5.14 shows the trajectory of C parallel to the abscissa and the trajectory of D, an inclined line, for the + sign in front of the radical, having $q = 2$. For both signs the trajectories in Fig. 5.15 are given.

It is found that increasing q extends the distances from the abscissa for the trajectory of C, and the lines representing the trajectory of D approach the ordinate (Fig. 5.16).

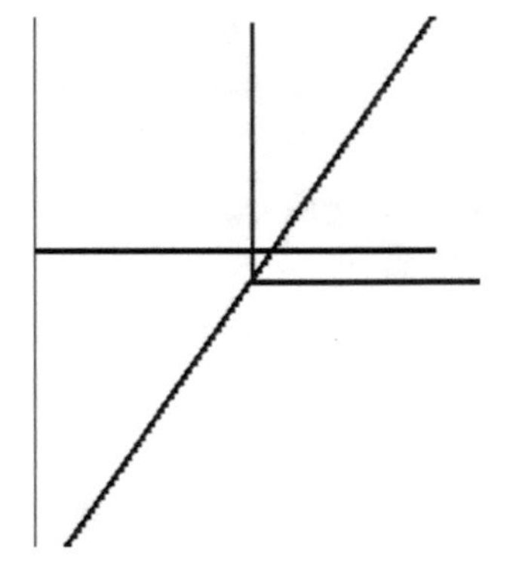

Fig. 5.14 The trajectories of C and D for the sign + and $q = 2$

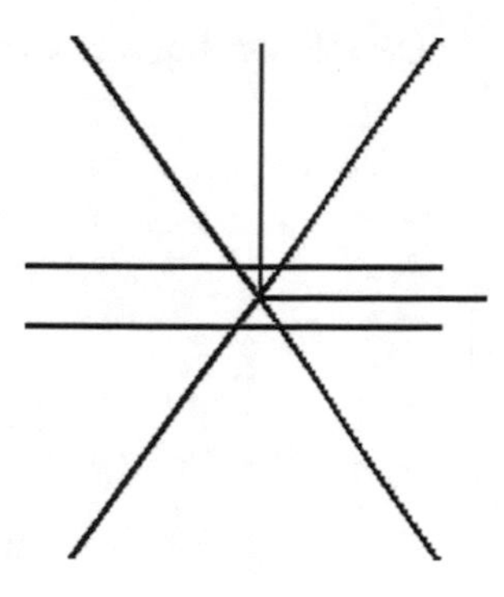

Fig. 5.15 The trajectories of C and D for both signs at q = 2

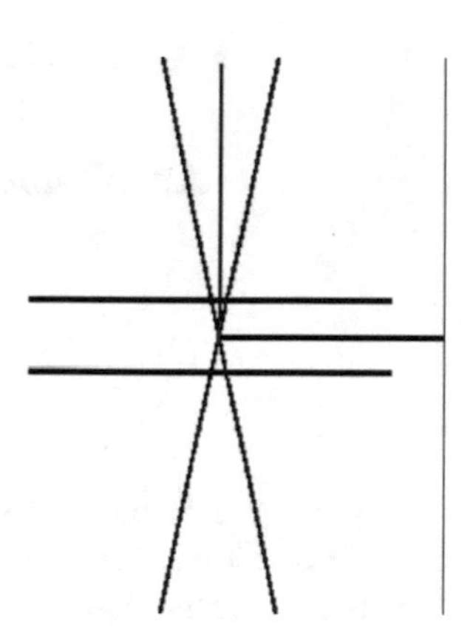

Fig. 5.16 The trajectories of C and D for both signs at q = 5

5.2 The Straight Lines Are Arbitrary

Consider the non-parallel straight lines OA and EB, fixed, on which the ends of the line AB slide (Fig. 5.17).

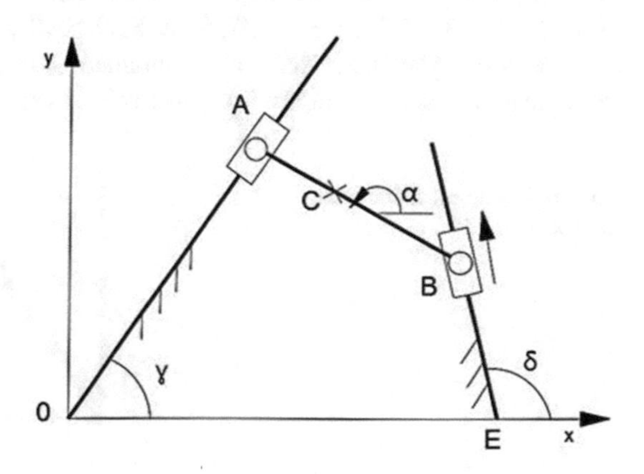

Fig. 5.17 AB slides non-parallel straight lines

The relationships are written:

$$x_B = x_E + EB\cos\delta \tag{5.5}$$

$$y_B = y_E + EB\sin\delta \tag{5.6}$$

$$x_A = x_B + AB\cos\alpha = OA\cos\gamma \tag{5.7}$$

$$y_A = y_B + AB\sin\alpha = OA\sin\gamma \tag{5.8}$$

$$OA = (y_B + AB\sin\alpha)/sin\gamma = (x_B + AB\cos\alpha)/\cos\gamma \tag{5.9}$$

$$ABcos\gamma\sin\alpha + y_B\cos\gamma = x_B\sin\gamma + ABsin\gamma\cos\alpha \tag{5.10}$$

$$x_C = x_B + BC\cos\alpha \tag{5.11}$$

$$y_C = y_B + BC\sin\alpha \tag{5.12}$$

From Eqs. 5.5 and 5.6 result the coordinates of B at the cycling of the EB stroke. From Eqs. 5.7 and 5.8 we obtain Eq. 5.9, reaching the trigonometric Eq. 5.10 from which it results the angle α. With Eqs. 5.11 and 5.12 the coordinates of the generating point are obtained.

For the control position the mechanism of Fig. 5.18 was obtained.

For the initial data: AB = 70; BC = 40; XE = 105;$\gamma = 60$;$\delta = 120$ resulted in the trajectory of Fig. 5.19.

Next, the inclinations of the two fixed lines were modified, resulting in many trajectories given below, each specifying the values of the angles of inclination for these lines (Figs. 5.20, 5.21, 5.22, 5.23, 5.24, 5.25, 5.26, 5.27, 5.28, 5.29, 5.30, 5.31, 5.32, 5.33, 5.34, 5.35, 5.36, 5.37, 5.38, 5.39, 5.40, 5.41, 5.42 and 5.43).

The same system of axes was maintained. Many types of curves have resulted, of a great diversity of shapes. For some sets of values, no curves were obtained, the

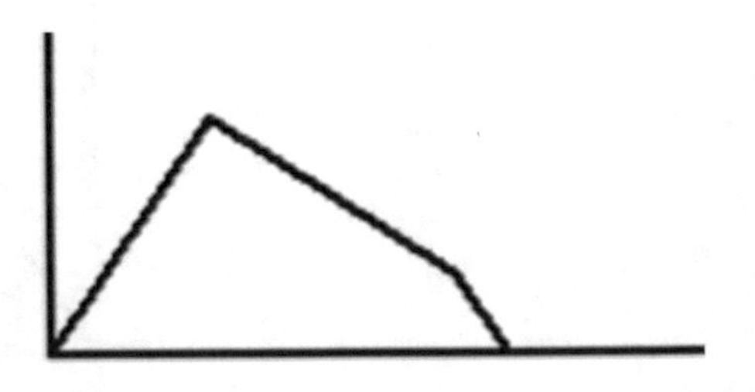

Fig. 5.18 The mechanism in one position

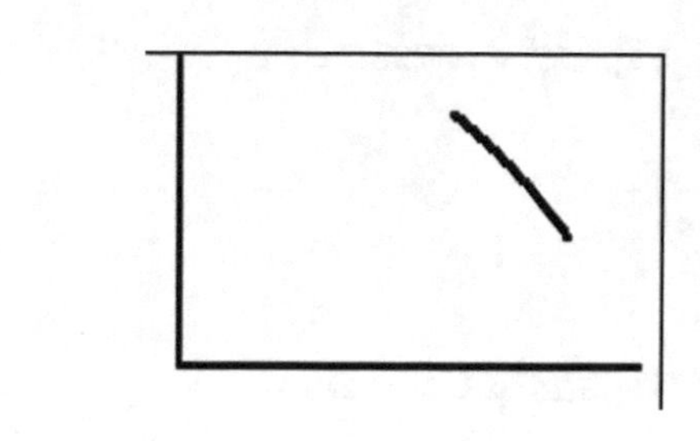

Fig. 5.19 The trajectory for the initial data

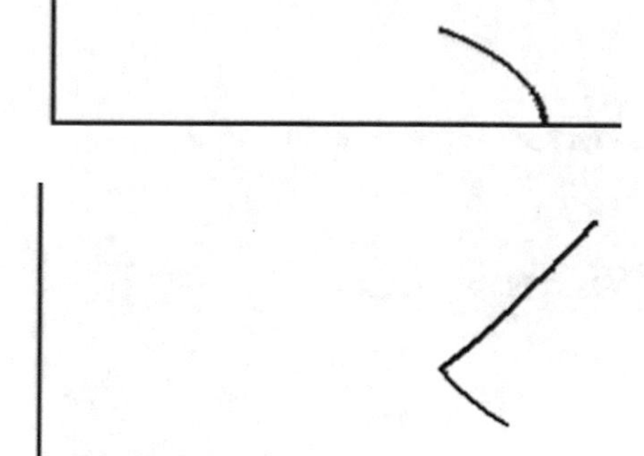

Fig. 5.20 $\gamma = 90; \delta = 0$

Fig. 5.21 $\gamma = 30; \delta = 80$

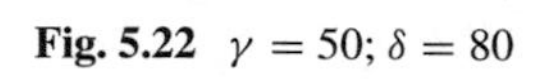

Fig. 5.22 $\gamma = 50; \delta = 80$

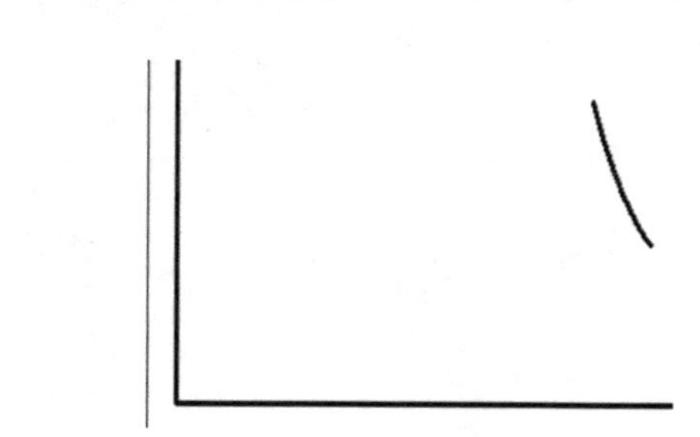

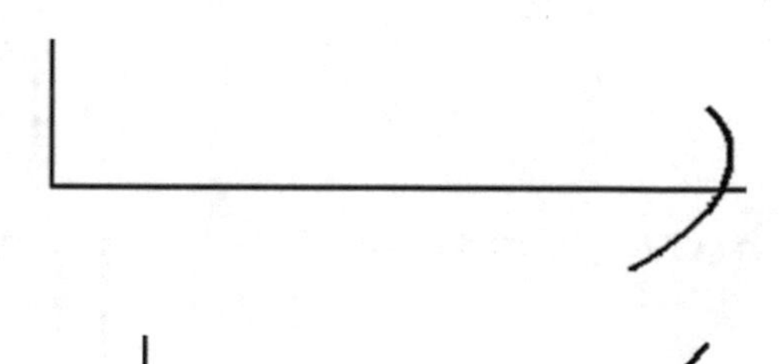

Fig. 5.23 $\gamma = 180; \delta = 80$

Fig. 5.24 $\gamma = 220; \delta = 80$
80

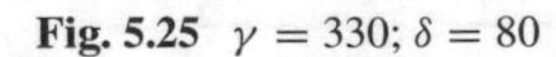

Fig. 5.25 $\gamma = 330; \delta = 80$

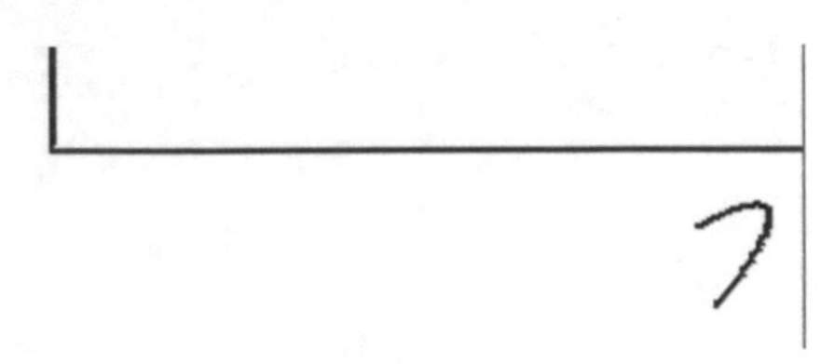

Fig. 5.26 $\gamma = 60; \delta = 10$

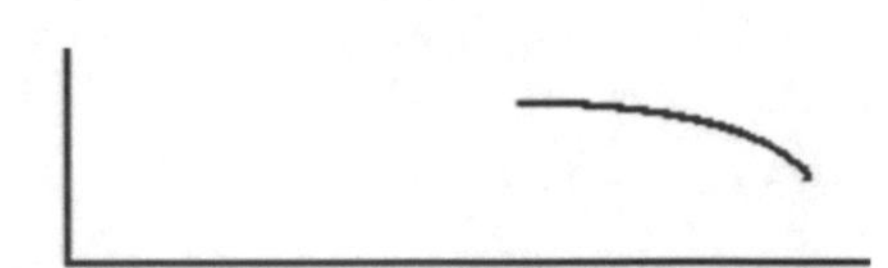

Fig. 5.27 $\gamma = 60; \delta = 30$

Fig. 5.28 $\gamma = 60; \delta = 100$

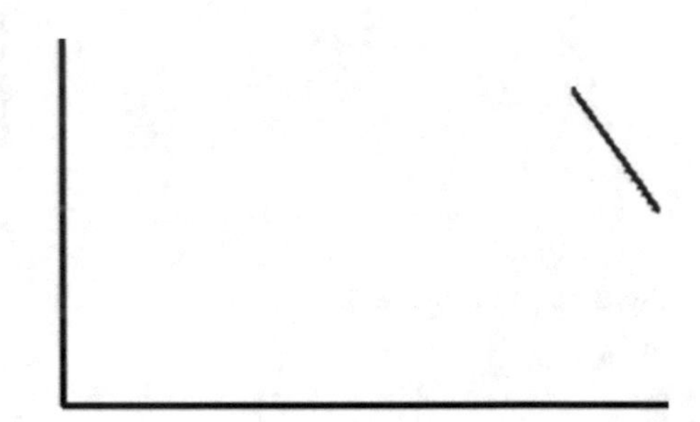

Fig. 5.29 $\gamma = 60; \delta = 130$

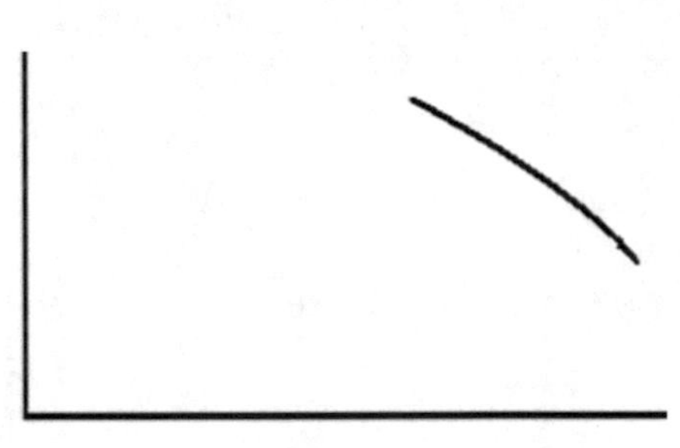

Fig. 5.30 $\gamma = 60; \delta = 180$

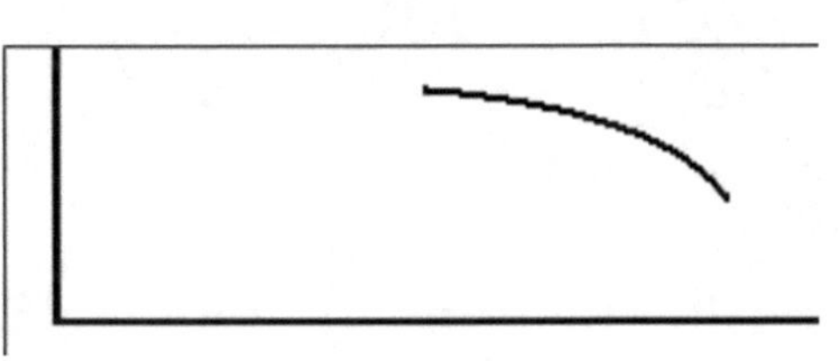

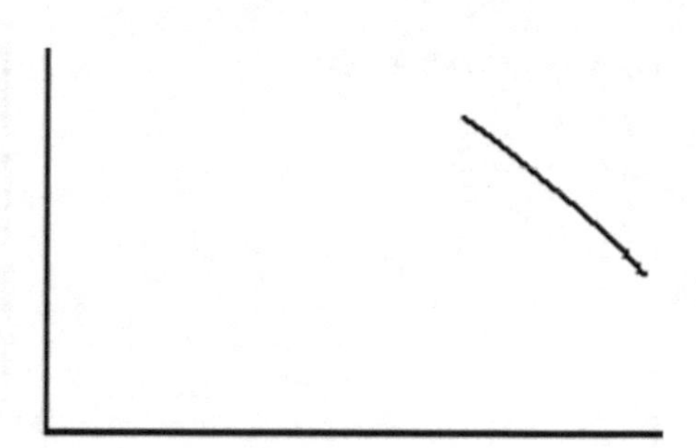

Fig. 5.31 $\gamma = 60; \delta = 120$

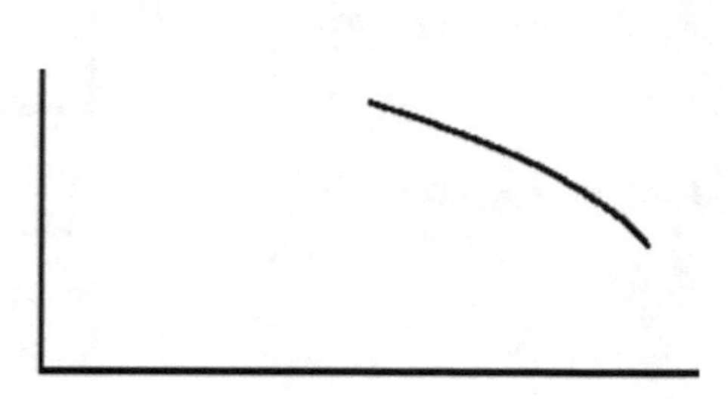

Fig. 5.32 $\gamma = 60; \delta = 330$

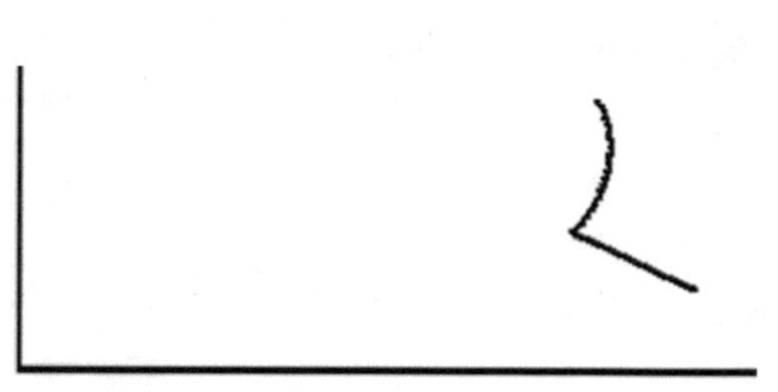

Fig. 5.33 $\gamma = 30; \delta = 120$

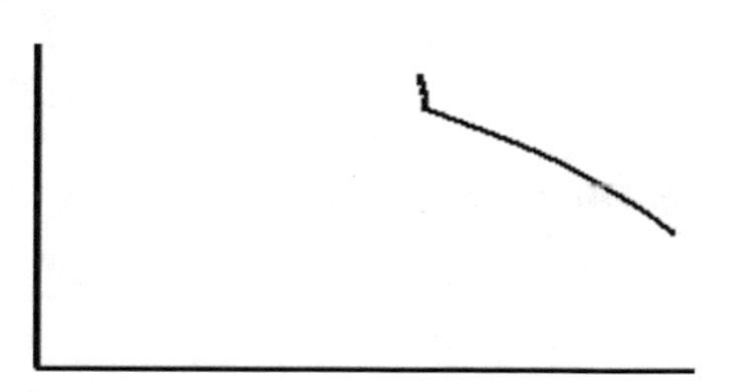

Fig. 5.34 $\gamma = 50; \delta = 140$

mechanism not working with those data due to the constant length of the element AB.

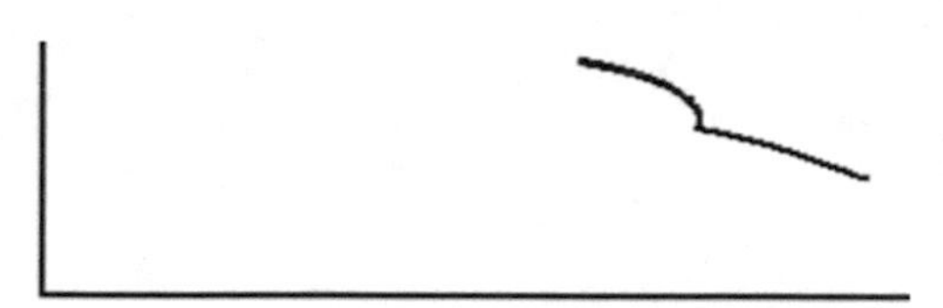

Fig. 5.35 $\gamma = 30; \delta = 150$

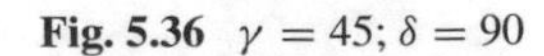

Fig. 5.36 $\gamma = 45; \delta = 90$

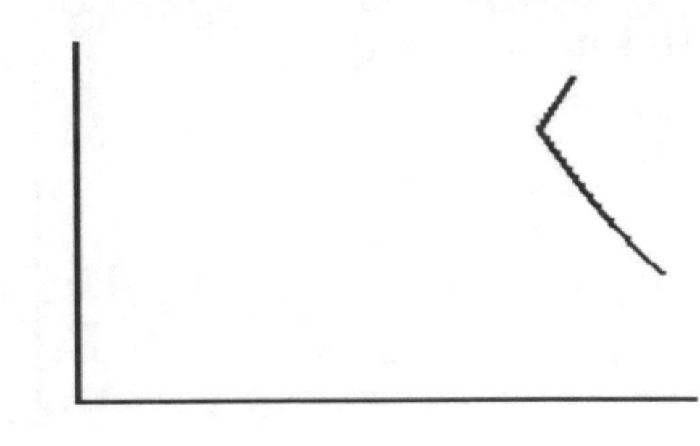

Fig. 5.37 $\gamma = 70; \delta = 200$

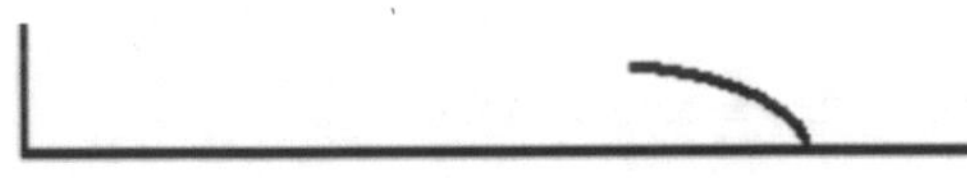

Fig. 5.38 $\gamma = 120; \delta = 45$

Fig. 5.39 $\gamma = 130; \delta = 55$

Fig. 5.40 $\gamma = 300; \delta = 50$

Fig. 5.41 $\gamma = 55; \delta = 330$

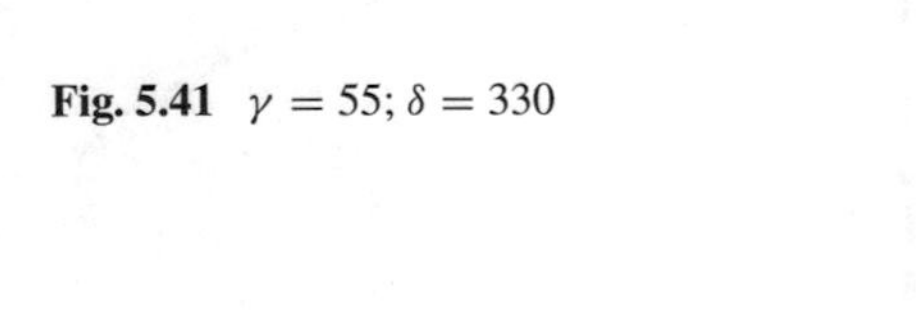

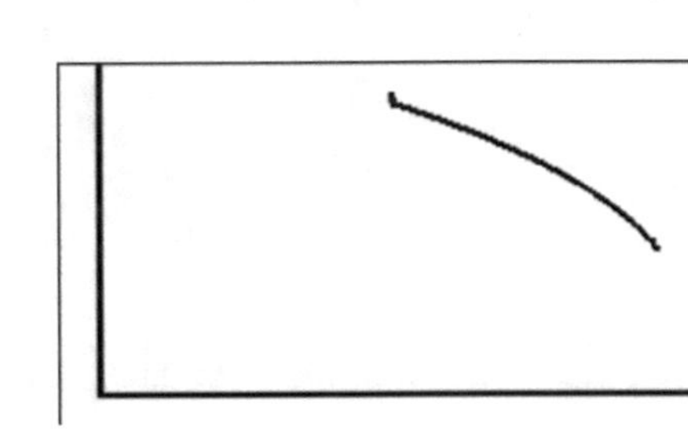

Fig. 5.42 $\gamma = 300; \delta = 20$

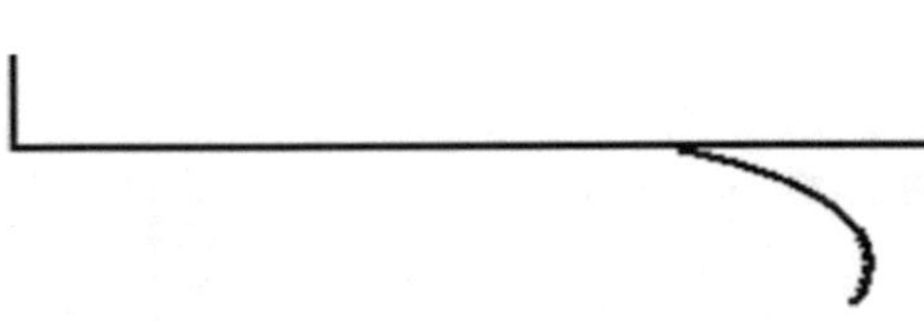

Fig. 5.43 $\gamma = 300; \delta = 180$

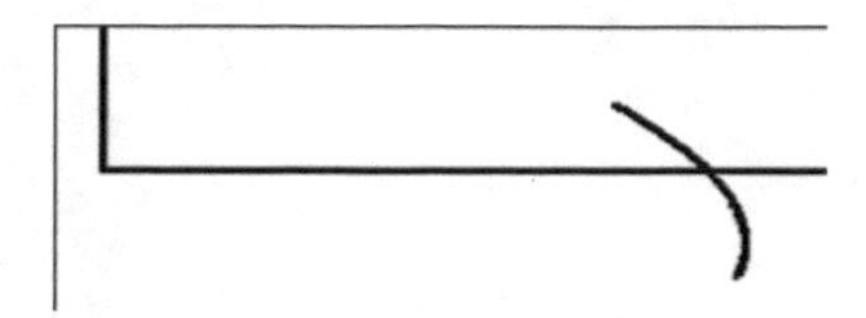

References

1. Popescu I, Sass L, Duţă A, Romanescu AE (2019) Double-heart curve generated by original mechanism. În: J Ind Des Eng Graph. Papers of the International Conference on Engineering Graphics and Design, IGEGD, 15–17 May 14 Issue 1 2019, Craiova, pp 41–46
2. Sass, Ludmila; Duţă, Alina; Popescu, Iulian—Nature gives us beauty, insoiring the technical creativity. J Ind Des Eng Graph. Papers of the International Conference on Engineering Graphics and Design, Section 3: Engineering Computer Graphics

Chapter 6
Loci Generated by Two Segment Lines Bound Between Them

Abstract It is considered a straight line that rotates around a fixed point, and at the free end it connects with another straight line that rotates around the same point. The mechanism has two leading elements [1], obtaining the trajectories of the end of the second line, with shapes dependent on the correlation of the two movements. Successive positions form aesthetic surfaces [2]. In another case, the first straight line slides on the abscissa, resulting in numerous trajectories and associated positions, also these one being aesthetic surfaces. In another case, the first line slides on the abscissa and the second on the first line, the trajectories being straight lines.

6.1 Case 1

The line segment AB rotates around the fixed-point A, and the segment BC rotates around the point B (Fig. 6.1). The locus of point C. is required.

The equivalent mechanism is given in Fig. 6.2. Rotation couplings have been fitted in A and B, their rotation angles being correlated. The following relations are written:

$$x_C - AB\cos\varphi + BC\cos\alpha \tag{6.1}$$

$$y_C = AB\sin\varphi + BC\sin\alpha \tag{6.2}$$

The mechanism consists of two successive steering elements, with correlated movements, thus of type RR. The linear correlation was considered, q being realized either with the electromagnetic processes or with assistive mechanisms (gears, belts). The geometrical places made depend on the values set for q.

Initial data: a = 40 mm, b = 50 mm, $\varphi = 0 \ldots 360°$.

For q = 0 a circle was obtained (Fig. 6.3).

Hereinafter the loci resulting for different values of q are given (Figs. 6.4 and 6.5).

I. Popescu et al., *Problems of Locus Solved by Mechanisms Theory*,
Springer Tracts in Mechanical Engineering,
https://doi.org/10.1007/978-3-030-63079-9_6

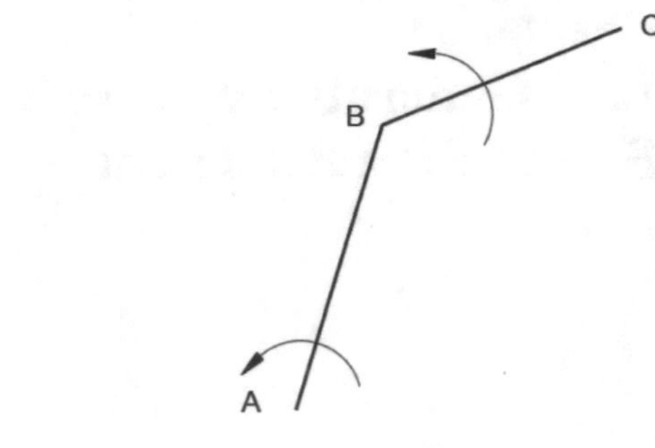

Fig. 6.1 The geometric construction

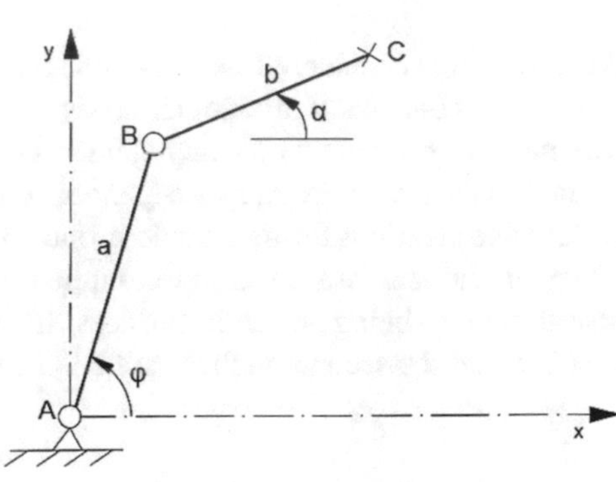

Fig. 6.2 The equivalent mechanism

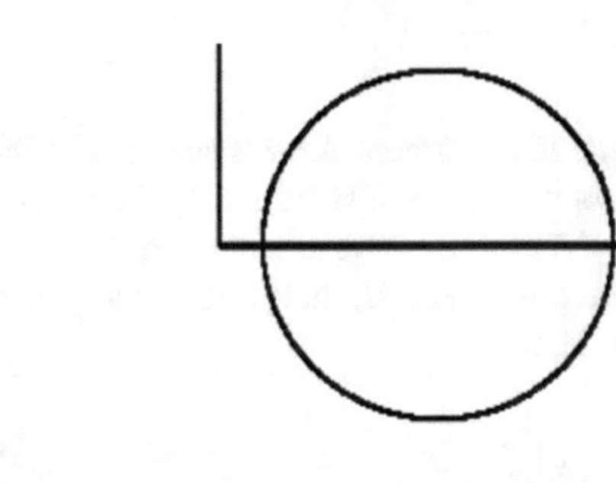

Fig. 6.3 $q = 0$

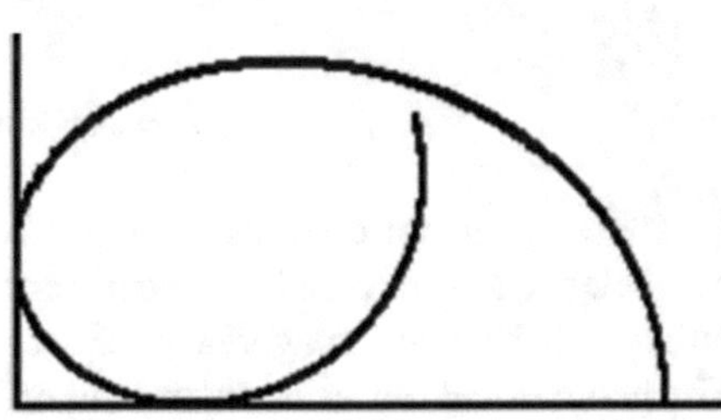

Fig. 6.4 $q = 0.2$

Fig. 6.5 $q = -0.2$

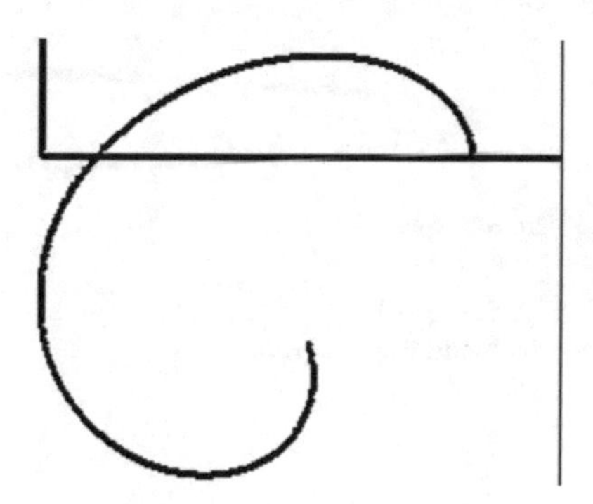

At $q = -0.2$ it means that AB rotates in the trigonometric direction and BC in the clockwise direction. The loci are different (Fig. 6.6 and 6.7).

For $q = 1$ we obtain a circle with radius AB + BC. For $q = -1$ an ellipse results (Figs. 6.8, 6.9, 6.10, 6.11 and 6.12).

For q with decimals, the curve is incomplete (Figs. 6.13, 6.14, 6.15, 6.16, 6.17, 6.18, 6.19, 6.20, 6.21, 6.22, 6.23, 6.24, 6.25, 6.26 and 6.27).

Many curves have resulted as loci, of great diversity, depending on the correlation between the movements of the two line segments.

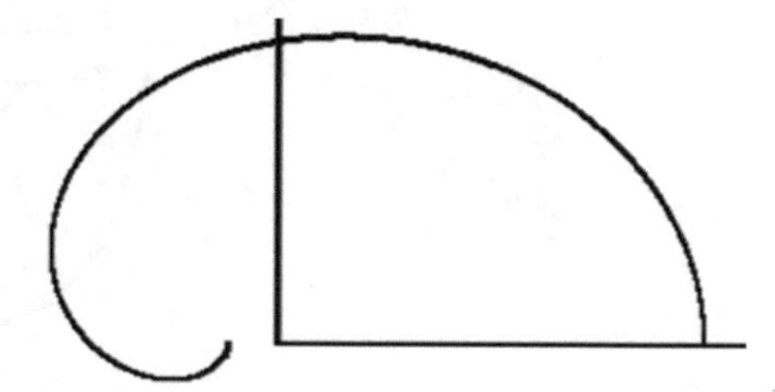

Fig. 6.6 $q = 0.5$

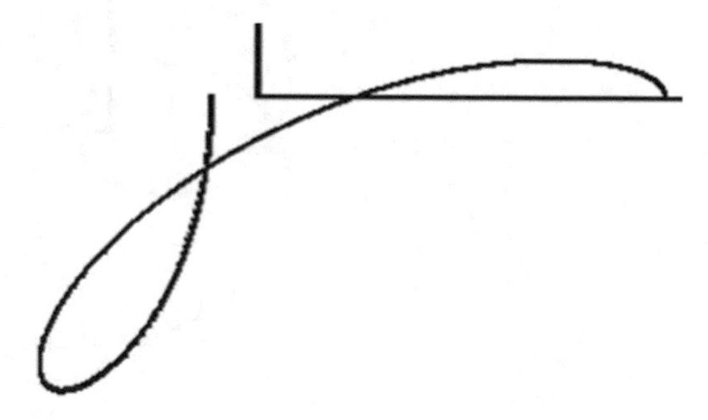

Fig. 6.7 $q = -0.5$

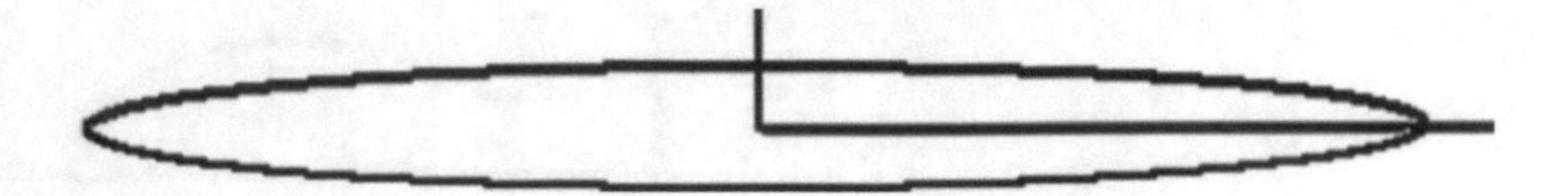

Fig. 6.8 q = −1

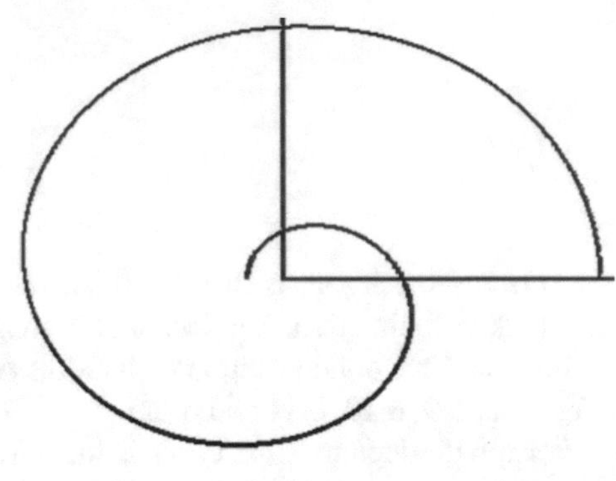

Fig. 6.9 q = 1.5 archimedical spiral type

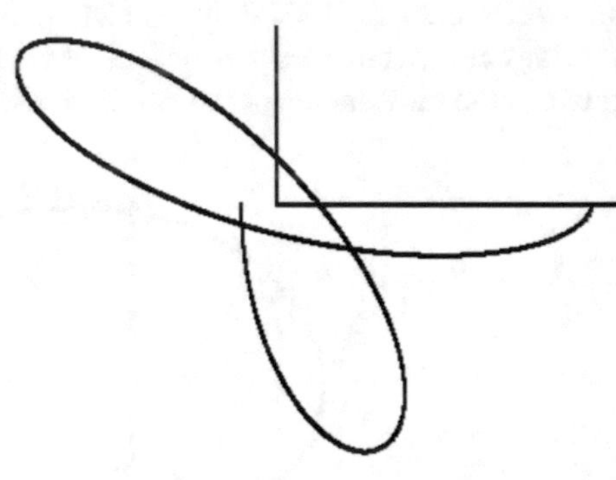

Fig. 6.10 q = −1.5

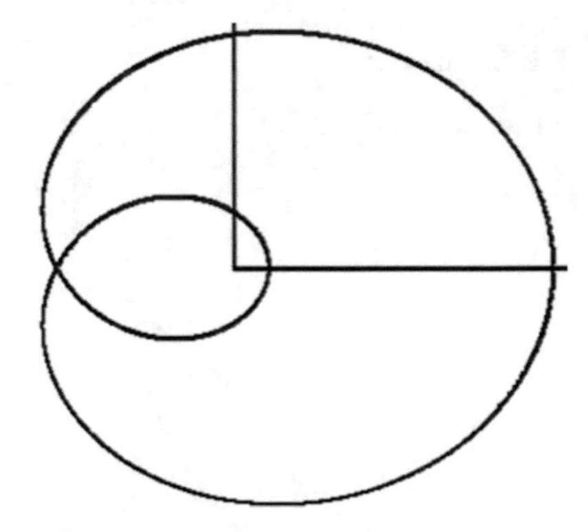

Fig. 6.11 q = 2 Pa's snail type

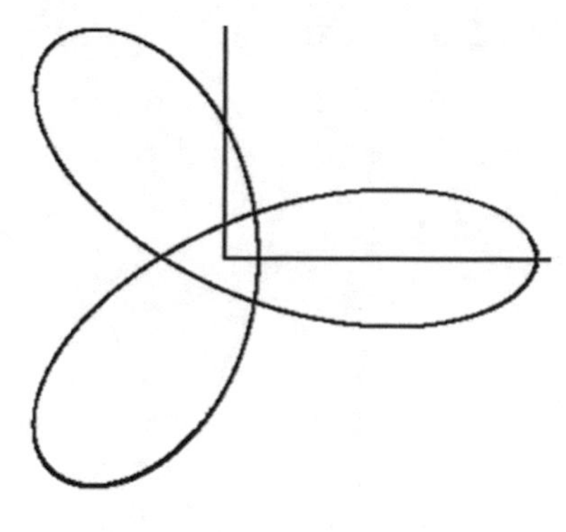

Fig. 6.12 q = −2

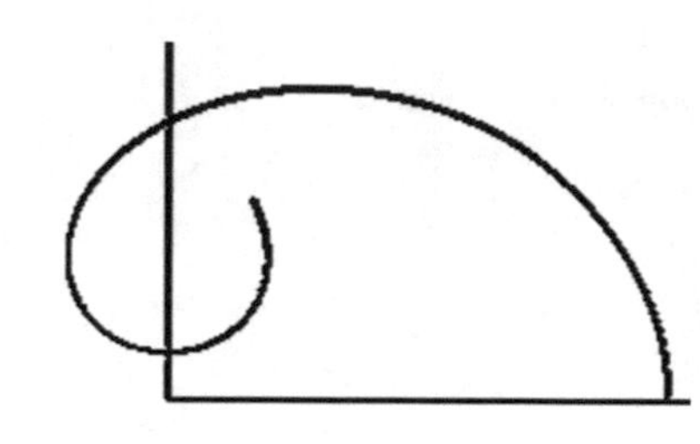

Fig. 6.13 q = 0.333

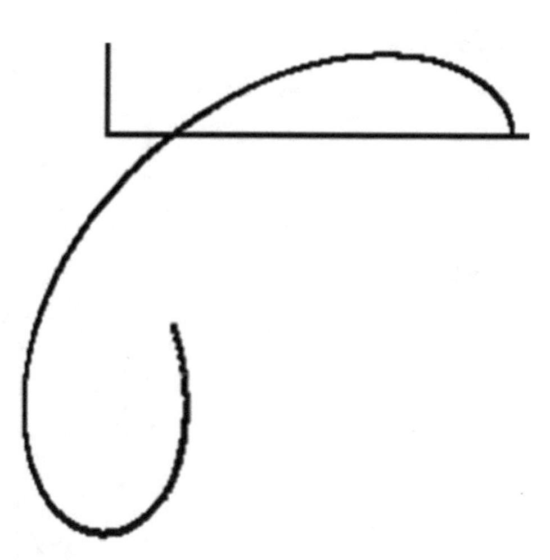

Fig. 6.14 q = −0.333

The successive positions of the bars AB and BC are given in the following images (Figs. 6.28, 6.29, 6.30, 6.31, 6.32, 6.33, 6.34, 6.35, 6.36, 6.37, 6.38, 6.39, 6.40 and 6.41).

It is found that depending on the correlation coefficient q between the two movements, the successive positions are completely different from one case to another. It is also noted that many of these images have a special aesthetics.

Fig. 6.15 q = 2.33

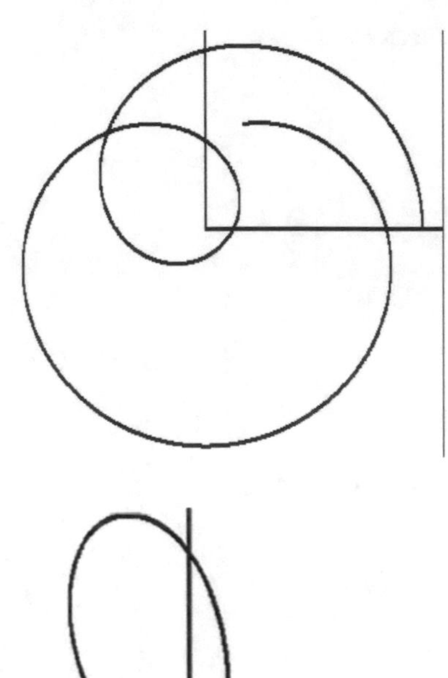

Fig. 6.16 q = −2.33

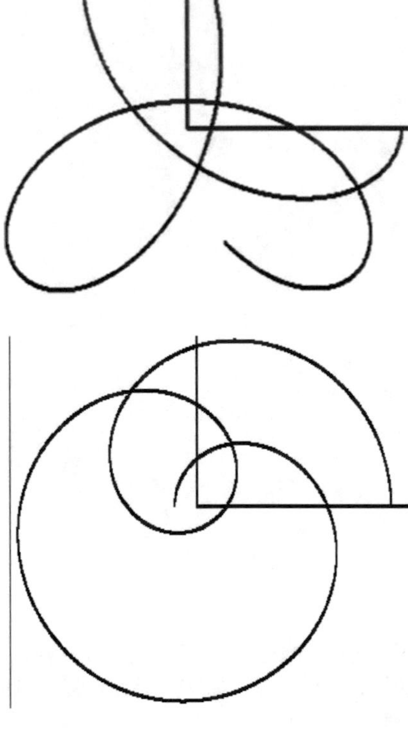

Fig. 6.17 q = 2.5

Fig. 6.18 q = −2.5

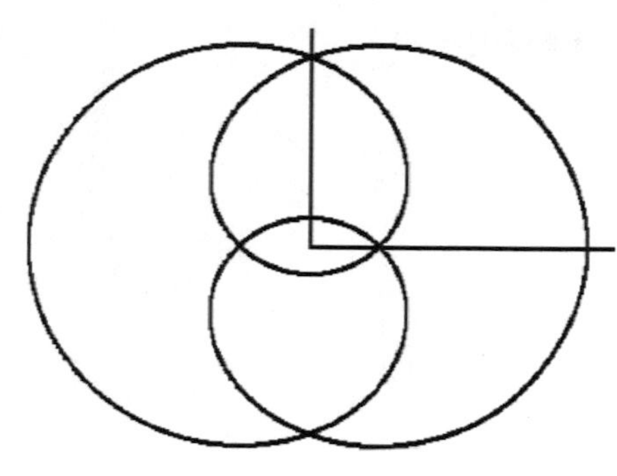

Fig. 6.19 q = 3

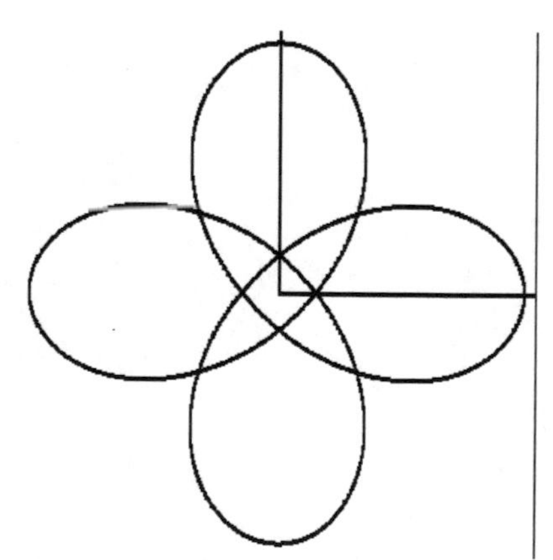

Fig. 6.20 q = −3

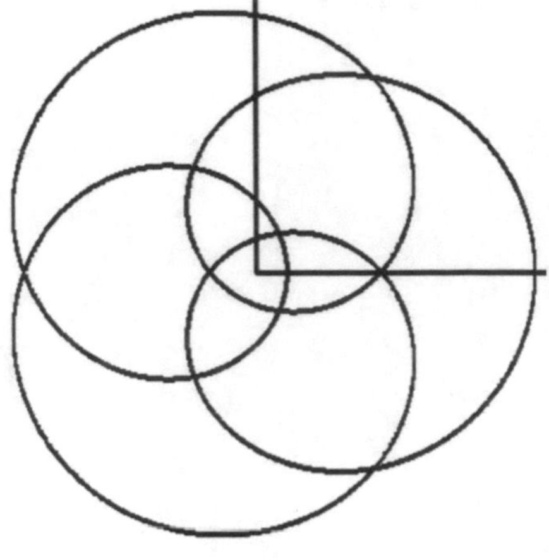

Fig. 6.21 q = 4

Fig. 6.22 q = −4

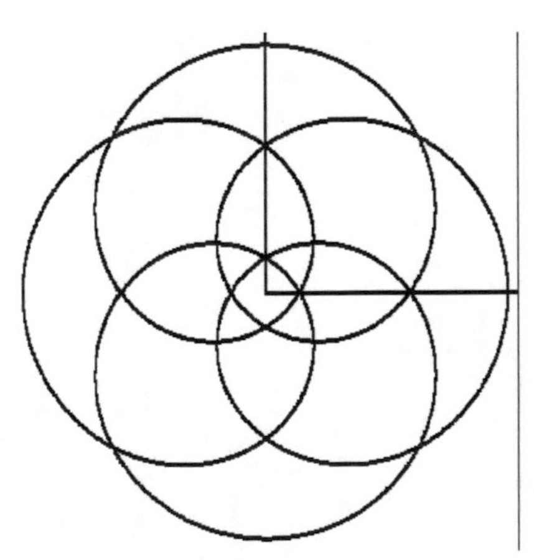

Fig. 6.23 q = 5

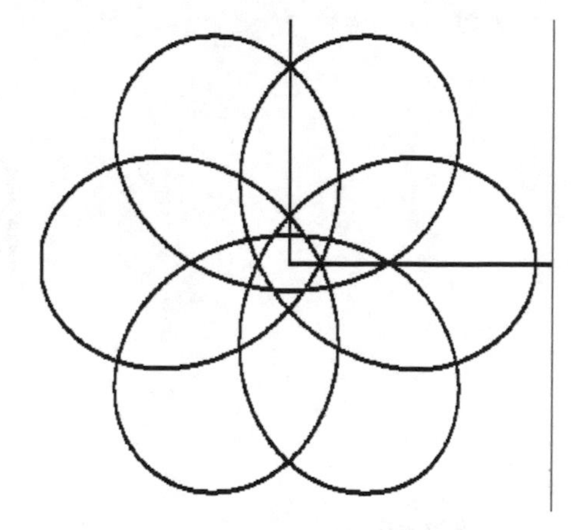

Fig. 6.24 q = −5

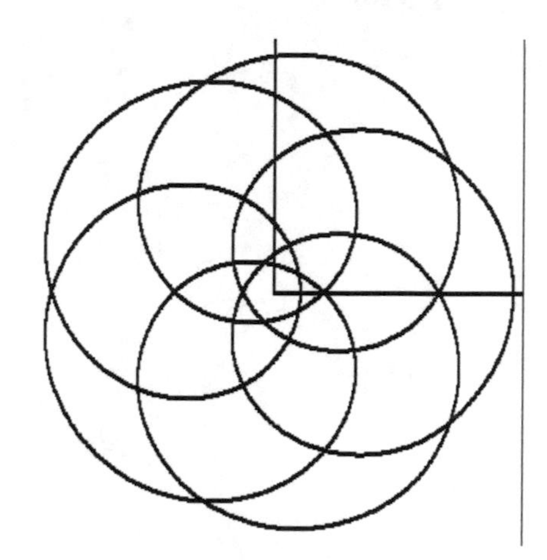

Fig. 6.25 q = 6

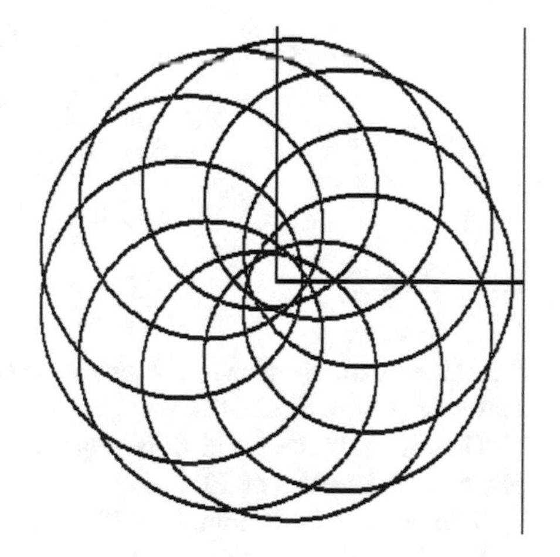

Fig. 6.26 q = 10

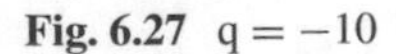

Fig. 6.27 q = −10

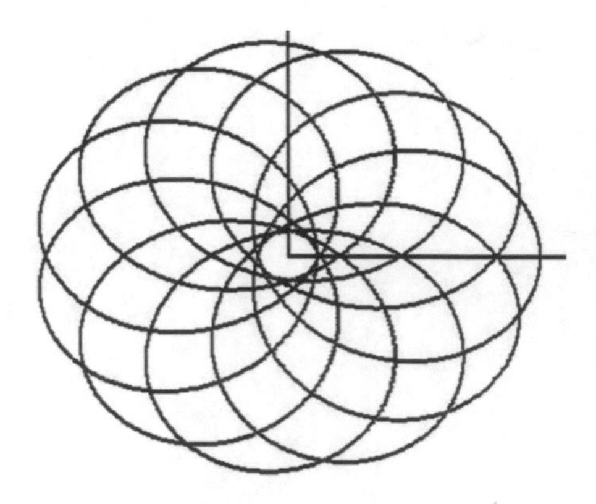

Fig. 6.28 q = 0

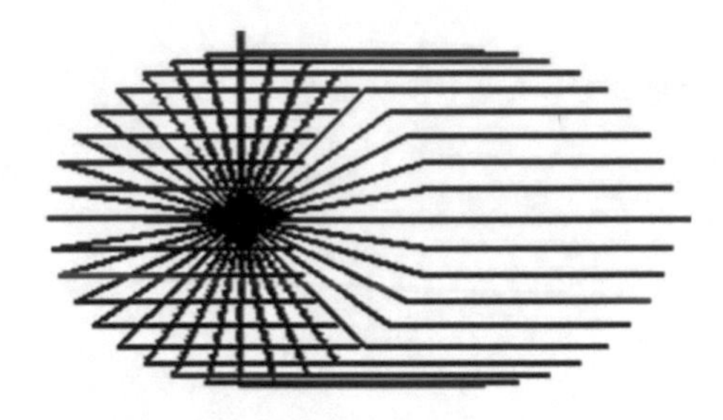

Fig. 6.29 q = 0.2

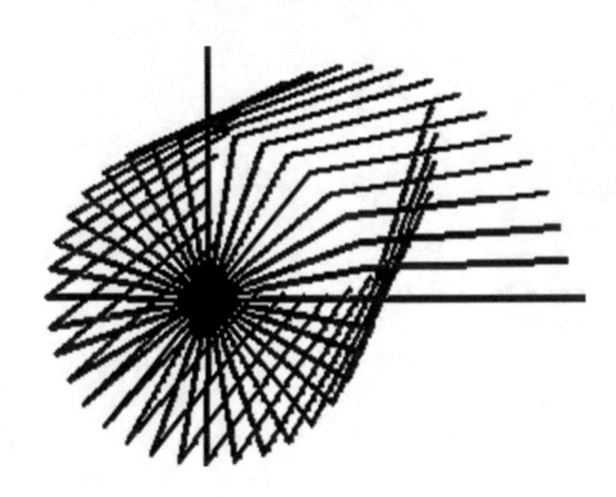

6.2 Case 2

It is started from the previous problem, except that the point A is no longer fixed, but moves on a line (Fig. 6.42).

The locus of the point C is requested. The equivalent mechanism is given in Fig. 6.43, being of type PR.

Initial data: a = 40 mm, b = 30 mm; $\gamma = 50°$; S = - 100… + 100 mm.

Based on Fig. 6.43 the following relations are written:

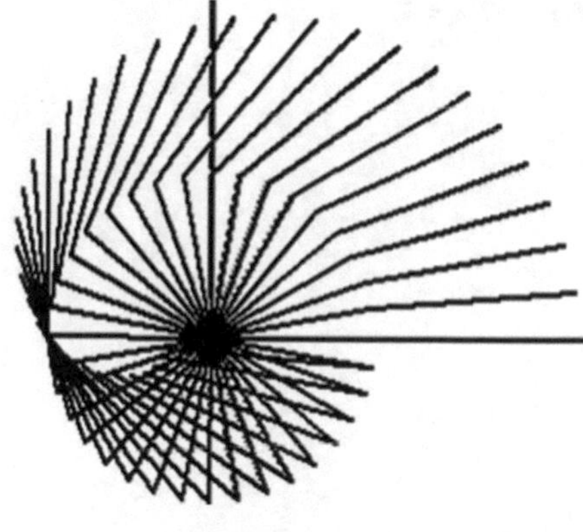

Fig. 6.30 q = 0.5

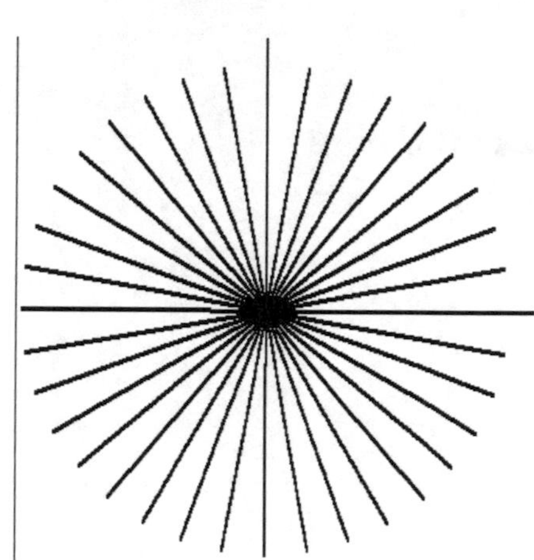

Fig. 6.31 q = 1

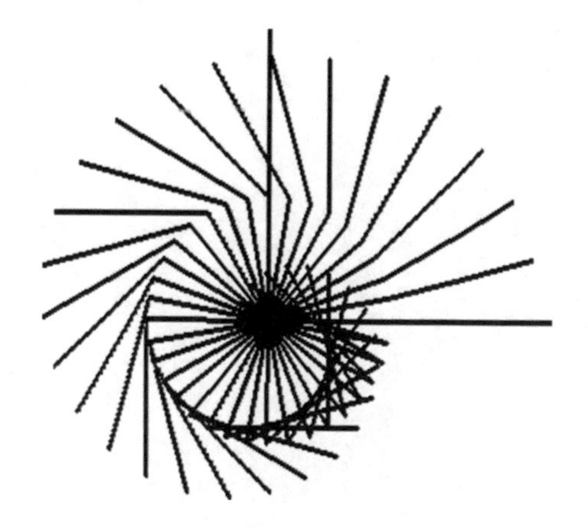

Fig. 6.32 q = 1.5

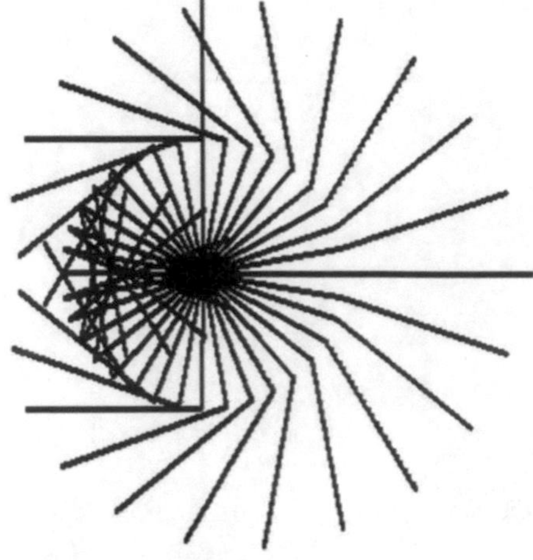

Fig. 6.33 q = 2

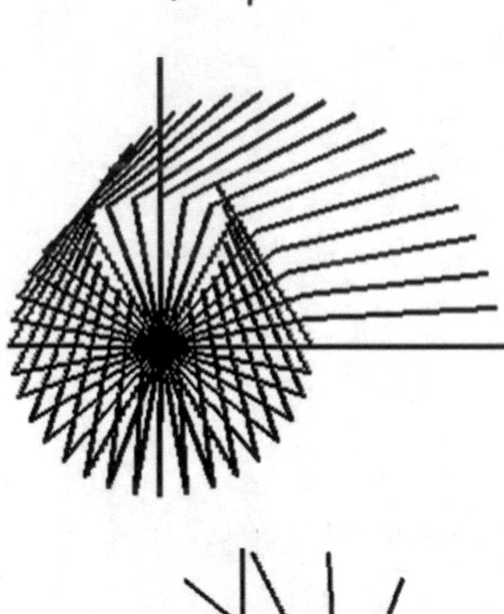

Fig. 6.34 q = 0.333

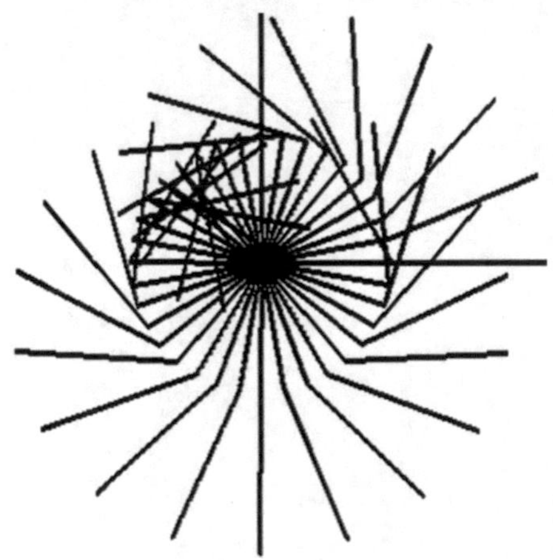

Fig. 6.35 q = 2.33

Fig. 6.36 q = 2.5

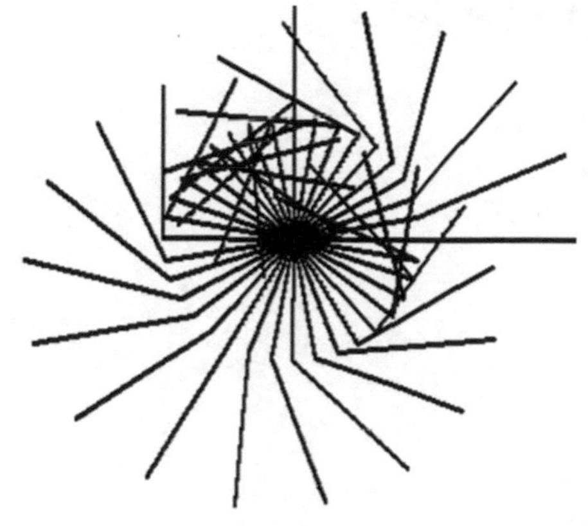

Fig. 6.37 q = 3

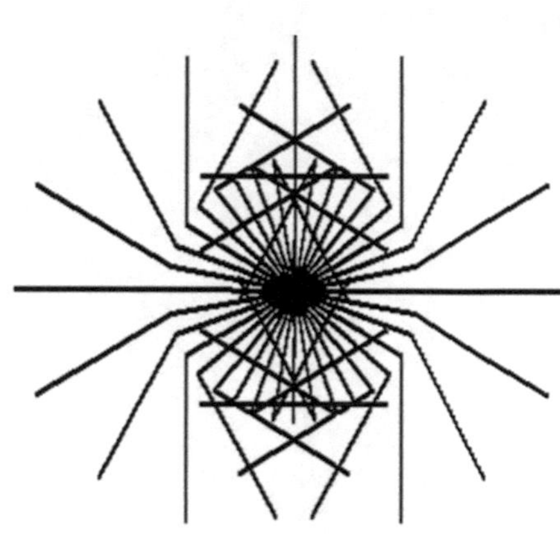

Fig. 6.38 q = 4

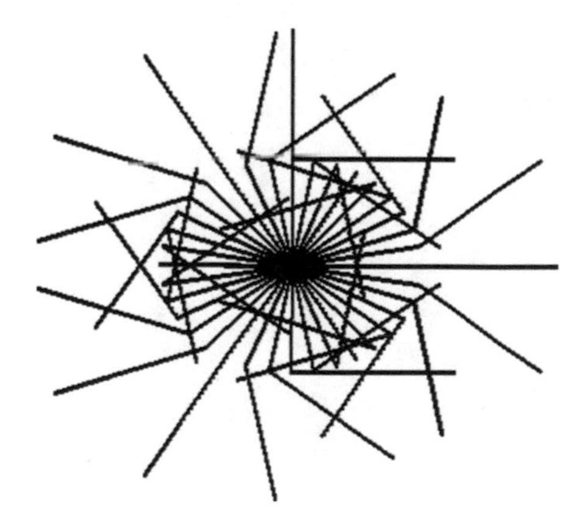

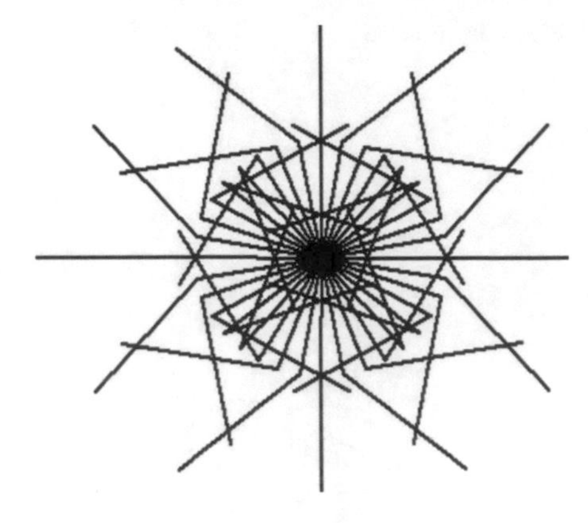

Fig. 6.39 q = 5

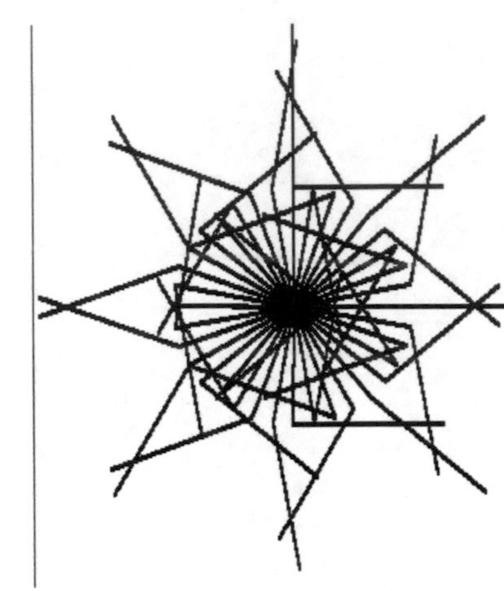

Fig. 6.40 q = 8

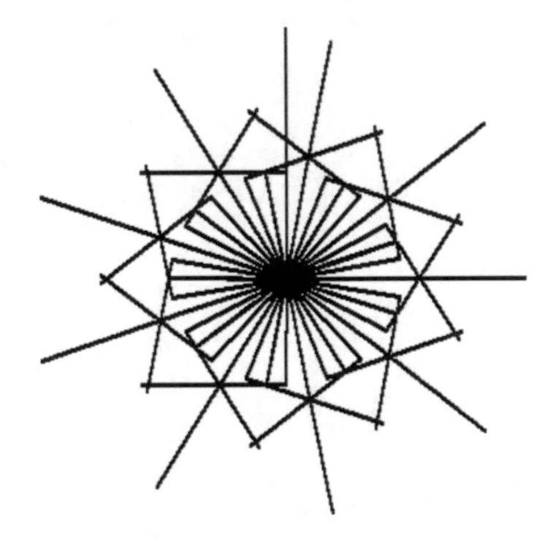

Fig. 6.41 q = 10

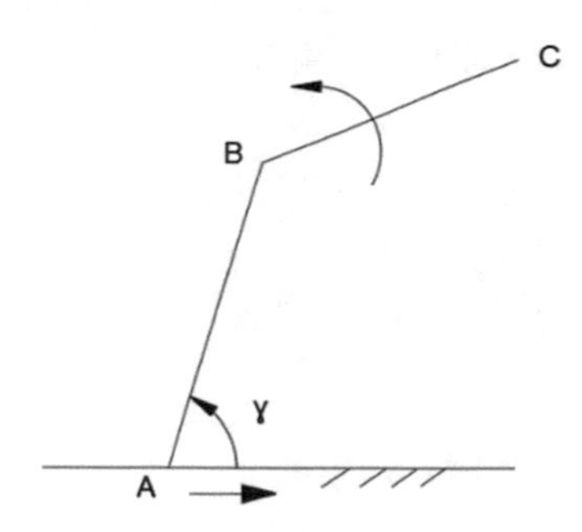

Fig. 6.42 The geometric construction

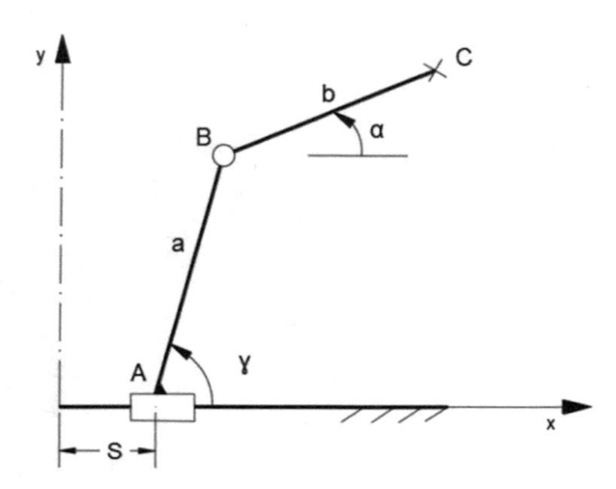

Fig. 6.43 The equivalent mechanism

$$x_C = s + a\cos\gamma + b\cos\alpha \tag{6.3}$$

$$y_C = a\sin\gamma + b\sin\alpha \tag{6.4}$$

$$\alpha = q.s \tag{6.5}$$

This time having high values for the S stroke, the angle α, measured in radians, must be small, so that the values of q will be low. Below are the loci for different values of q, positive and negative (Figs. 6.44, 6.45, 6.46, 6.47, 6.48, 6.49, 6.50, 6.51, 6.52, 6.53, 6.54, 6.55, 6.56, 6.57, 6.58, 6.59 and 6.60).

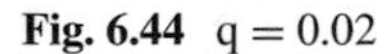

Fig. 6.44 q = 0.02

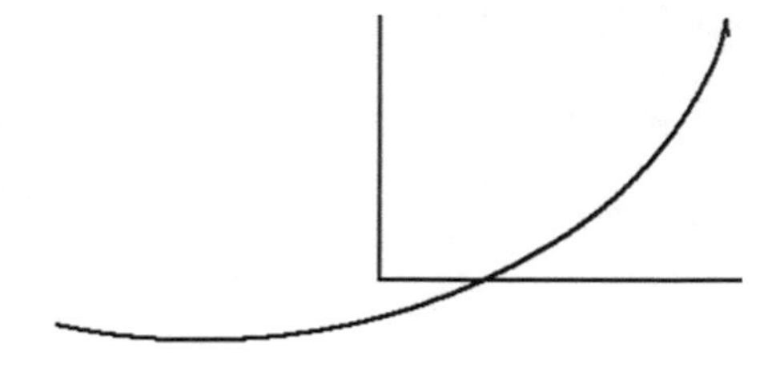

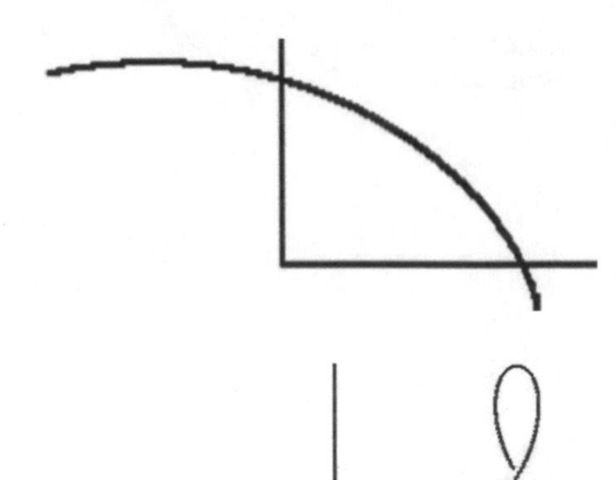

Fig. 6.45 q = −0.02

Fig. 6.46 q = 0.03

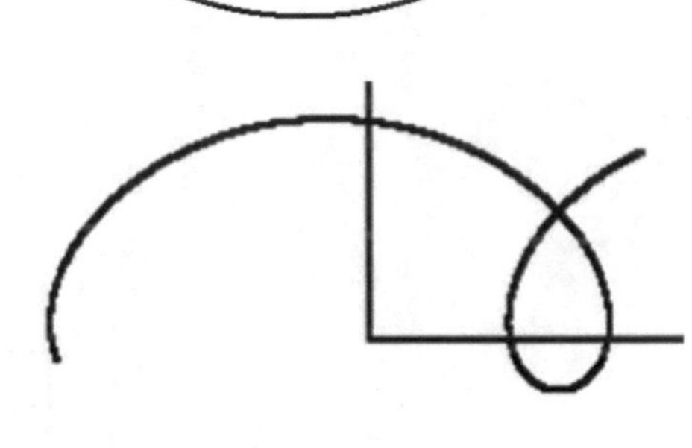

Fig. 6.47 q = −0.04

Fig. 6.48 q = 0.05

Fig. 6.49 q = 0.06

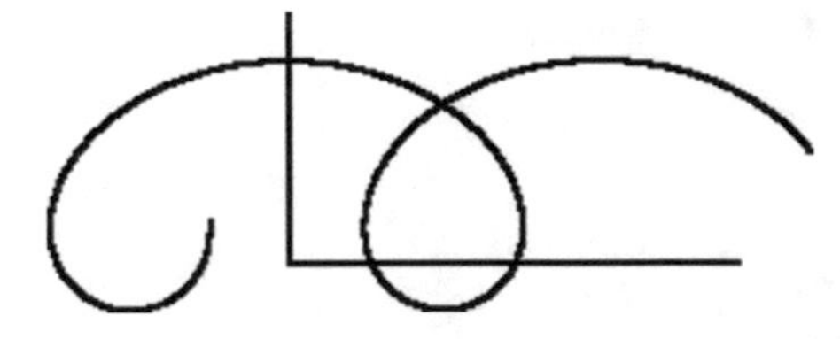

Fig. 6.50 q = −0.06

Fig. 6.51 $q = 0.07$

Fig. 6.52 $q = -0.08$

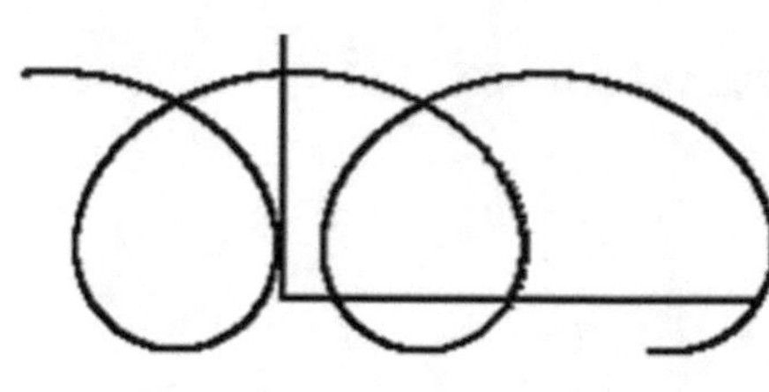

Fig. 6.53 $q = 0.09$

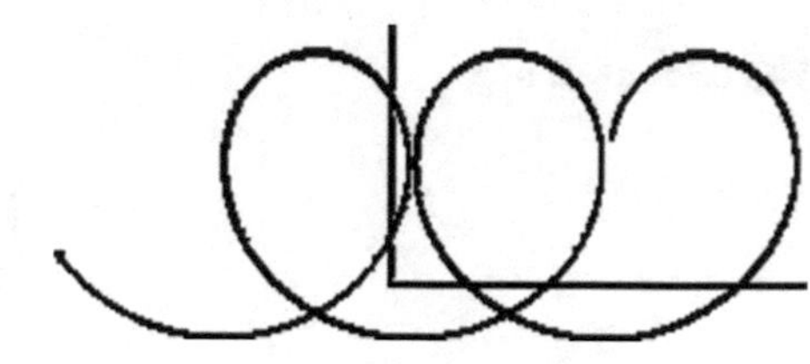

Fig. 6.54 $q = 0.1$

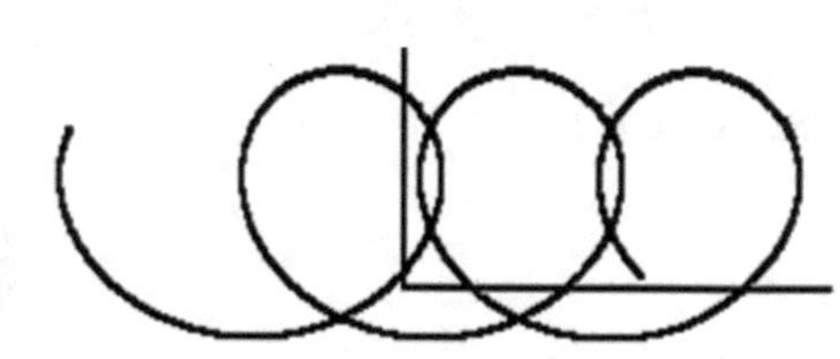

Fig. 6.55 $q = -0.1$

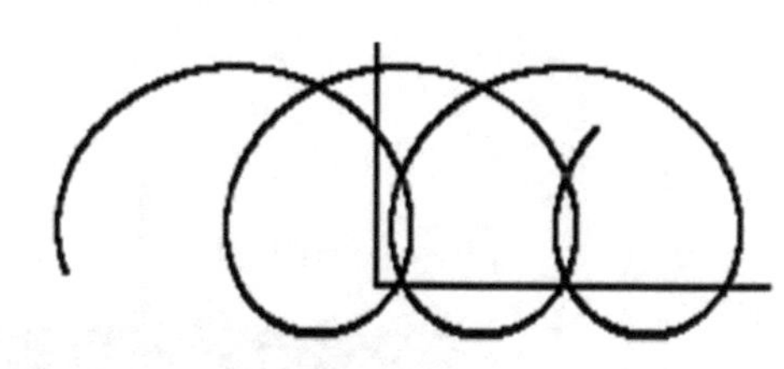

Fig. 6.56 $q = 0.15$

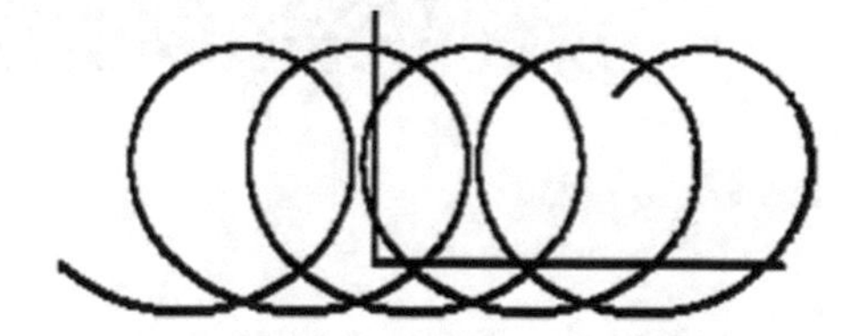

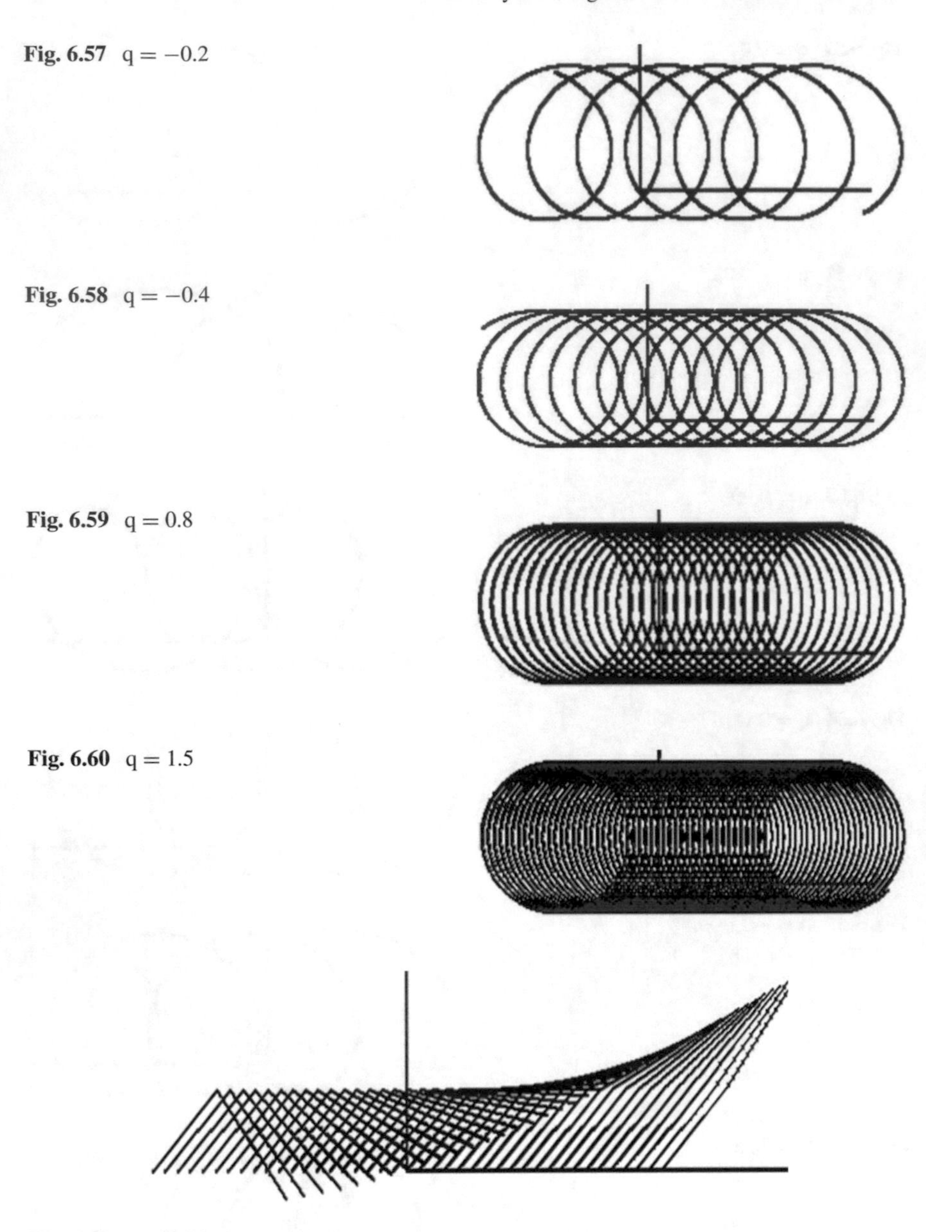

Fig. 6.57 q = −0.2

Fig. 6.58 q = −0.4

Fig. 6.59 q = 0.8

Fig. 6.60 q = 1.5

Fig. 6.61 q = 0.01

The resulting loci are orthocycloid-like curves with different numbers of loops at different values of q, with the upper peaks for q > 0 and the lower loops for q < 0. As q increases, a sequence of orthocycloid-type curves is observed.

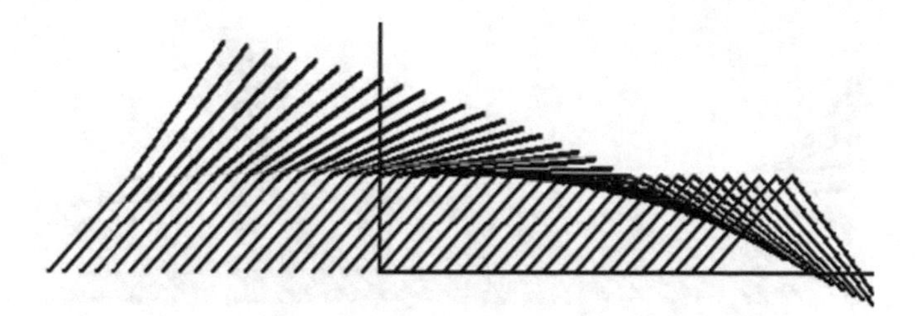

Fig. 6.62 $q = -0.01$

The successive positions of the elements for different values of q are presented below (Figs. 6.61, 6.62, 6.63, 6.64, 6.65, 6.66, 6.67, 6.68, 6.69, 6.70, 6.71, 6.72, 6.73 and 6.74).

These figures with successive positions of the line segments are very interesting and some are even aesthetic. We observe the symmetries in some areas for the same values of q, one positive and the other negative.

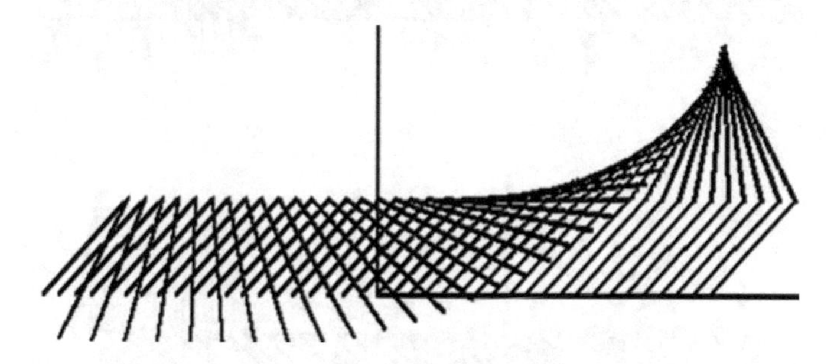

Fig. 6.63 $q = 0.02$

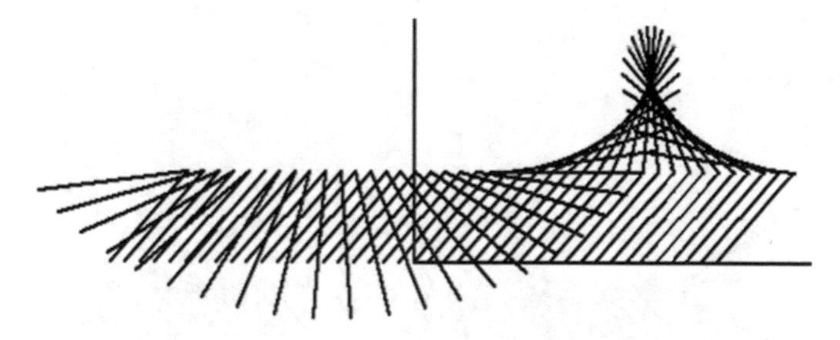

Fig. 6.64 $q = 0.03$

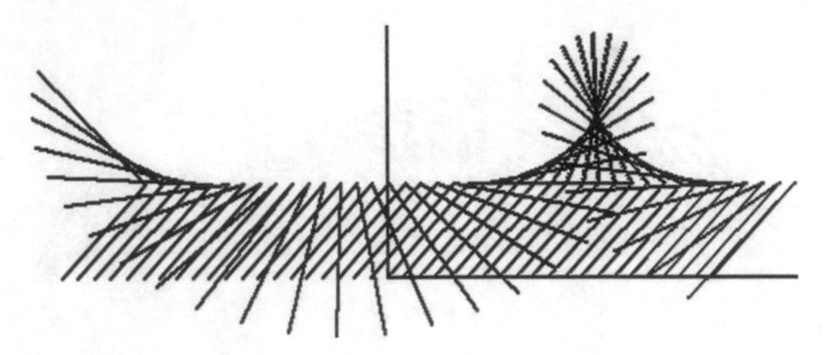

Fig. 6.65 q = 0.04

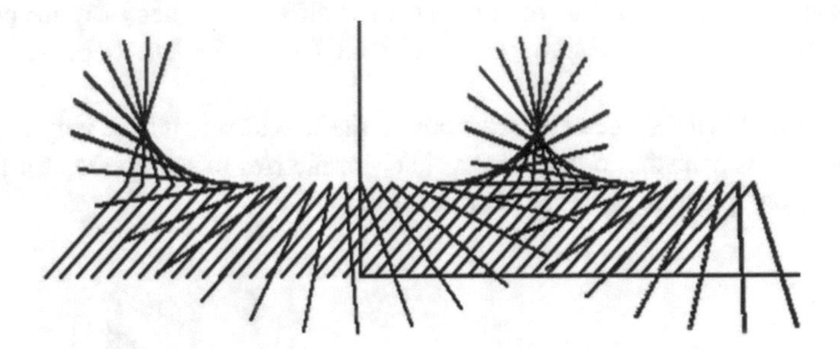

Fig. 6.66 q = 0.05

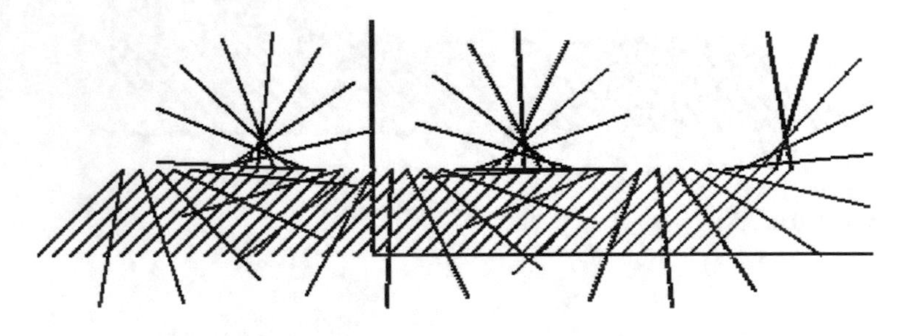

Fig. 6.67 q = 0.08

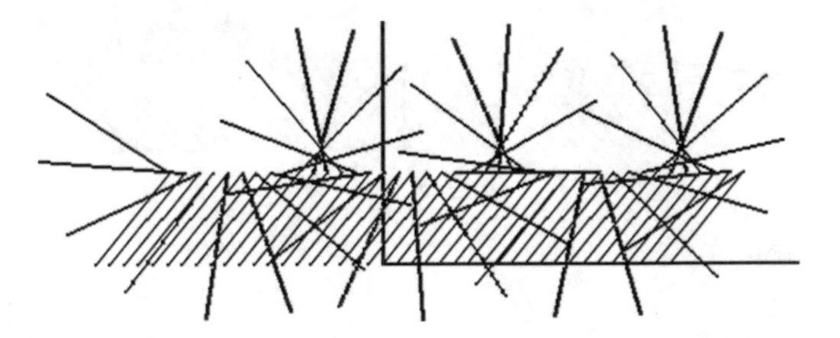

Fig. 6.68 q = 0.1

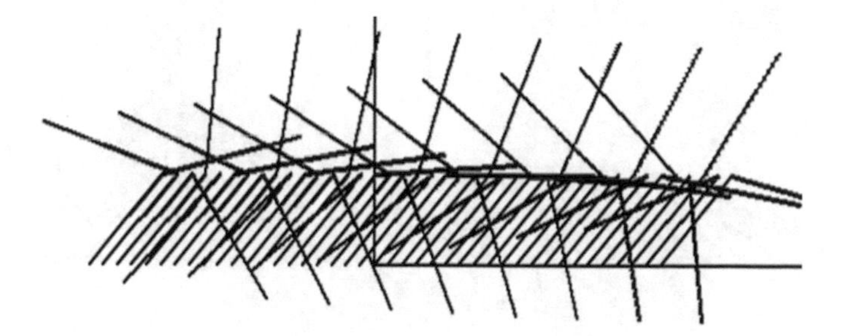

Fig. 6.69 q = 0.5

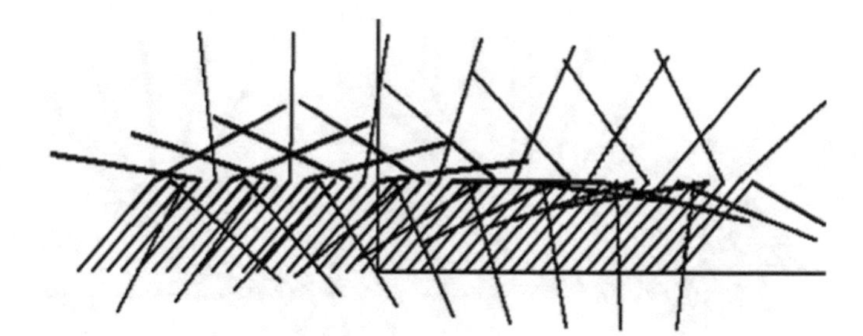

Fig. 6.70 q = 1

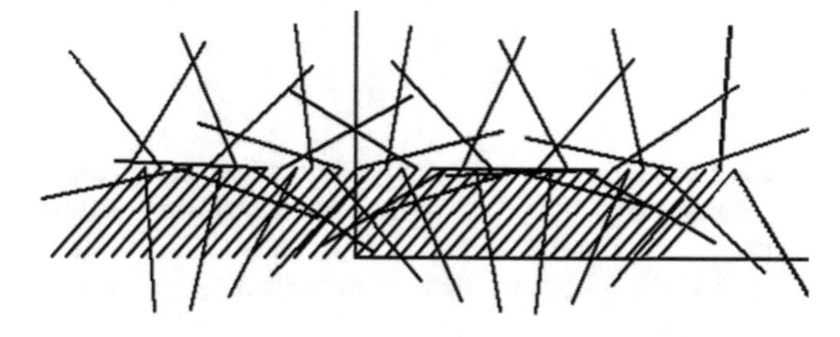

Fig. 6.71 q = 2

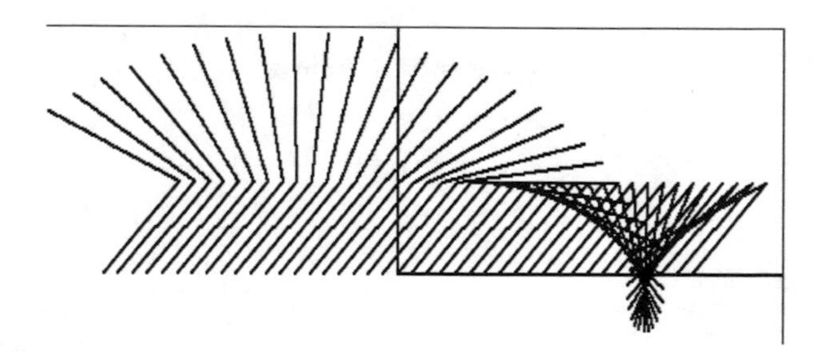

Fig. 6.72 q = 5

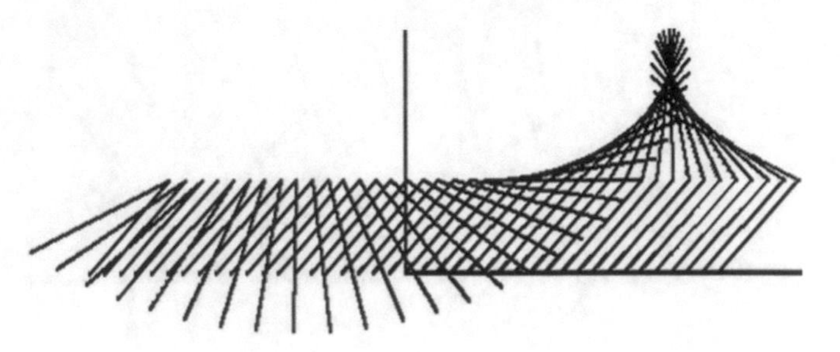

Fig. 6.73 q = −5

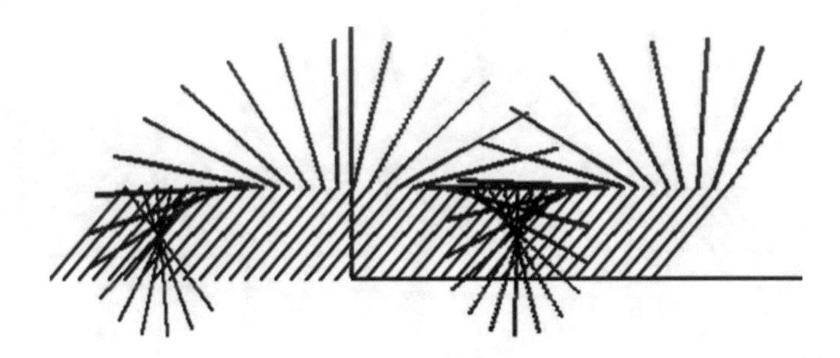

Fig. 6.74 q = 10

6.3 Case 3

In this case, point A slides on a fixed line, and point B slides on AB, the obliquities of the lines being constant (Fig. 6.75). The equivalent mechanism is given in Fig. 6.76, being of PP type.

Based on Fig. 6.47 the following relations are written:

$$x_C = s + a\cos\gamma + b\cos\alpha \tag{6.6}$$

$$y_C = a\sin\gamma + b\sin\alpha \tag{6.7}$$

$$a = q.s \tag{6.8}$$

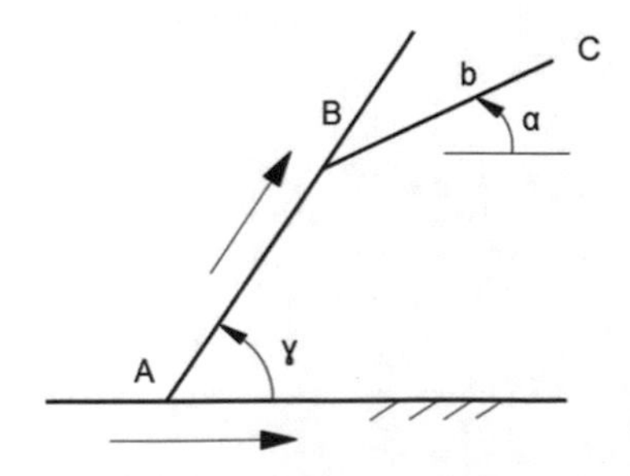

Fig. 6.75 The geometric construction

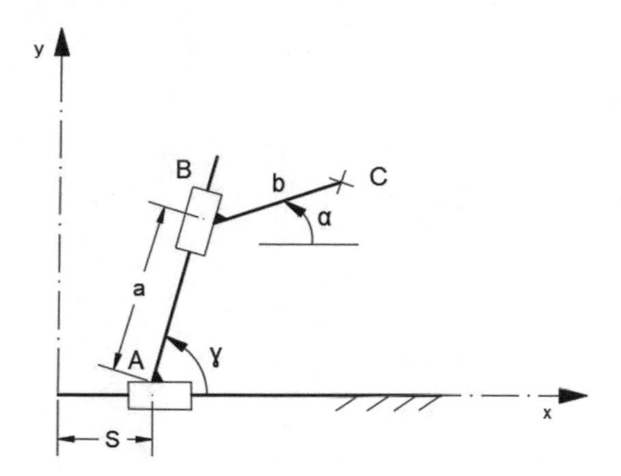

Fig. 6.76 The equivalent mechanism

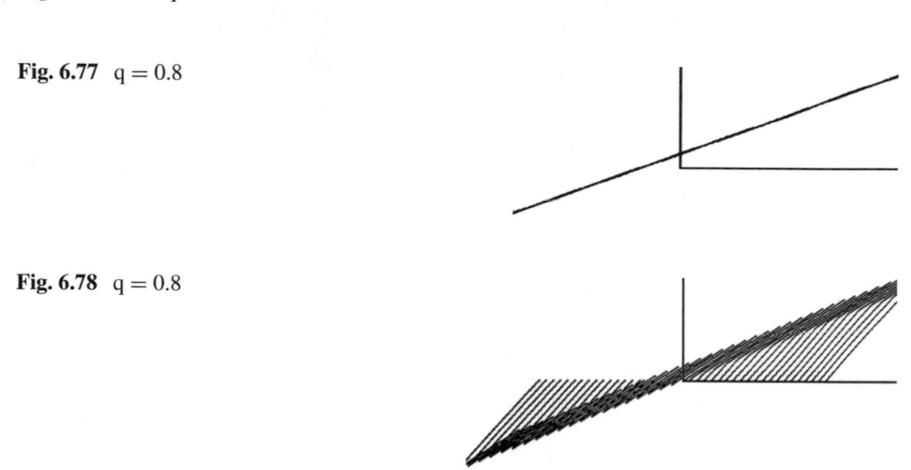

Fig. 6.77 q = 0.8

Fig. 6.78 q = 0.8

It is found that the length a is variable. Initial data: b = 50 mm; $\gamma = 45$; $\alpha = 30°$.

In this case the loci are straight, with different obliquities of the assembly depending on the value of q (Fig. 6.77).

The successive positions are similar to the case of Fig. 6.78, with different inclinations as in the case of Fig. 6.77.

References

1. Popescu I, LucaL (2015) Geometric locus generated by a mechanism with three dyads. In: Analele Universităţii «Constantin Brâncuşi» din Târgu-Jiu, Seria Inginerie, nr. 3/2015, pp 29–37
2. Popescu I, Romanescu AE, Sass L (2016) A locus problem solved through methods from the theory of mechanisms. In: International conference: innovative manufacturing engineering & energy, IManE, 23–25.09.2016, Kallithea Chalkidiki Greece, p 7

Chapter 7
Problem of a Locus with Four Intercut Lines

Abstract A straight line rotates around a fixed point, and at the free end it connects with another straight line that rotates around the same point, and the other end slides on the abscissa. Another line FB rotates around F point, the end—C point sliding to the line DE, requiring the trajectory of point C. Many different curves are obtained, with quite complicated shapes [2], depending on the correlation of the movements of the two leading elements. Another range of curves resulted for the trajectory of the point of intersection of the lines AD with BC, where two slides were introduced that slide on the elements AD and BC and have a torque of rotation between them. In both situations, both directions of rotation of the conductive elements were taken into account: in the same direction or in opposite directions.

7.1 The Trajectory of C Point

A straight line DE rests on the circle with radius AD, (with A fixed at the origin of the system), and on the abscissa. Another BC line rests on point B on a circle with radius FB, and with point C on line DE. The locus of point C is required, while FB and AD rotate (Fig. 7.1).

The mechanism that fulfills these conditions is shown in Fig. 7.2, where there was a slide in E and one in C, so that these points can be moved.

The following relations are written:

$$\mathrm{x}_F = AF \cos\varphi \tag{7.1}$$

$$\mathrm{y}_F = AF \sin\varphi \tag{7.2}$$

$$\mathrm{x}_B = \mathrm{x}_F + AF \cos\psi \tag{7.3}$$

$$\mathrm{Y}_B = \mathrm{y}_F + AF \sin\psi \tag{7.4}$$

I. Popescu et al., *Problems of Locus Solved by Mechanisms Theory*,
Springer Tracts in Mechanical Engineering,
https://doi.org/10.1007/978-3-030-63079-9_7

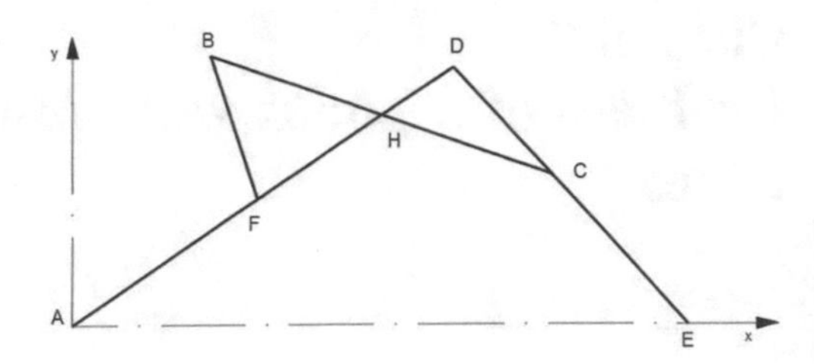

Fig. 7.1 Loci problem

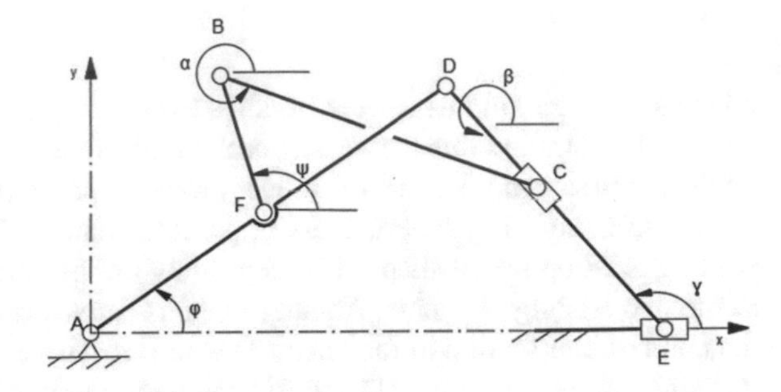

Fig. 7.2 The equivalent mechanism

$$x_C = x_B + BC\cos\alpha = x_E + EC\cos\gamma \tag{7.5}$$

$$y_C = y_B + BC\sin\alpha = EC\sin\gamma \tag{7.6}$$

$$tg\gamma = \left[y_B + BC\sin\alpha\right]/[x_B + BC\cos\alpha - x_E] \tag{7.7}$$

$$(x_B + BC\ \cos\ \alpha - x_E)tg\gamma = y_B + BC\ \sin\alpha \tag{7.8}$$

$$x_D = AD\cos\varphi = x_E + DE\cos\gamma \tag{7.9}$$

$$y_D = AD\sin\varphi = DE\sin\gamma \tag{7.10}$$

It is established that $\psi = q\varphi$, to determine the locus of point C for different values of q. AD = 25; DE = 48; FB = 28; BC = 67; AF = 30 were taken as initial data.

Figure 7.3 shows the mechanism generated for a position, for q = 1, and Fig. 7.4 shows the successive positions of the mechanism.

The searched locus is given in Fig. 7.5 for q = 0.

Figures from 7.6, 7.7, 7.8, 7.9, 7.10, 7.11, 7.12, 7.13, 7.14 and 7.15 shows the loci for other values of q.

Fig. 7.3 The mechanism in one position

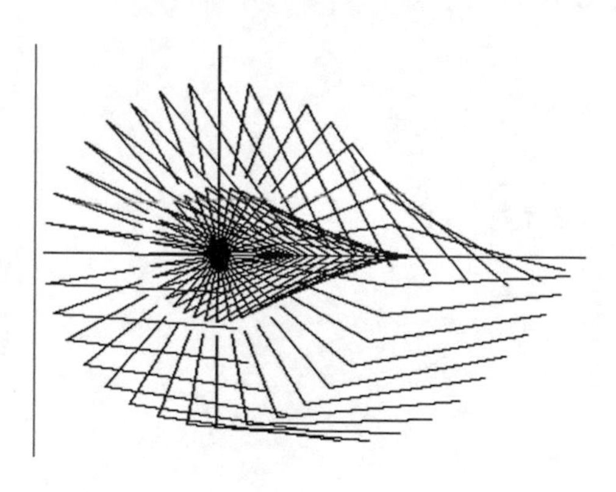

Fig. 7.4 Successive positions

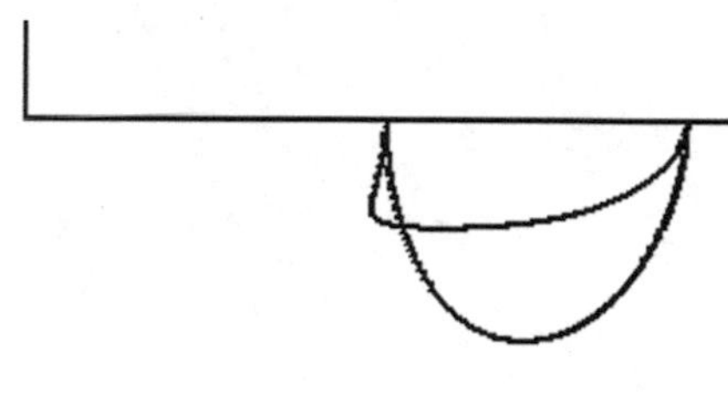

Fig. 7.5 Locus $q = 0$

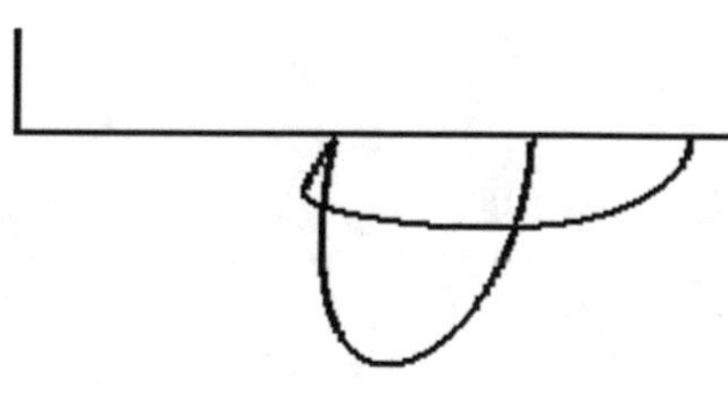

Fig. 7.6 $q = 0.2$

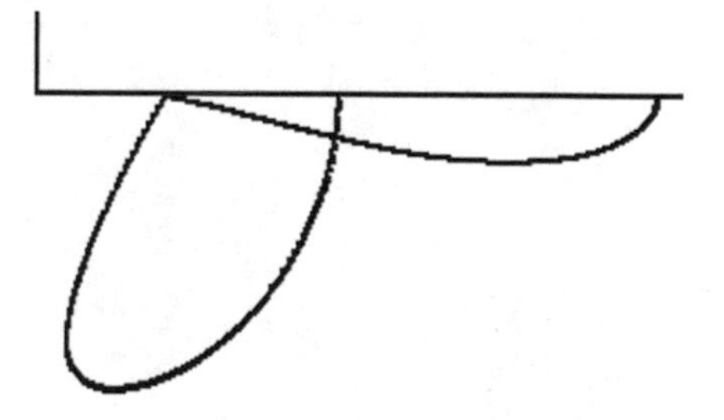

Fig. 7.7 $q = 0.5$

Fig. 7.8 q = 0.75

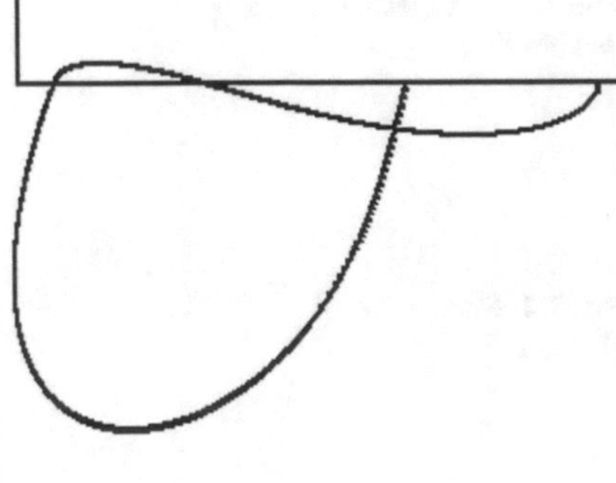

Fig. 7.9 q = 1

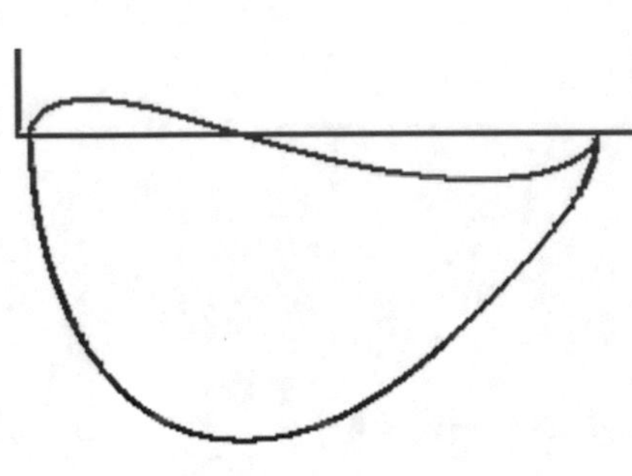

Fig. 7.10 q = 1.5

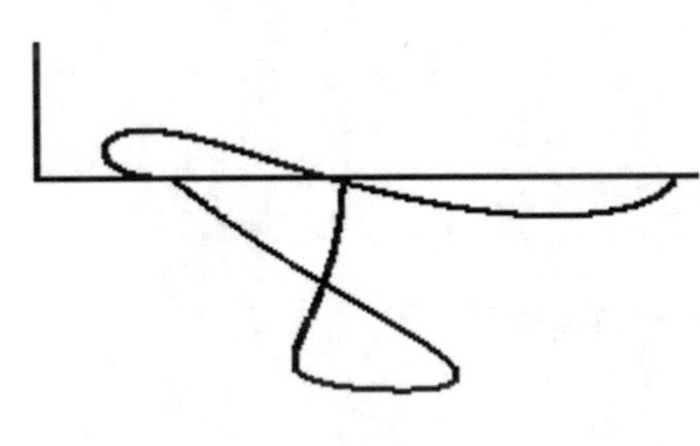

Fig. 7.11 q = 2

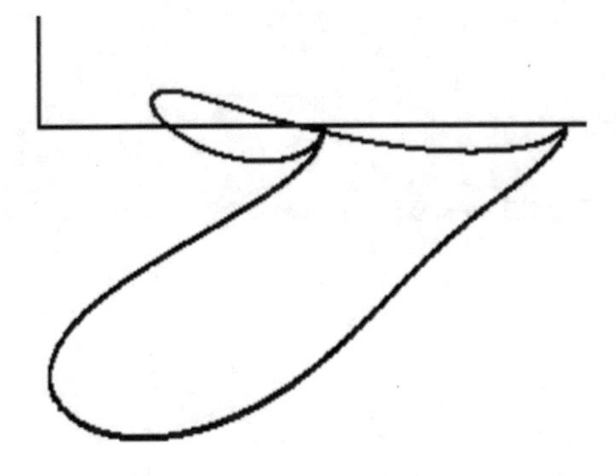

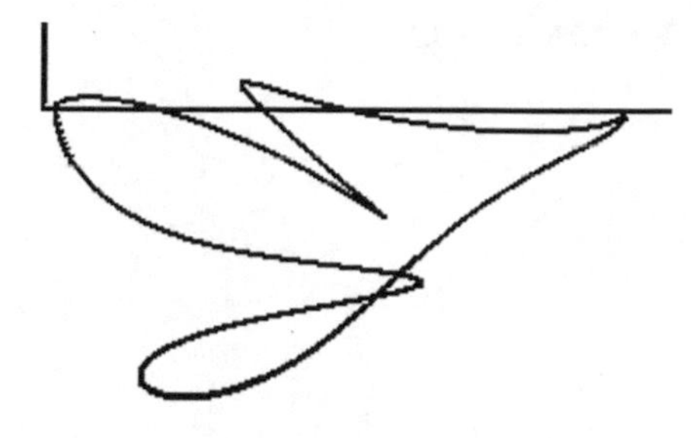

Fig. 7.12 q = 3

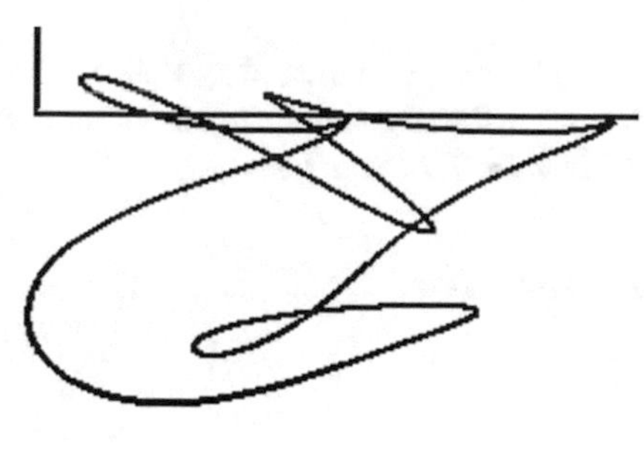

Fig. 7.13 q = 4

Fig. 7.14 q = 5

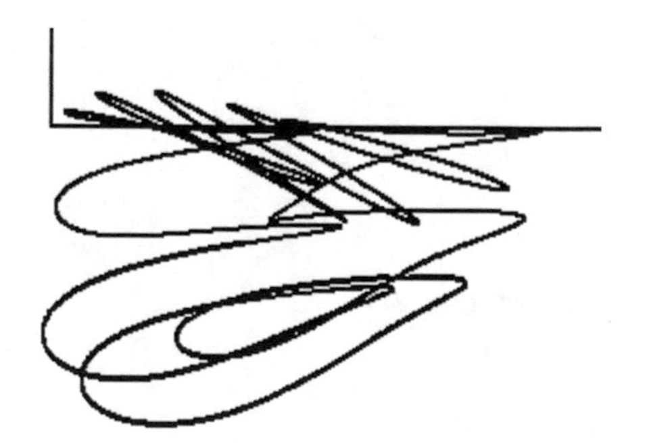

Fig. 7.15 q = 10

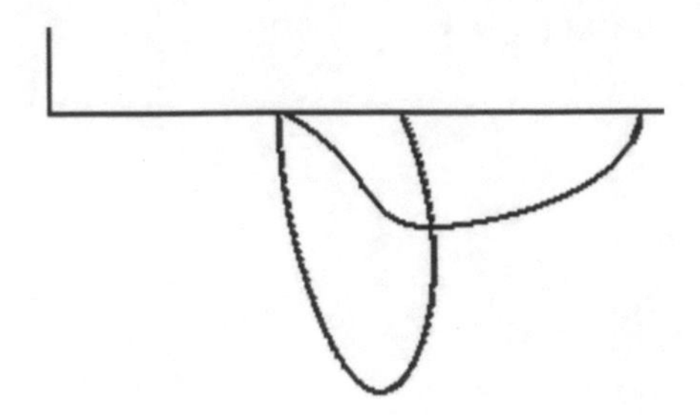

Fig. 7.16 q = –0.3

For negative values of q, that is upon the clockwise rotation of FB, while AD rotates in the trigonometric direction, Figures from 7.16, 7.17, 7.18, 7.19, 7.20, 7.21, 7.22, 7.23, 7.24 and 7.25.

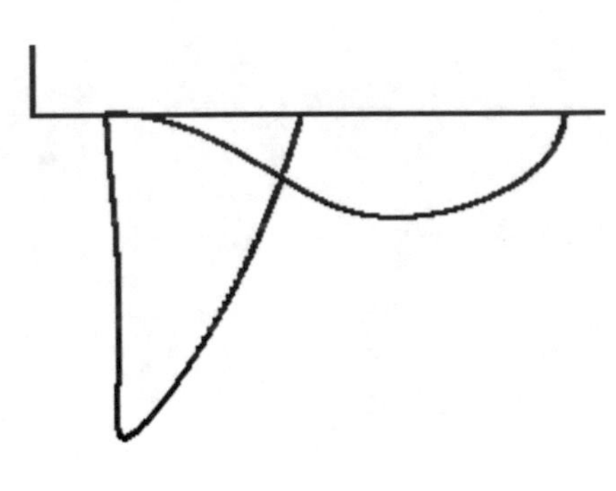

Fig. 7.17 q = –0.6

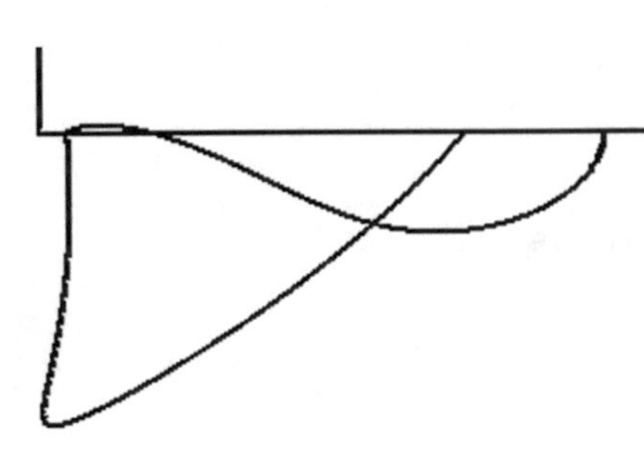

Fig. 7.18 q = –0.8

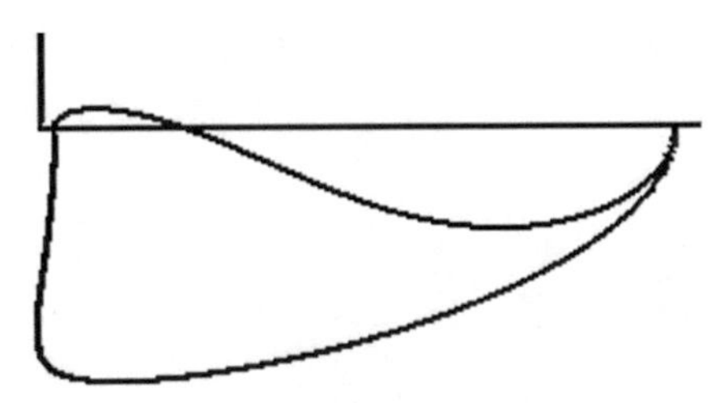

Fig. 7.19 q = –1

Fig. 7.20 q = –1.5

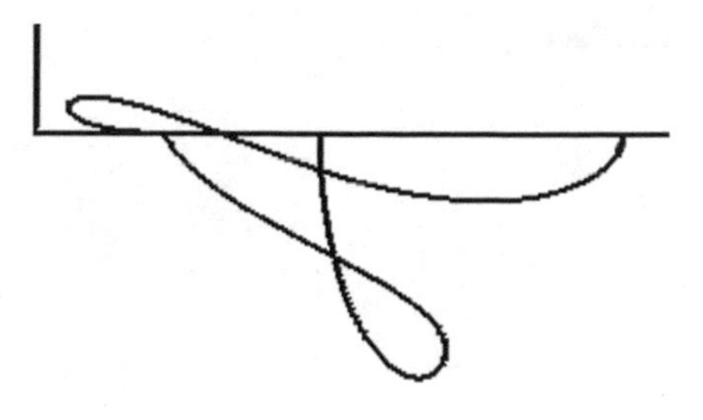

Fig. 7.21 q = –2

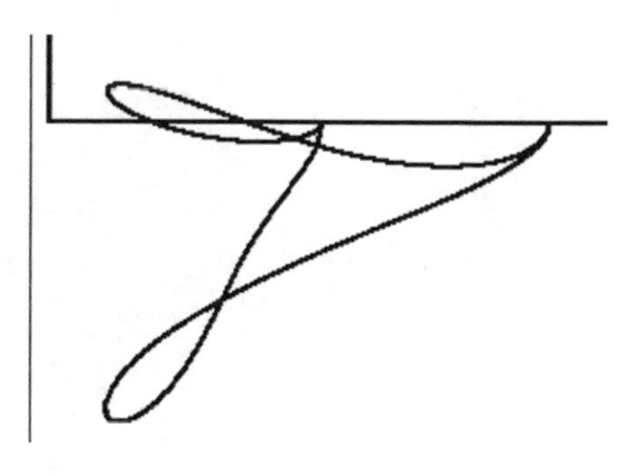

Fig. 7.22 q = –3

Fig. 7.23 q = –4

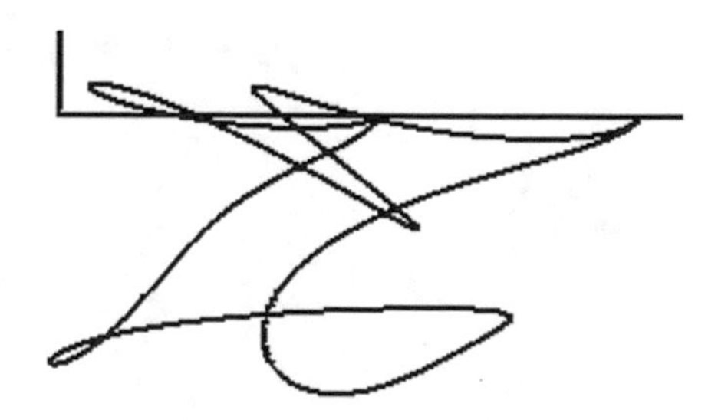

Different shapes are observed for the geometric place, more and more complicated when increasing q.

Fig. 7.24 q = –5

Fig. 7.25 q = –10

7.2 The Trajectory of H Point

The trajectory of the H point of intersection between AD and BC is also sought.

The new mechanism appears in Fig. 7.26, where two elements with zero lengths were introduced, i.e. two slides connected by a rotating torque.

The relationships are written:

$$x_H = AH\cos\varphi = x_B + BH\cos\alpha \tag{7.11}$$

$$y_H = AH\sin\varphi = y_B + BH\sin\alpha \tag{7.12}$$

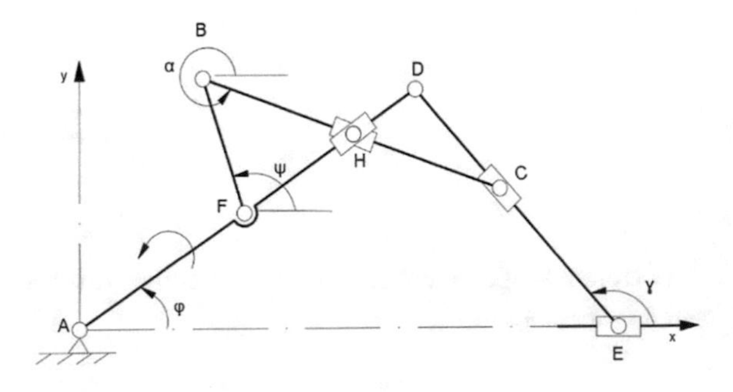

Fig. 7.26 Another mechanism

$$tg\,\varphi = [y_B + BH\sin\alpha]/[x_B + BH\cos\alpha] \tag{7.13}$$

$$BH = [y_B - x_B \mathrm{tg}\varphi]/[\cos\alpha tg\varphi - sin\alpha] \tag{7.14}$$

$$AH = [y_B + BH\sin\alpha]/\sin\varphi \tag{7.15}$$

The relation given by Eq. 7.13 is obtained by dividing Eq. 7.12 by Eq. 7.11. From Eq. 7.13 we deduce Eq. 7.14, then from Eq. 7.12 it results Eq. 7.15. From Eqs. 7.11 and 7.12 result the coordinates of the tracer point, H.

Using the data from the previous mcchanism, numerous figures given below resulted, for the values of q used in the trajectories of point C, having $\psi = q\varphi$. Below the figures are given the values of q (Figs. 7.27, 7.28, 7.29, 7.30, 7.31, 7.32, 7.33, 7.34, 7.35, 7.36, 7.37, 7.38, 7.39, 7.40, 7.41, 7.42, 7.43, 7.44 and 7.45).

Many forms of trajectories have resulted, some being closed curves, but most open, with several branches [1].

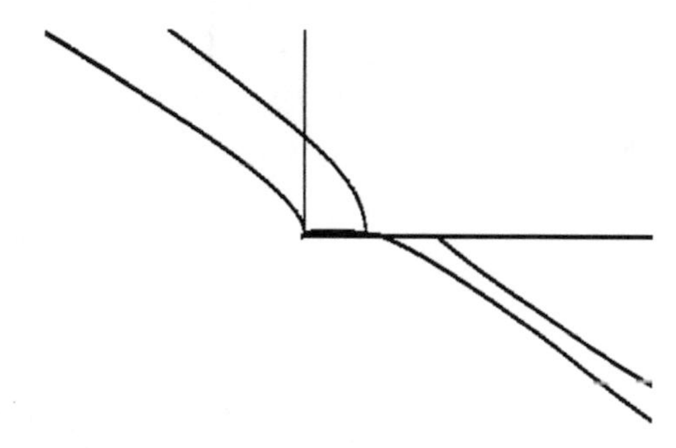

Fig. 7.27 q = 0.2

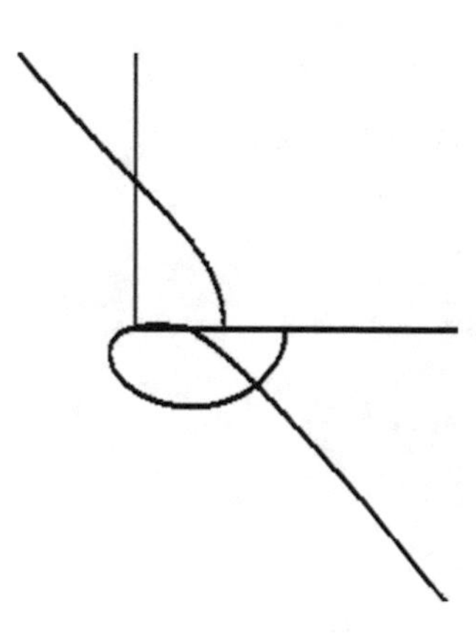

Fig. 7.28 q = 0.5

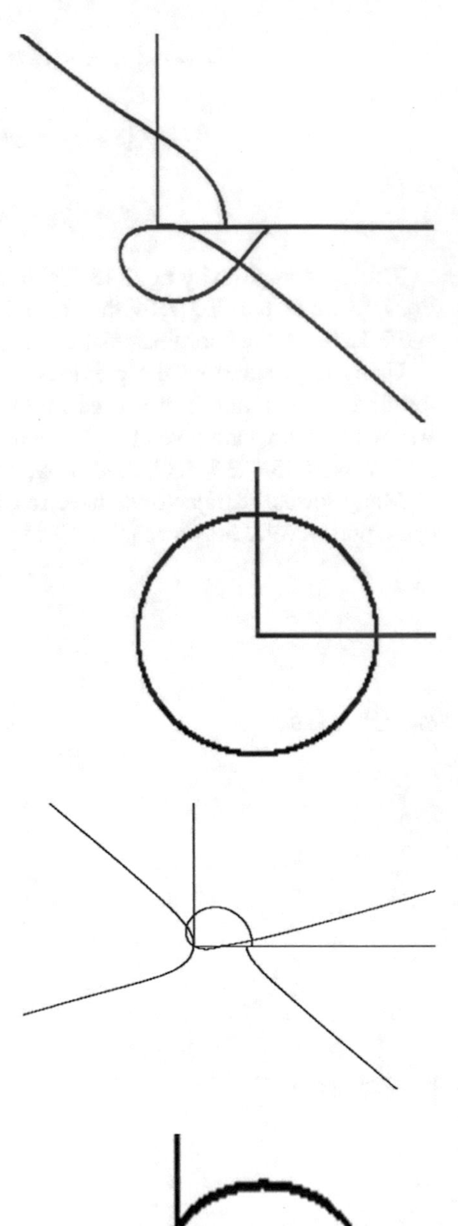

Fig. 7.29 q = 0.75

Fig. 7.30 q = 1

Fig. 7.31 q = 1.5

Fig. 7.32 q = 2

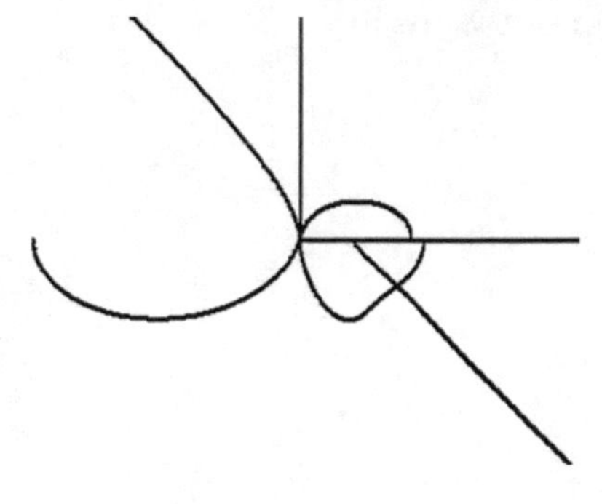

Fig. 7.33 q = 3

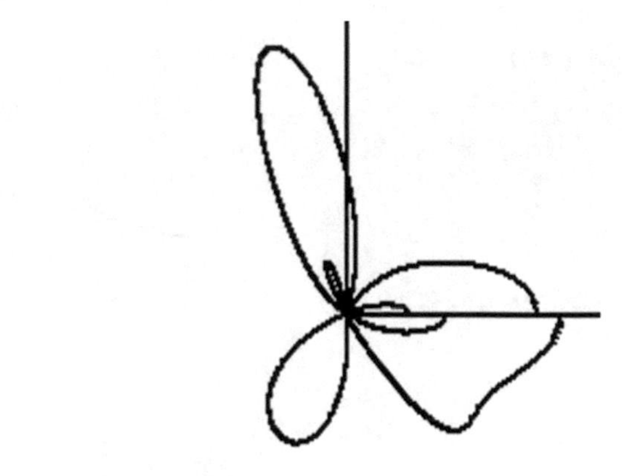

Fig. 7.34 q = 4

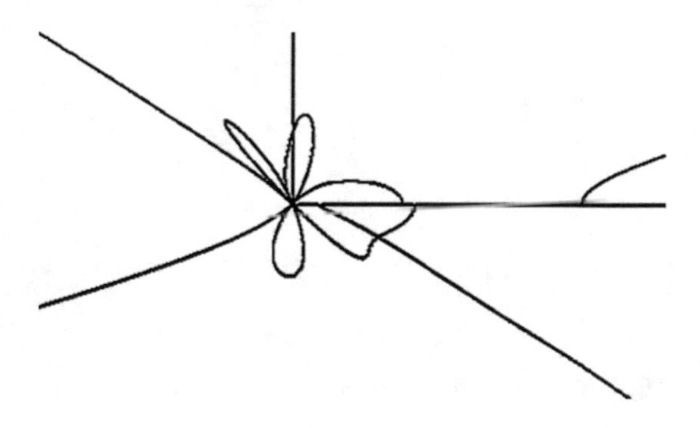

Fig. 7.35 q = 5

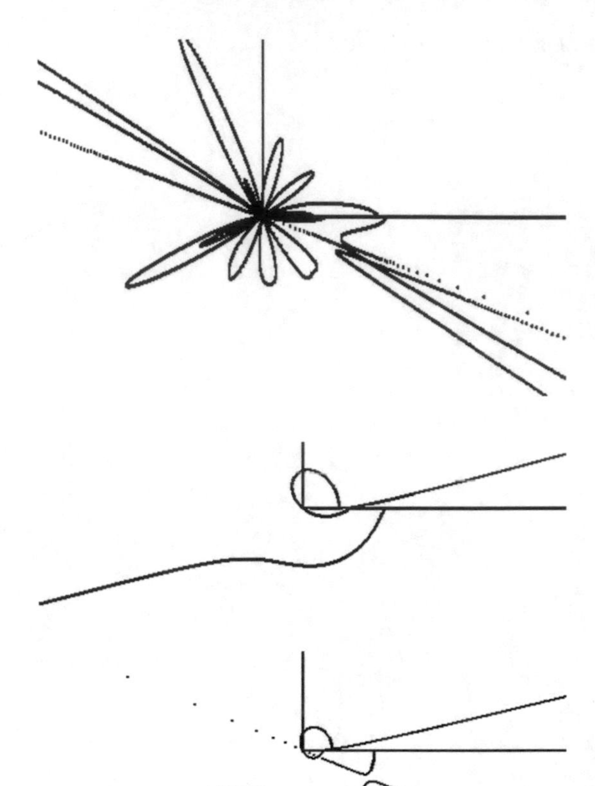

Fig. 7.36 q = 10

Fig. 7.37 q = −0.3

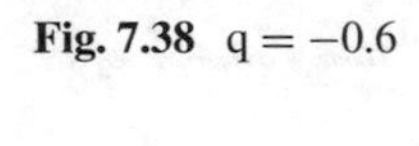

Fig. 7.38 q = −0.6

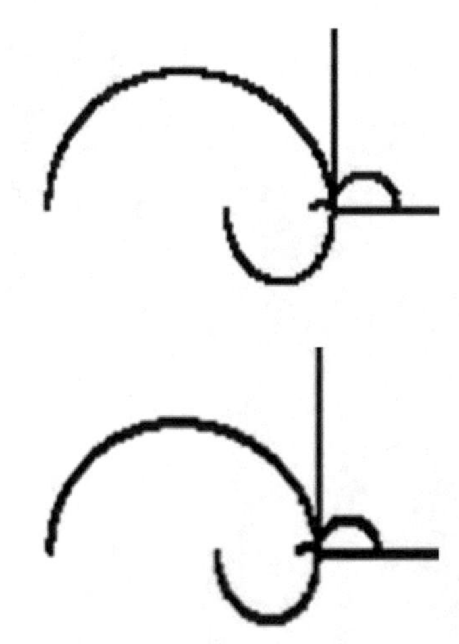

Fig. 7.39 q = −1

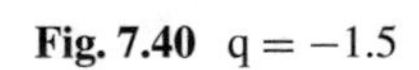

Fig. 7.40 q = −1.5

Fig. 7.41 q = −2

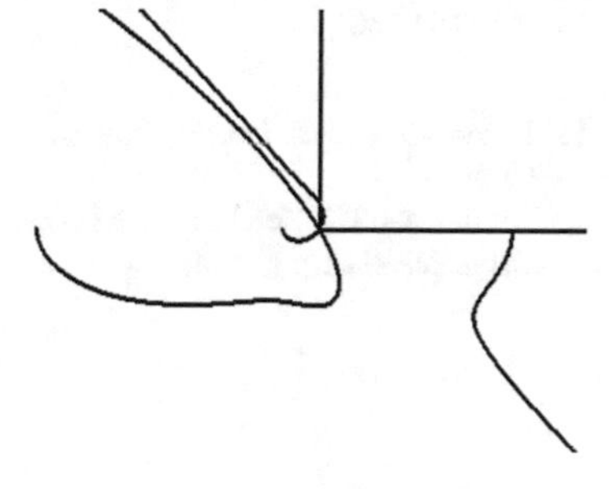

Fig. 7.42 q = −3

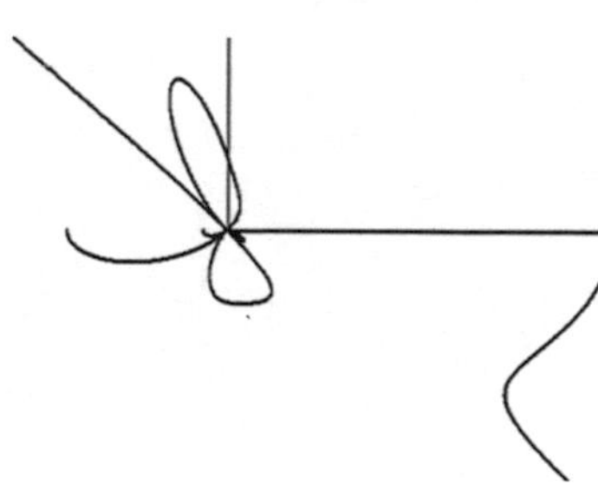

Fig. 7.43 q = −4

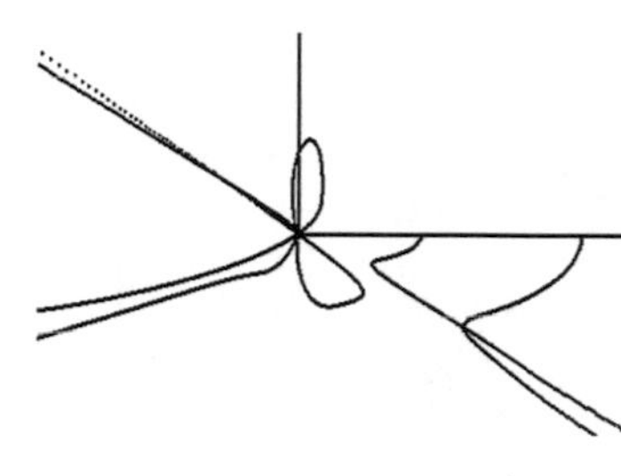

Fig. 7.44 q = −5

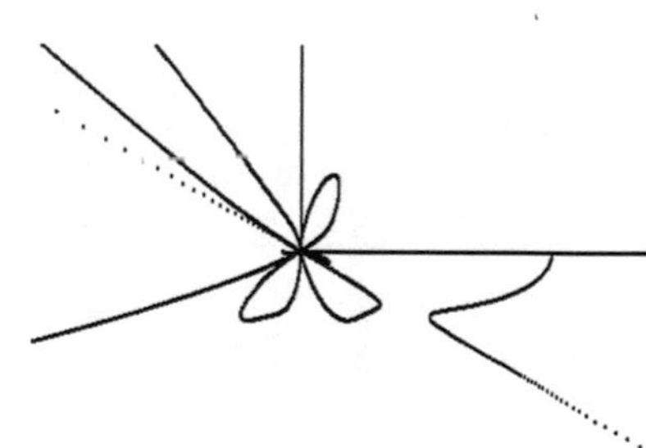

Fig. 7.45 q = −10

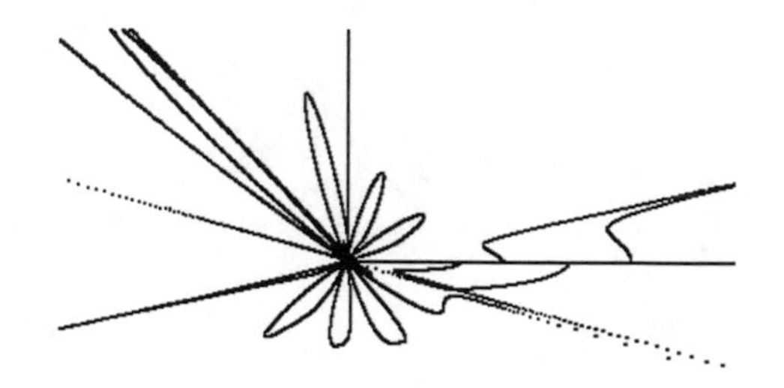

References

1. I. Popescu (2016) Locuri geometrice şi imagini estetice generate cu mecanisme. Editura Sitech, Craiova
2. Teodorescu ID, Teodorescu SD (1975) Culegere de probleme de geometrie superioară. Editura Didactică şi Pedagigică, Bucureşti

Chapter 8
"Kappa" and "Kieroid" Curves Resulted as Loci

Abstract A crooked line with an angle of 90°, has one end that goes on the abscissa, and the other rotates around the origin of the system and slides through that point. The trajectory of points B and D on the bent line is required. The result is the "Kappa" curve, in several variants depending on the length of the long arm of the bent line. Next we also start from a crooked line but with the bending angle of different values, resulting in a parallelogram and generating curves similar to the "Kappa" curve but with deformed branches. Successive positions are also given, which have the shapes of curves. We study also the "Kieroid" curve obtained when EC line has the ends moving on two fixed lines parallel to the ordinate. Point E is also in rotation around B moving on the conducting element AC. The trajectories of B and D are required, i.e. the "Kieroid" curve, which has two branches, one drawn by D and the other by B. By changing the distance between the fixed vertical lines, other curves result, some of the same kind, others different.

8.1 The "Kappa' Curve

The "Kappa" curve was studied by Gutschoven, Barrov and Sluse, as a tangent spiral. The name comes from the small hand letter, the kappa of the Greek alphabet. It is shown in [1, 2] how to obtain this curve as a locus (Fig. 8.1): moving a set square with the point B on an axis, the line MO passing permanently through O. The points M and z describe the two-branch "Kappa" curve.

The equivalent mechanism is shown in Fig. 8.2. It is built as follows:

- The CBA element is placed with the BAD side passing through the fixed A point, where we positioned a slider rotating around the fixed A point;
- The measure of B angle is 90 degrees;
- Point C moves on the Y-coordinate with the slider 1.

We search for the locus of the point D when C moves on the abscissa.

I. Popescu et al., *Problems of Locus Solved by Mechanisms Theory*,
Springer Tracts in Mechanical Engineering,
https://doi.org/10.1007/978-3-030-63079-9_8

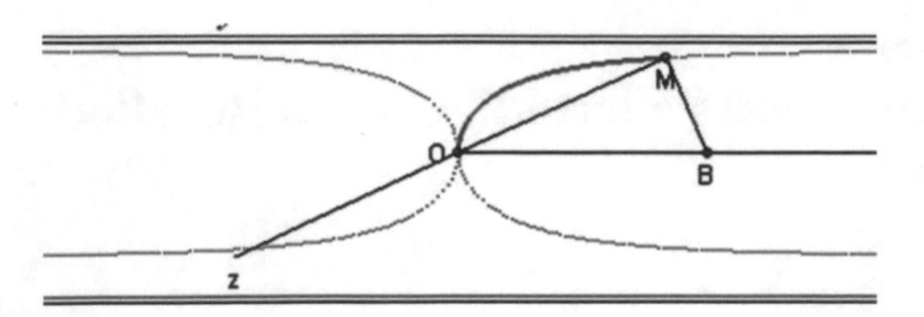

Fig. 8.1 Geometric conditions [3]

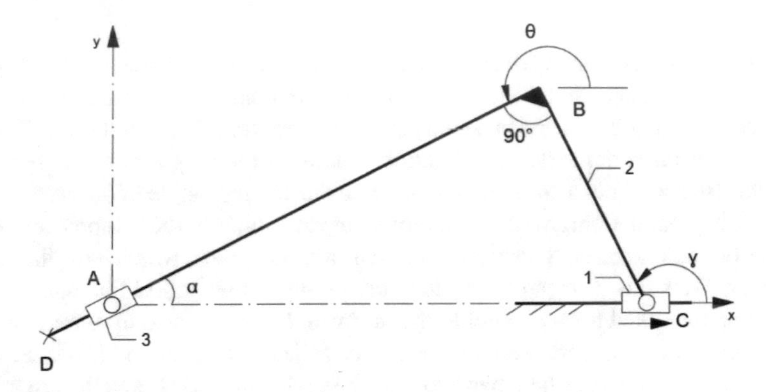

Fig. 8.2 The equivalent mechanism

The following are written:

$$x_B = AC + CB\cos\gamma = AB\cos\alpha \tag{8.1}$$

$$y_B = CB\sin\gamma = AB\sin\alpha \tag{8.2}$$

$$\gamma = \alpha + 90 \tag{8.3}$$

$$x_D = x_B + BD\cos\theta \tag{8.4}$$

$$y_D = y_B + BD\sin\theta \tag{8.5}$$

$$\theta = \gamma + 90 = \alpha + \pi \tag{8.6}$$

We considered as initial data: CB = 50; BD = 120. In Fig. 8.3 it is shown the mechanism in one position and the drawn curve.

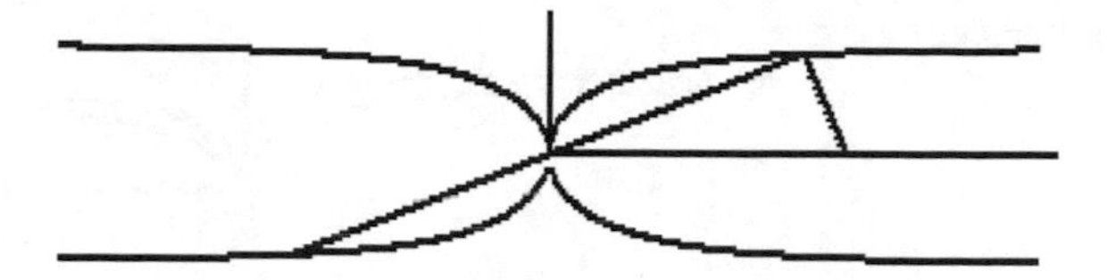

Fig. 8.3 The locus

The successive positions of the mechanism are shown in Fig. 8.4. In Fig. 8.5 it is shown which zone of the curve is drawn by the point D, and in Fig. 8.6 it is presented which zone is drawn by the point B.

Next, the lenght of BD line was modified, resulting in "Kappa" curves, too, but with other dimensions, with branches differently positioned in relation to the normal curve (Figs. 8.7 and 8.8).

In the same site another graphic construction appears in Fig. 8.9.

Here, the "Kappa" curve is described by the point M of the right-angle triangle POM, when P turns on a circle with the radius OP and the PM line stays parallel to the fixed line OB [1, 2].

The equivalent mechanism is shown in Fig. 8.10.

The following expressions are written:

$$x_B = x_A + AB \cos \varphi \tag{8.7}$$

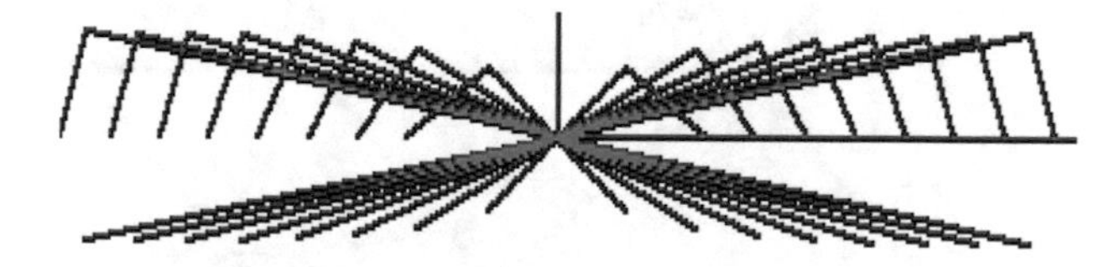

Fig. 8.4 Successive positions

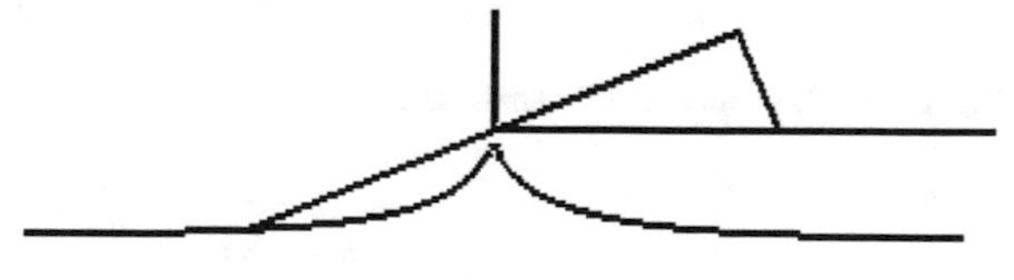

Fig. 8.5 Curve branch

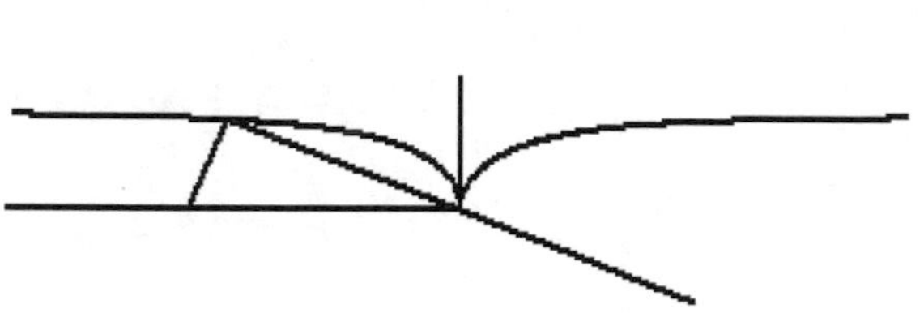

Fig. 8.6 Zone drawn by B point

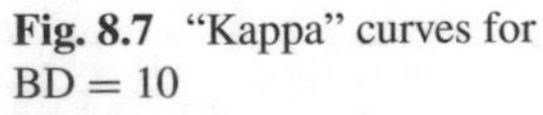

Fig. 8.7 "Kappa" curves for BD = 10

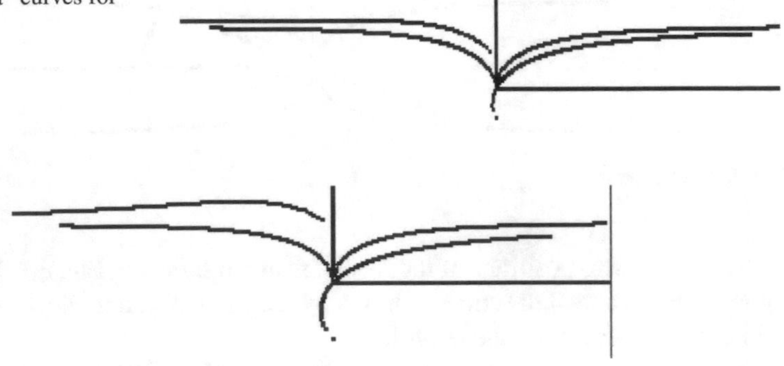

Fig. 8.8 "Kappa" curves for BD = 20

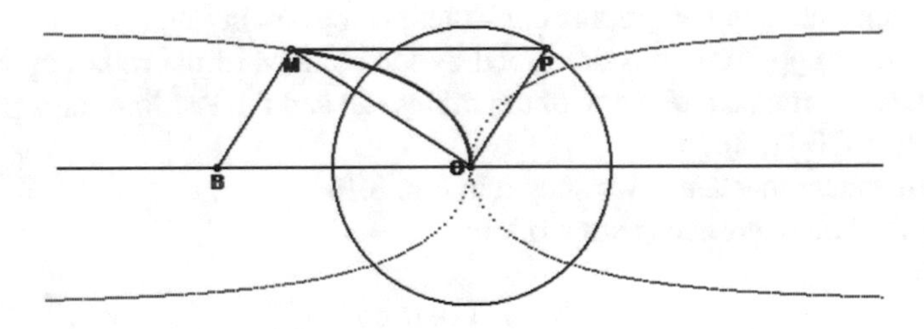

Fig. 8.9 Other geometrical conditions

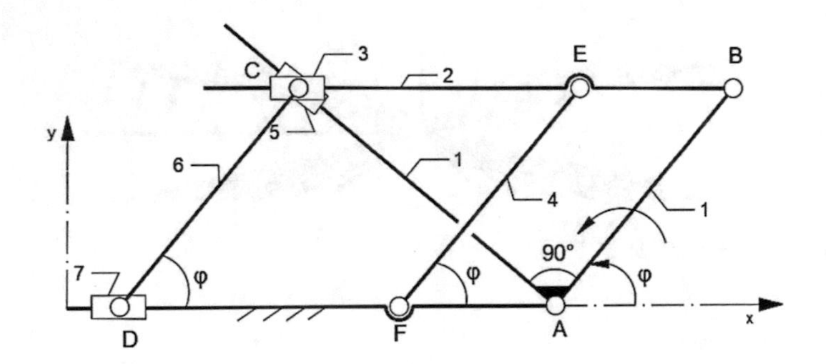

Fig. 8.10 The equivalent mechanism

$$y_B = AB \sin \varphi \tag{8.8}$$

$$x_E = x_F + AB \cos \varphi \tag{8.9}$$

$$y_E = AB \sin \varphi \tag{8.10}$$

$$x_C = x_D + AB\ \cos\ \varphi \tag{8.11}$$

$$y_C = AB\ \sin\ \varphi \tag{8.12}$$

$$x_C = x_A + AC\ \cos(\varphi + 90) \tag{8.13}$$

$$\frac{AC}{\sin\ \varphi} = \frac{AB}{\sin(90 - \varphi)} \tag{8.14}$$

in ACD triangle.

$$(x_C - x_D)\text{tg}\,\varphi = y_C. \tag{8.15}$$

In Fig. 8.11, the generated mechanism is shown in one position, and the Fig. 8.12 shows the successive positions of the ACD triangle. The mechanism positions can be seen in Fig. 8.13.

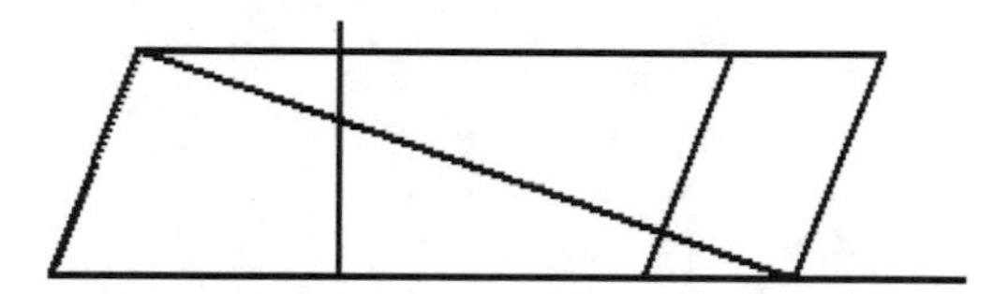

Fig. 8.11 The mechanism in a position

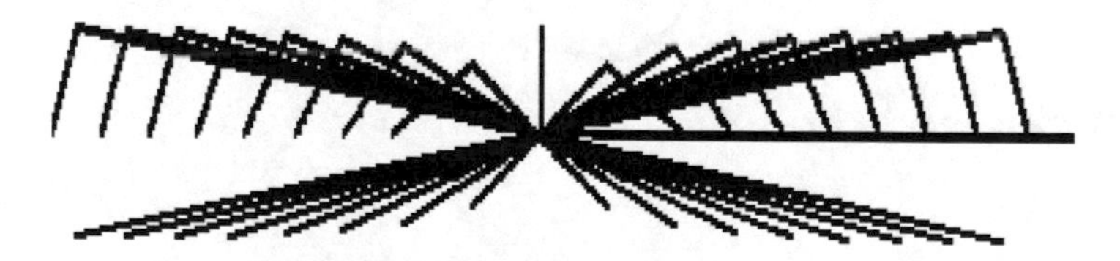

Fig. 8.12 Successive positions of the triangle ACD

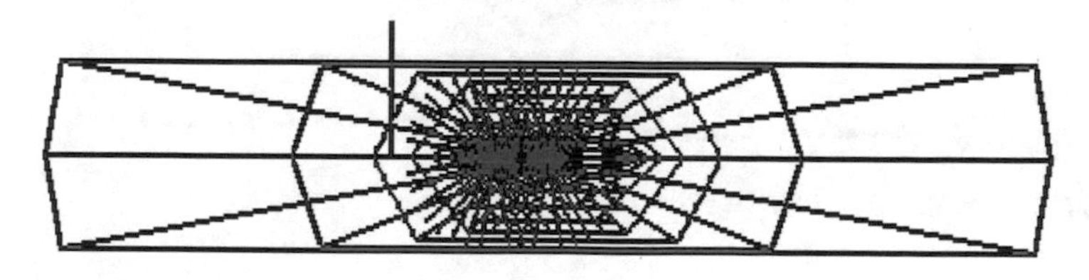

Fig. 8.13 Successive positions of the mechanism

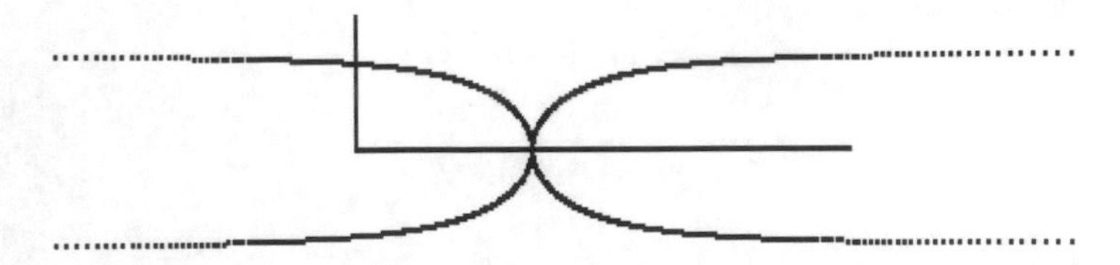

Fig. 8.14 The Locus

The following initial values were considered: XF = 48, XA = 72; AB = 40.

In Fig. 8.14 the drawn curve is shown (Figs. 8.15, 8.16 and 8.17).

Next, we wanted to know what curves are obtained if the BAC angle = α is 90 degrees no more. The achieved curves are shown below (Figs. 8.15, 8.16 and 8.17).

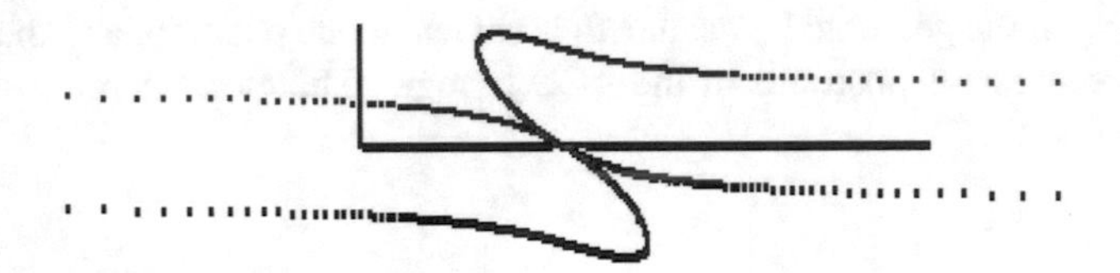

Fig. 8.15 Curves obtained for $\alpha = 30$

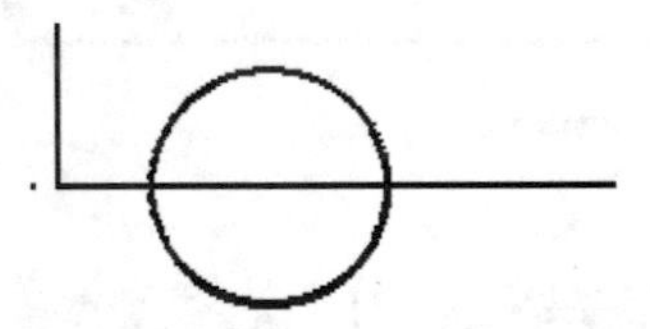

Fig. 8.16 Curves obtained for $\alpha = 0$

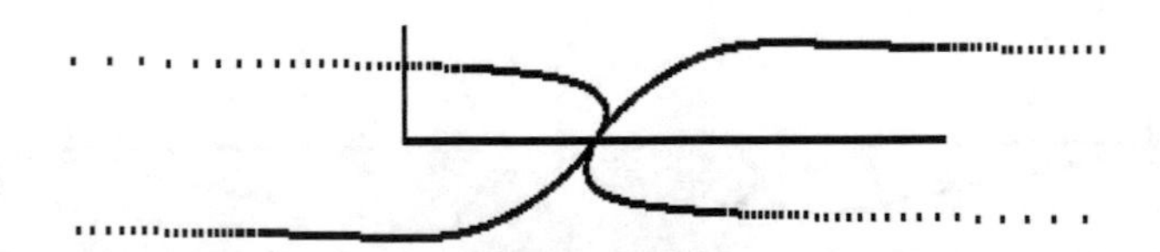

Fig. 8.17 Curves obtained for $\alpha = 120$

8.2 The "Kieroid" Curve

Kieroid is a rational quartz flat curve, created by P. J. Kiernan in 1945. In [1, 4] it is shown that having the N and P points with equal orders placed respectively on two straight lines parallel to the Y-coordinate, the locus of the intersection points between the circle having the center in P point with the ON line is the curve called"Kieroid" (Fig. 8.18).

The synthesis of the mechanism was made as follows (Fig. 8.19):

- Fixed lines were drawn through F and G points, parallel to the Y-coordinate;
- Points E and C slip on these straight lines, so there were placed sliders in C and E points;
- The D and B trajectory points are located on the mobile circle having the center in E point, so that there were also placed sliders in these points to slip on the AC line;
- In C point, another slider was placed on the AC element.

By structurally analyzing the mechanism, it was found that it was not functional, so that the slider placed in E point was removed. This one was structurally parasitic, because the motion of the E point was determined by the element 5, the weld from point C assuring for point E the motion on a vertical imposed by the guide CG. And as CE = GF, the line EF is no longer required.

The angle φ hangs in a loop and determines the locus of D together with B.

The following expressions are written:

$$x_C = \text{AC}\cos = x_G = \text{constant}; \tag{8.16}$$

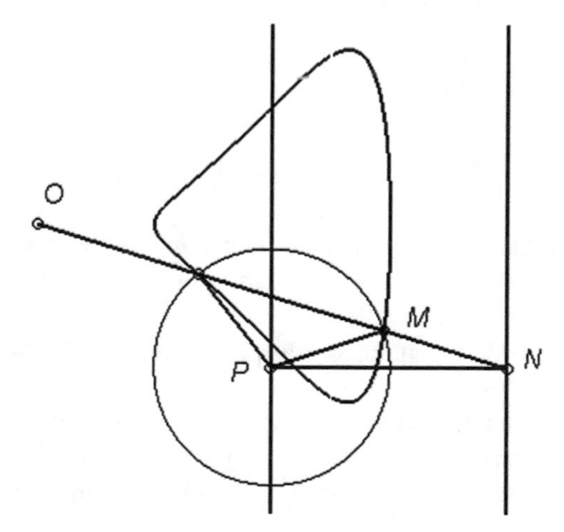

Fig. 8.18 Geometrical conditions [1, 4]

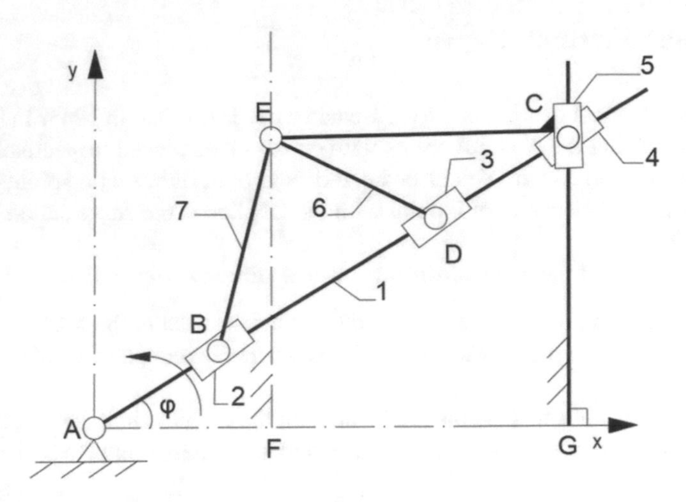

Fig. 8.19 The equivalent mechanism

$$y_C = \text{AC sin} = \text{GC} \tag{8.17}$$

$$x_E = x_F = \text{constant} \tag{8.18}$$

$$y_E = y_C \tag{8.19}$$

$$x_E = x_C - CE \tag{8.20}$$

$$\tan = y_B/x_B = y_D/x_D = y_C/x_C \tag{8.21}$$

$$(x_D - x_E)^2 + (y_D - y_E)^2 = \text{DE}^2 \tag{8.22}$$

$$(x_B - x_E)^2 + (y_B - y_E)^2 = \text{BE}^2 \tag{8.23}$$

In Fig. 8.20 it is shown the mechanism generated by the computer program for one position, and Fig. 8.21 gives the successive positions of the mechanism, for XF = 30; XG = 66; BE = 30; DE = 30.

Figure 8.22 shows the generated trajectory, i.e. the Kieroid curve and the mechanism in one position.

Interesting in this mechanism is that the D and B tracer points each generate a zone of the Kieroid curve, completing each other in order to obtain the complete curve. Thus, point B marks the left zone (Fig. 8.23) and point D marks the the right one (Fig. 8.24).

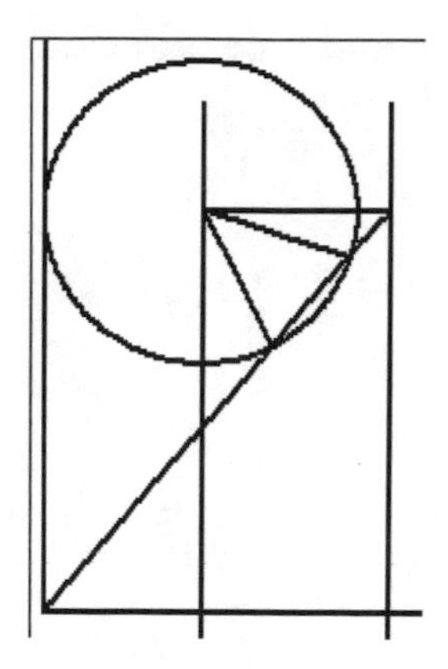

Fig. 8.20 The mechanism in one position

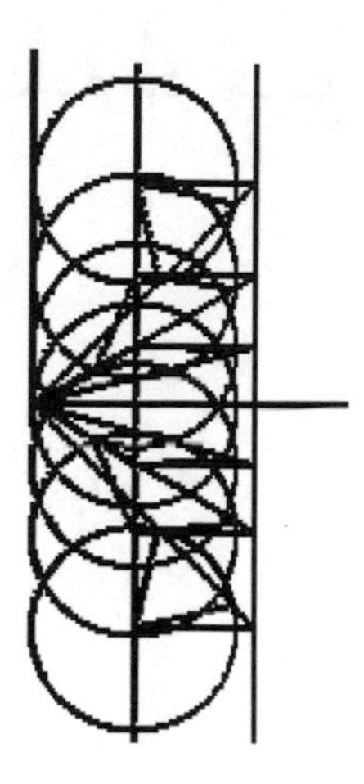

Fig. 8.21 Succesive positions

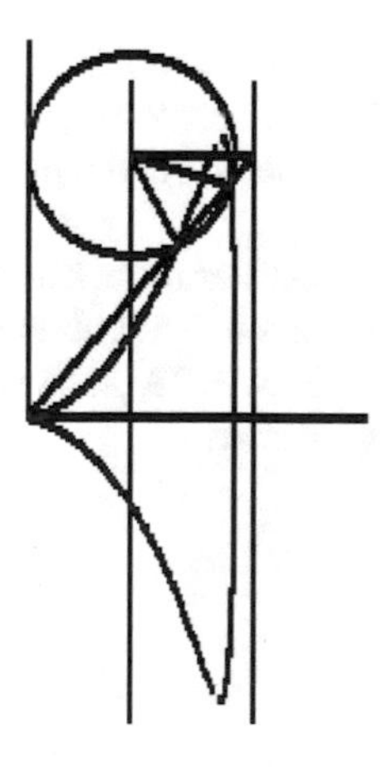

Fig. 8.22 The generated trajectory

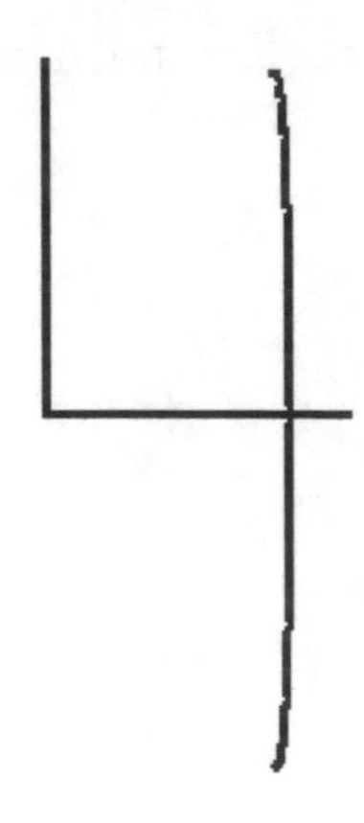

Fig. 8.23 The locus of point B

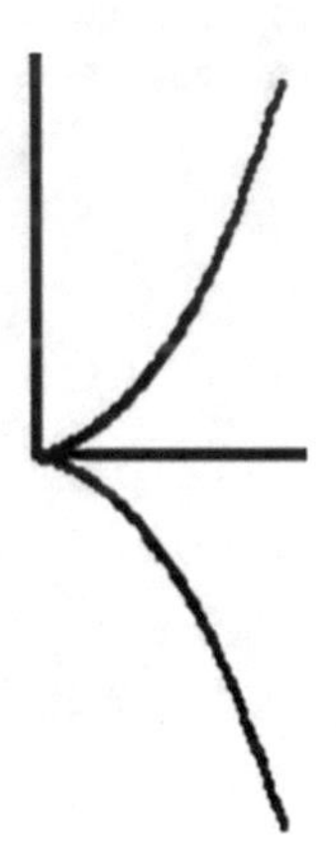

Fig. 8.24 The locus of point D

For the values: XF = 40: XG = 70: BE = 30: DE = 30, the kieroid curve shown in Fig. 8.25 was generated.

Therefore, the mechanism described above draws kieroid curves (Figs. 8.26 and 8.27).

Furthermore, other properties of this mechanism have been established, starting from a locus problem. Thus, XG was successively modified and the curves shown in Figs. 8.28, 8.29 and 8.30, to were obtained, some of them being known from geometry.

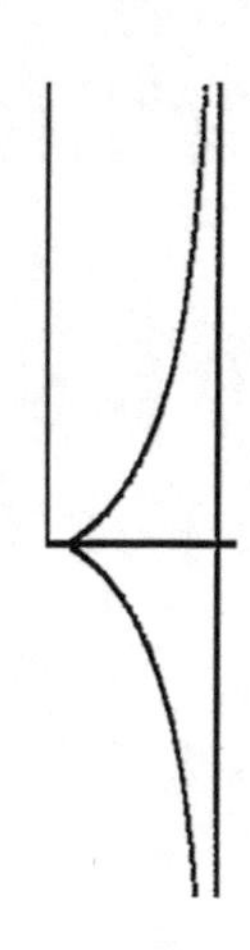

Fig. 8.25 Another Kieroid curve

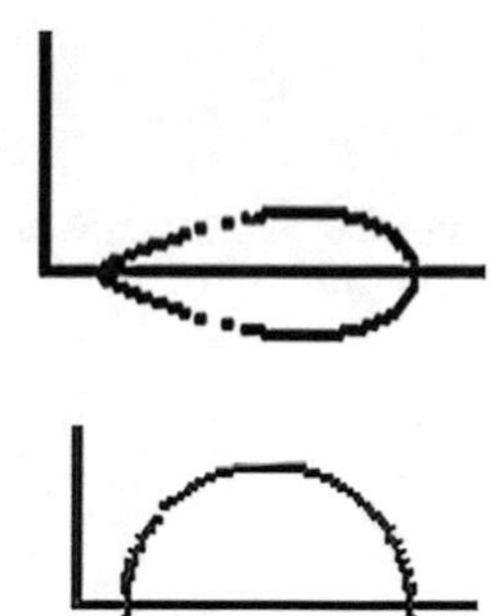

Fig. 8.26 Kieroid curve for XG = −70

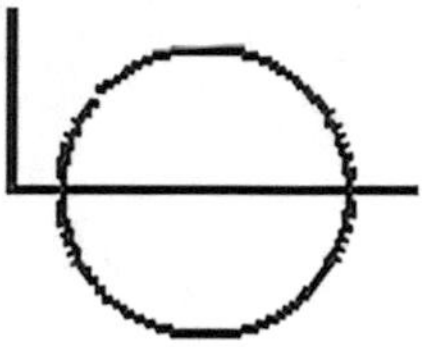

Fig. 8.27 Kieroid curve for XG = 0

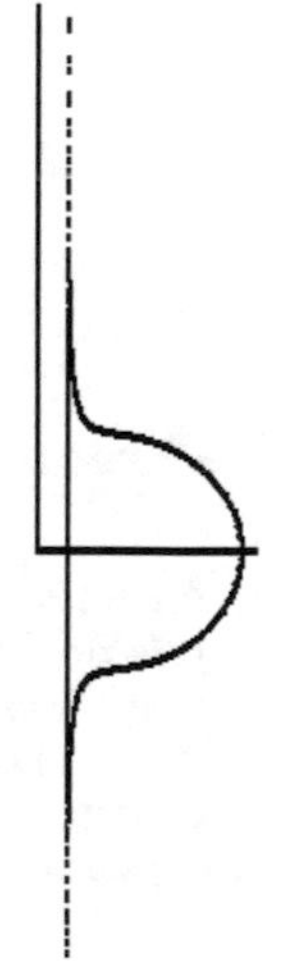

Fig. 8.28 Kieroid curve for xg = 10

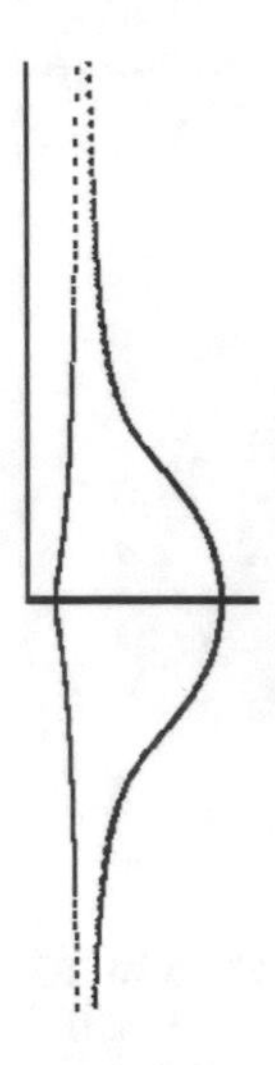

Fig. 8.29 Kieroid curve for XG = 20

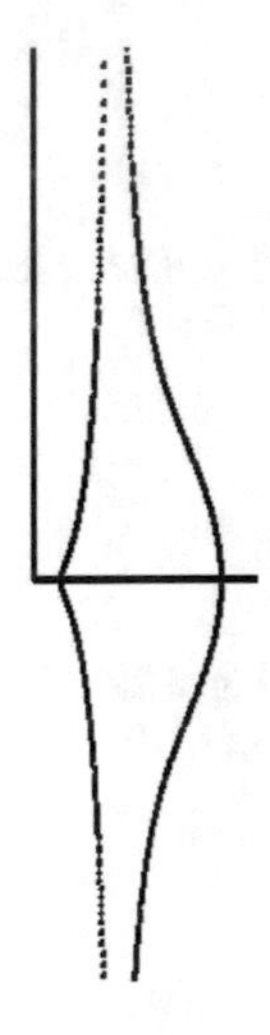

Fig. 8.30 Kieroid curve for XG = 30

References

1. Ferreol R (2006–2011) Encyclopedie des formes ramarquables courbes, surfaces, fractals, polyedre. https://www.mathcurve.com
2. https://www.mathcurve.com/courbes2d/kappa/kappa.shtml
3. Popescu I (2016) Locuri geometrice şi imagini estetice generate cu mecanisme. Editura Sitech, Craiova
4. https://www.mathcurve.com/courbes2d/kieroide/kieroide.shtml

Chapter 9
The 'Butterfly' Locus Type

Abstract Two concurrent fixed lines have at their point of intersection a torque of rotation, so that another line rotates around this point, this line intersecting with another line that goes with its heads on the fixed lines, looking for the geometric place of this point of intersection. This generates the "butterfly" curve. Many similar curves result in changing the angle between a fixed line and an abscissa.

We have the following problem (Fig. 9.1): there are given two fixed lines AB and AC, concurrent in A point. The segment that has a constant length, BC, slides with the heads on the fixed lines. A secant drawn through A point cuts the BC segment in point D, and the ADC angle is constant. The locus of point D is required [1].

The synthesis of the generating mechanism was done, as shown in Fig. 9.2.

The sides AB and AC are fixed, and the BC bar is gliding through the sliders from B and C points on the fixed lines [2].

In Fig. 9.2, we have BC, α, γ, ψ, that are constant and φ,λ, AB, AC, AD, XD, YD, that are variable. Based on Fig. 9.2, we can write the following expressions:

$$\varepsilon = \pi - (90 - \alpha - \gamma) - \delta = \pi - (\phi - \alpha) - \psi \tag{9.1}$$

$$\frac{BC}{\sin(90 - \alpha - \gamma)} = \frac{AB}{\sin\,\varepsilon} = \frac{AC}{\sin\,\delta} \tag{9.2}$$

$$x_D = AC\,\cos\,\alpha + CD\,\cos\,\lambda \tag{9.3}$$

$$y_D = AC\,\sin\,\alpha + CD\,\sin\,\lambda \tag{9.4}$$

$$\lambda = \pi + \alpha - \varepsilon \tag{9.5}$$

from where ε, δ, λ, AB, AC, XD, YD may be determined.

I. Popescu et al., *Problems of Locus Solved by Mechanisms Theory*,
Springer Tracts in Mechanical Engineering,
https://doi.org/10.1007/978-3-030-63079-9_9

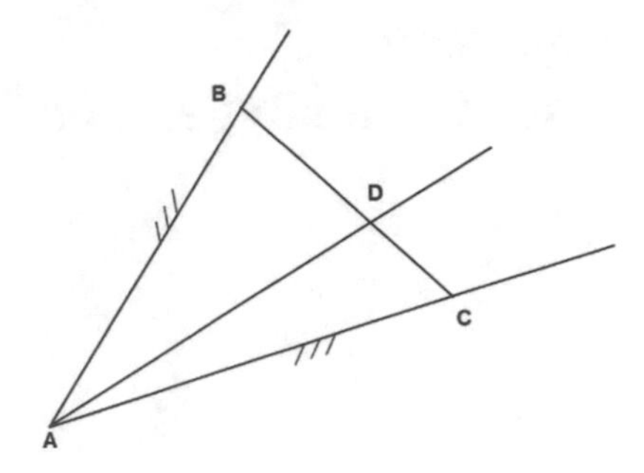

Fig. 9.1 Geometric conditions

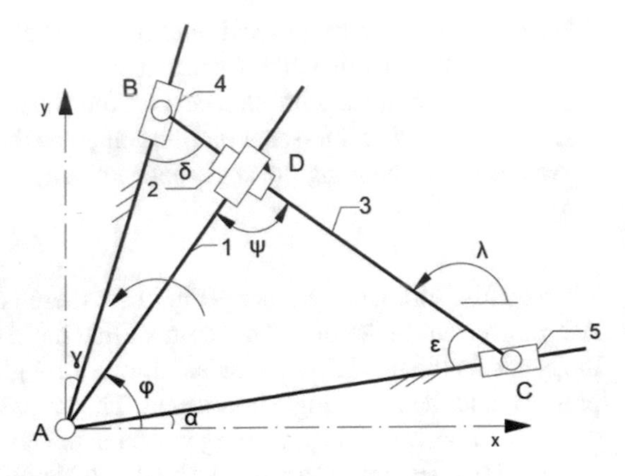

Fig. 9.2 Equivalent mechanism

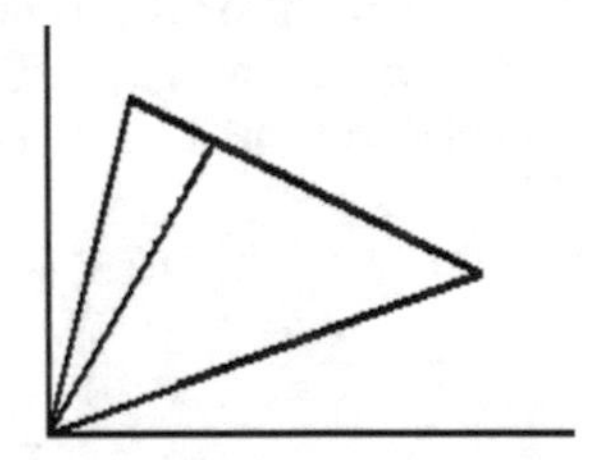

Fig. 9.3 The mechanism in one position

For $\varphi = 60$, BC $= 110$, $\alpha = 20$, $\gamma = 14$, $\psi = 93$, using a computer program based on the expressions above, the mechanism was drawn in Fig. 9.3, which was checked graphically, confirming the program. Figure 9.4 shows the successive positions of this mechanism.

All constants of the mechanism may receive other values, resulting in other curves. The data from the initial position (Fig. 9.3) was maintained and the angles and then the AC and BC lengths were cycled. The results are given in the following pictures, below which the values of the modified parameters were marked [3].

The curve resulting as locus, given in Figs. 9.5, 9.6 and 9.7, is a "butterfly", having four loops, that are equal two by two.

The curve in Fig. 9.8 may be obtained also for $\alpha = 180$.

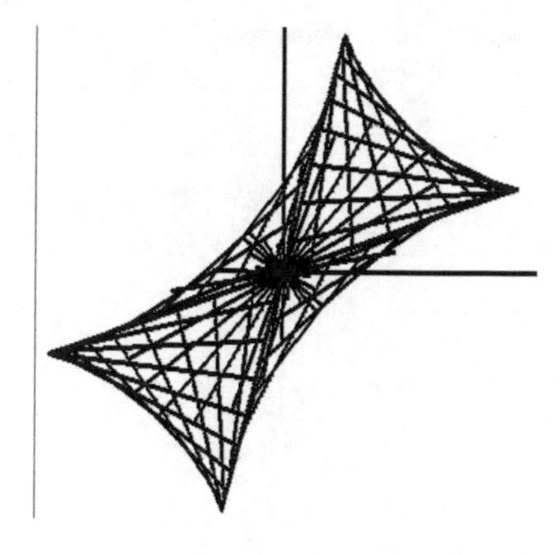

Fig. 9.4 The mechanism in successive positions

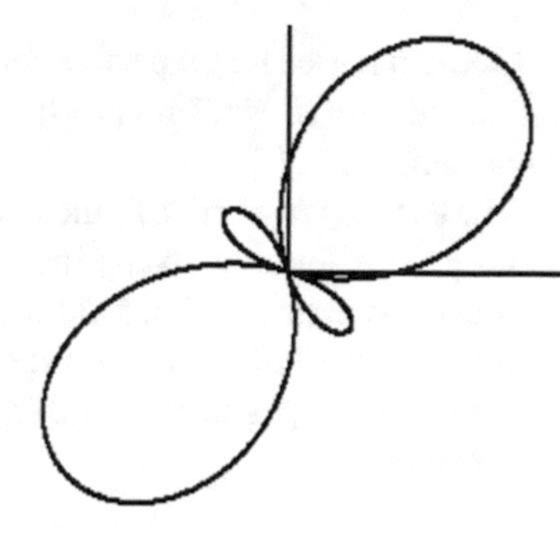

Fig. 9.5 The Butterfly locus

Fig. 9.6 The Butterfly locus for $\alpha = 0$

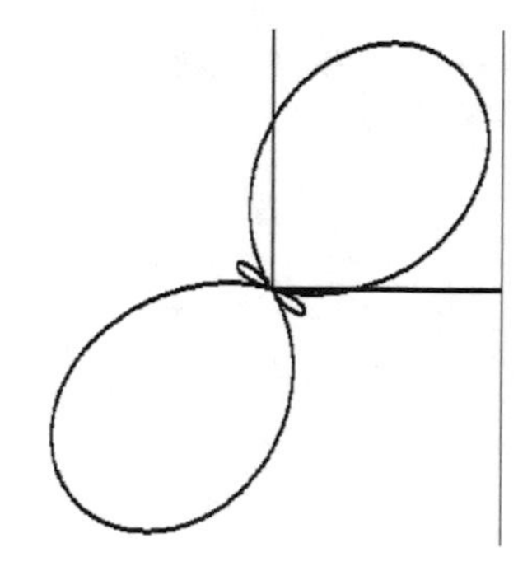

Fig. 9.7 The Butterfly locus for α =35

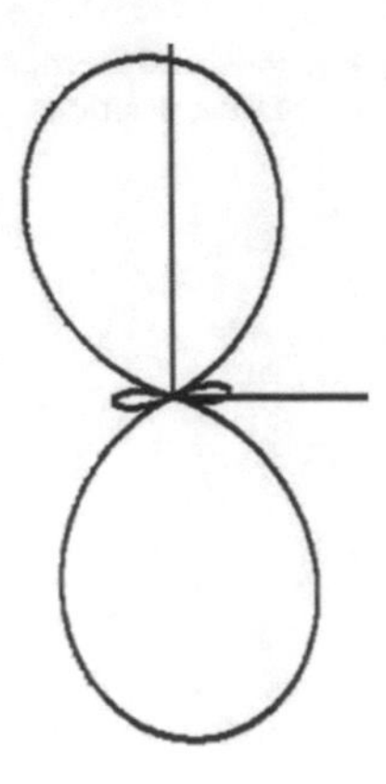

Fig. 9.8 The Butterfly locus for $\alpha = 300$

One can notice that the value for α does not interfere with the shape of the resulted curve, but only its size and position. For the following Figs. 9.9, 9.10, 9.11 and 9.12, γ was modified.

The curves are similar to the ones above, but they have different sizes. After that, α and γ were simultaneously modified (Fig. 9.13).

The curves are similar, but they are differently positioned in relation to the axes system. For the following pictures, ψ was modified (Figs. 9.14 and 9.15).

Here too the positions and dimensions of the curves were changed, but they are the same type.

Fig. 9.9 The Butterfly locus for $\gamma = 0$

Fig. 9.10 The Butterfly locus for $\gamma = 20$

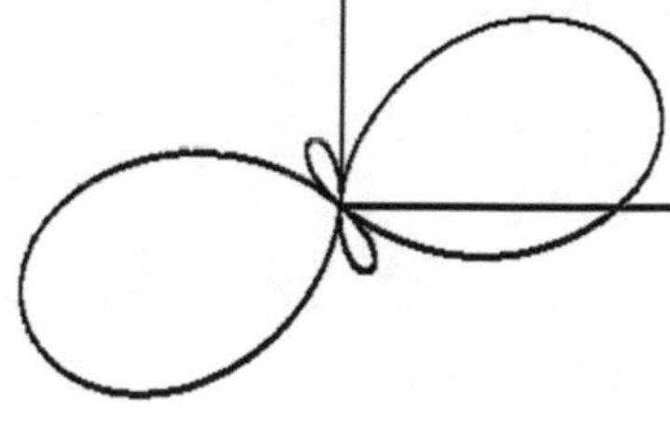

Fig. 9.11 The Butterfly locus for $\gamma = 40$

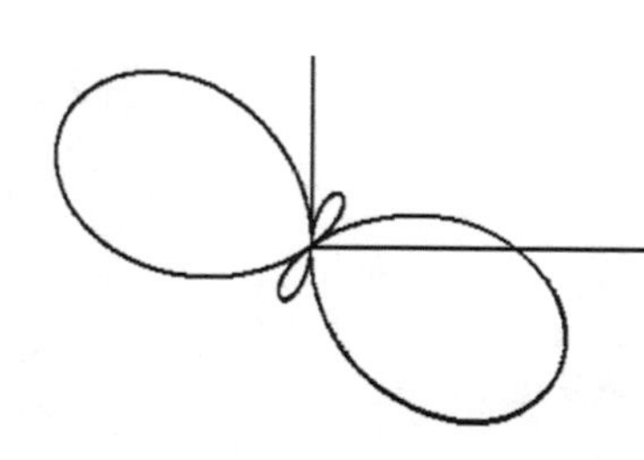

Fig. 9.12 The Butterfly locus for $\gamma = 320$

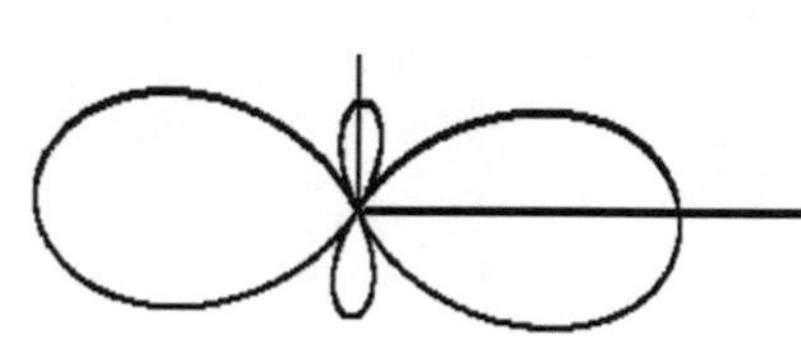

Fig. 9.13 The Butterfly locus for $\alpha = 210$, $\gamma = 120$

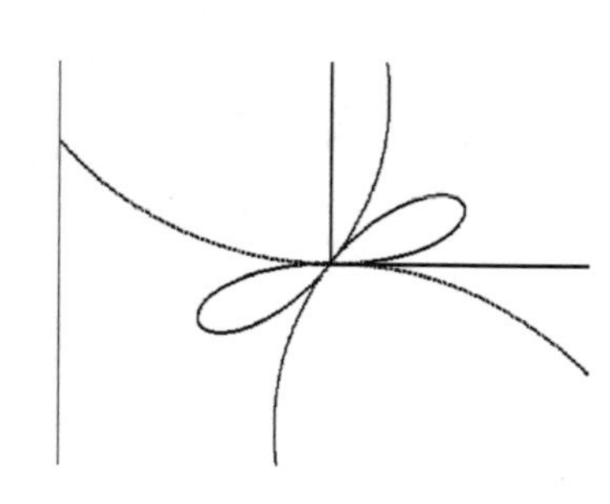

Fig. 9.14 The Butterfly locus for $\psi = 2$

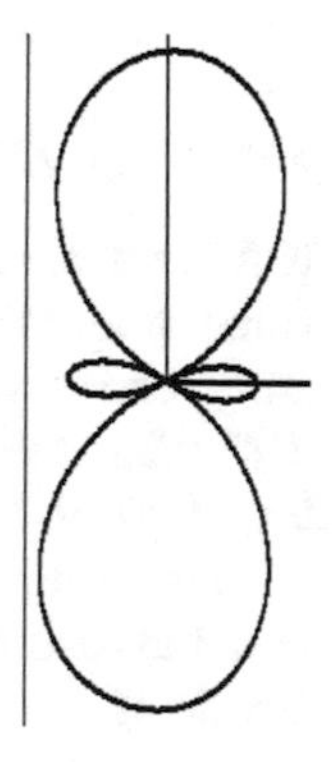

Fig. 9.15 The Butterfly locus for $\psi = 90$

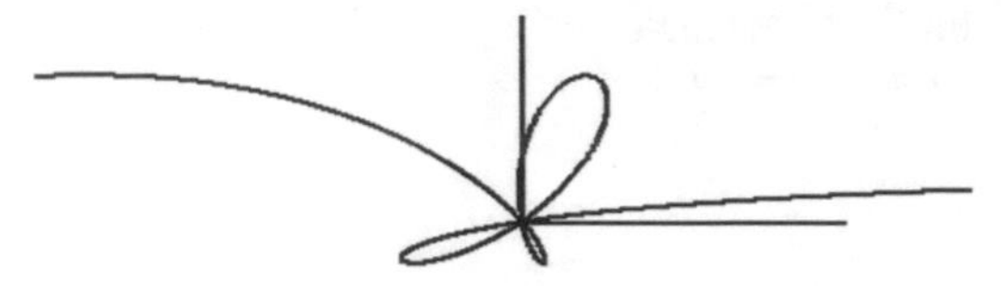

Fig. 9.16 q = 1

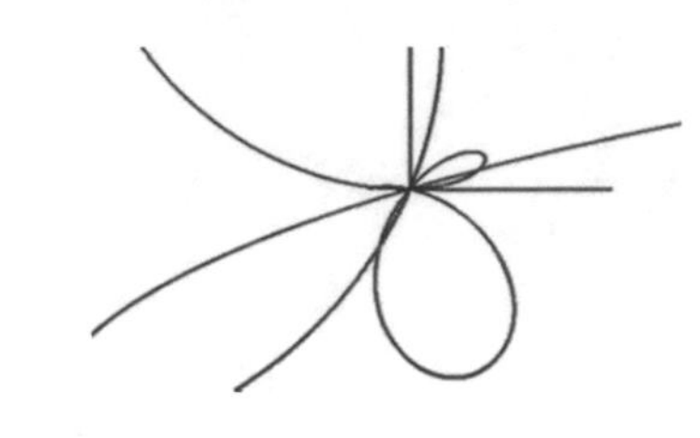

Fig. 9.17 q = 0.1

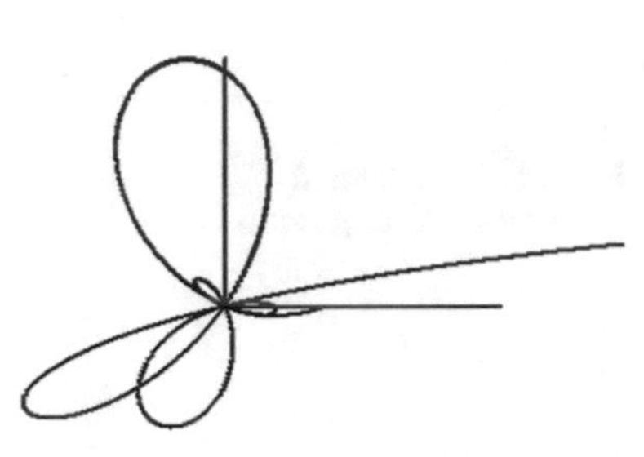

Fig. 9.18 q = 0.3

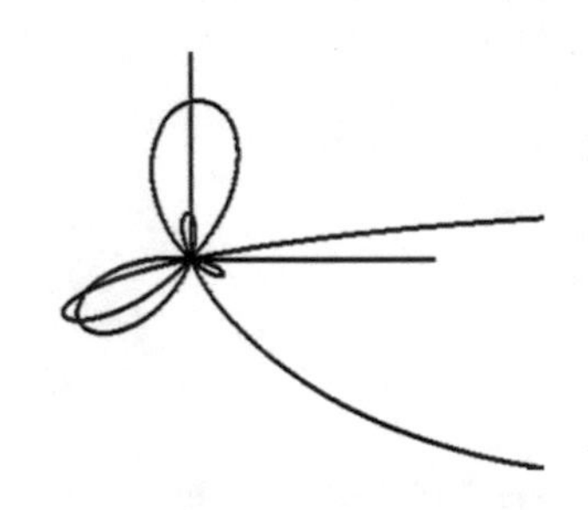

Fig. 9.19 q = 0.5

- **The line AC is an arbitrary one**.

If AC is arbitrary, but rotates around A with the angle $\psi = q.\varphi$, choosing different values of q, other curves result, having different from the above. Below the figures are given the values of q (Figs. 9.16, 9.17, 9.18, 9.19, 9.20, 9.21, 9.22, 9.23, 9.24, 9.25, 9.26, 9.27, 9.28, 9.29, 9.30, 9.31, 9.32, 9.33, 9.34, 9.35, 9.36, 9.37, 9.38, 9.39, 9.40, 9.41, 9.42, 9.43, 9.44, 9.45 and 9.46).

The result is a wide variety of curves, open, with several branches, different from the case when the leading elements rotate in the same direction or in opposite directions.

Fig. 9.20 q = 0.8

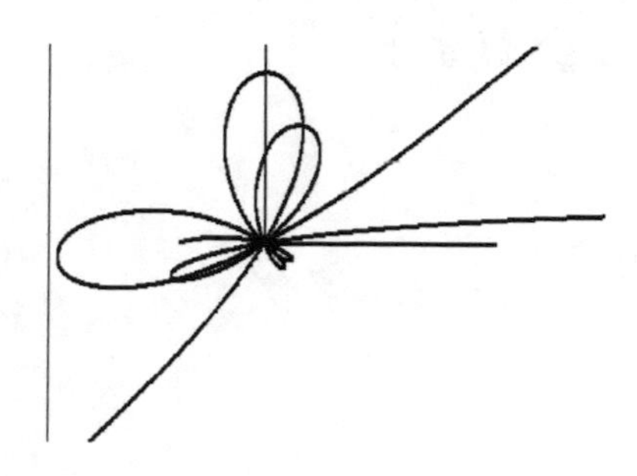

Fig. 9.21 q = 1.1

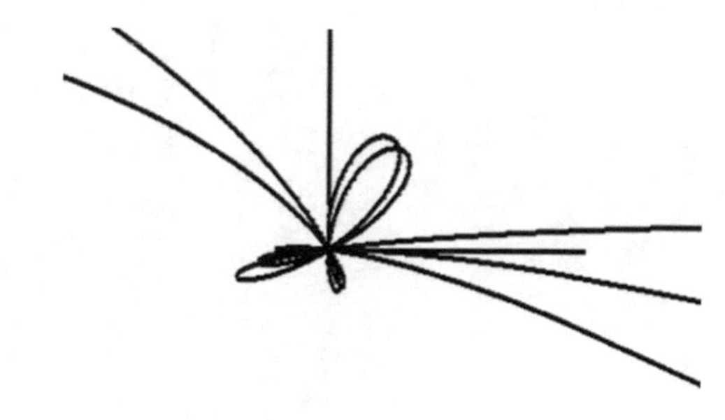

Fig. 9.22 q = 1.3

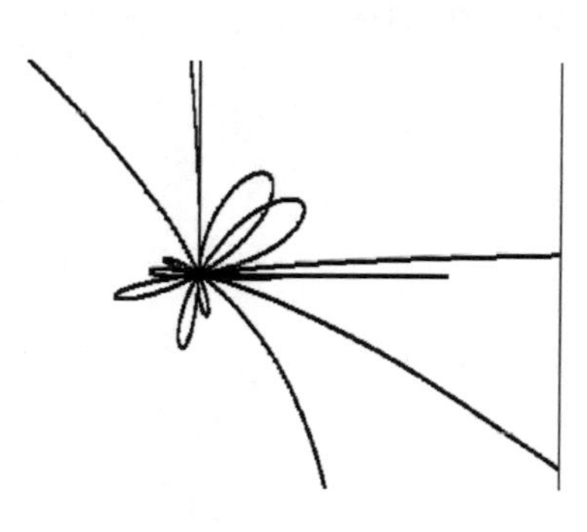

Fig. 9.23 q = 1.5

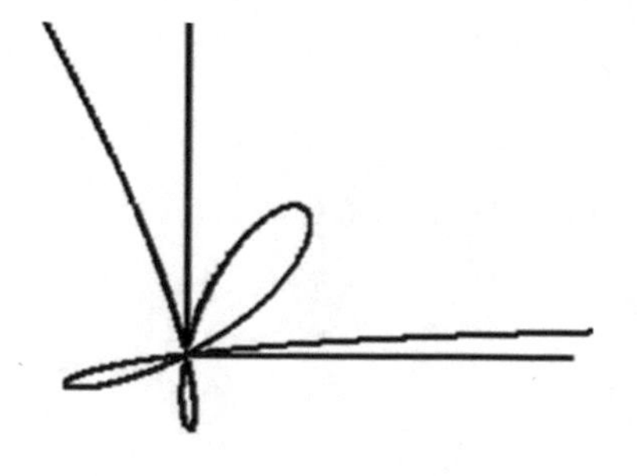

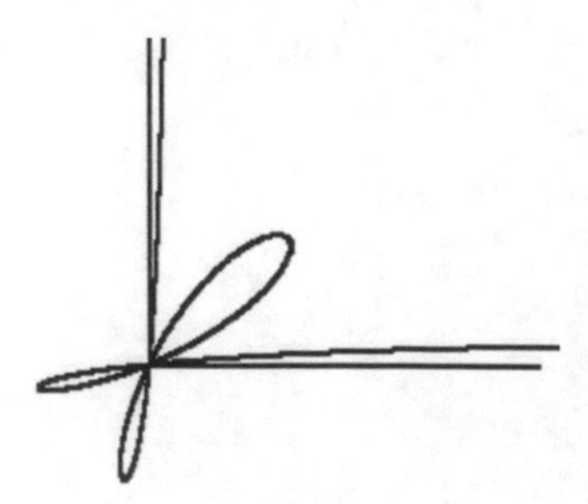

Fig. 9.24 q = 2

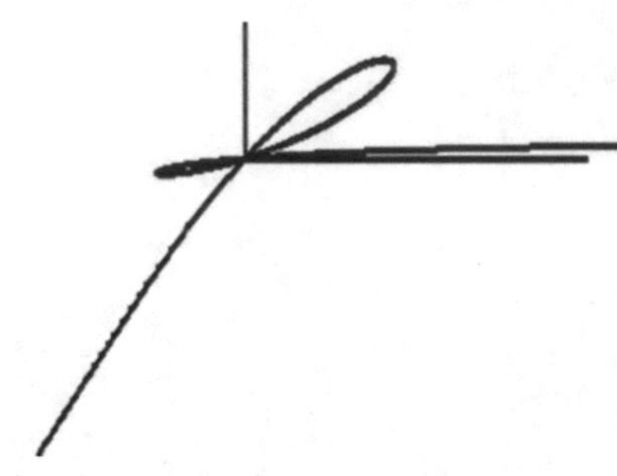

Fig. 9.25 q = 3

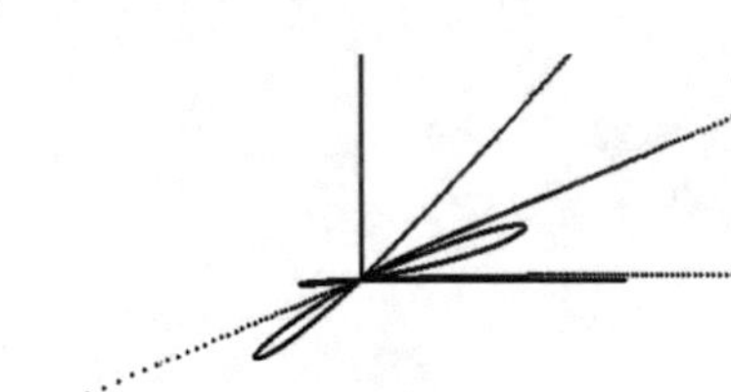

Fig. 9.26 q = 0.7

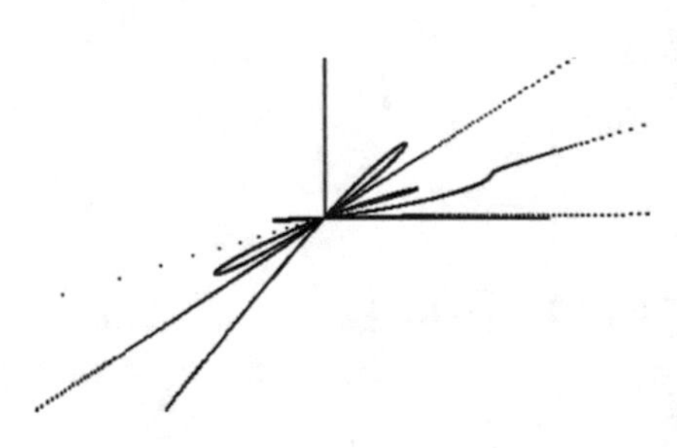

Fig. 9.27 q = 10

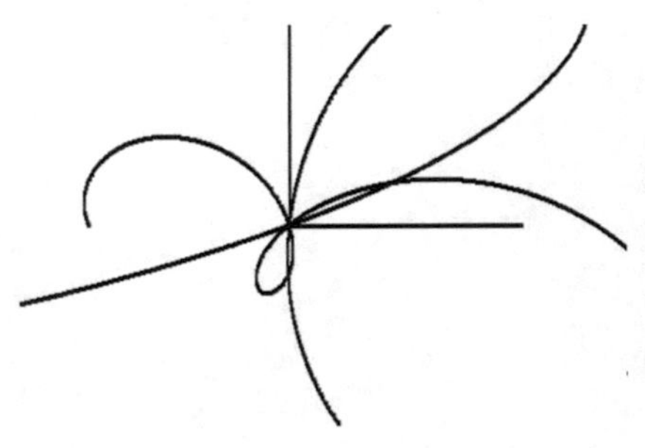

Fig. 9.28 q = −0.1

Fig. 9.29 q = −0.2

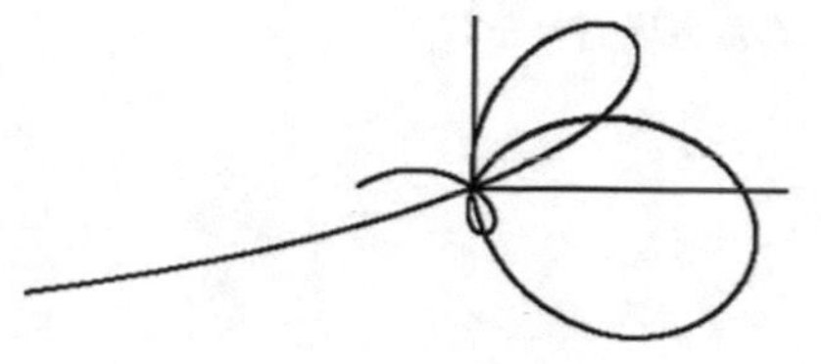

Fig. 9.30 q = −0.3

Fig. 9.31 q = −0.4

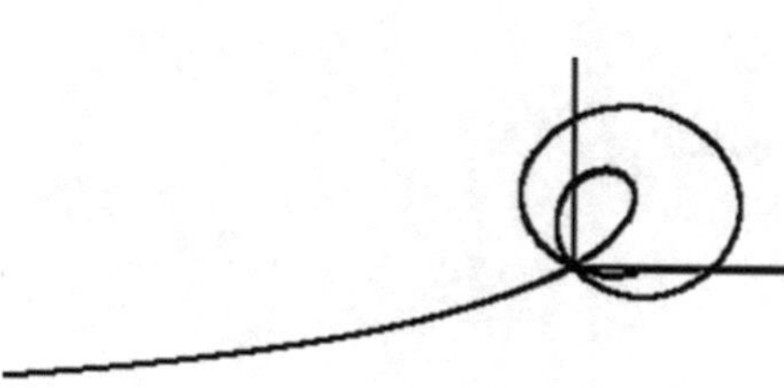

Fig. 9.32 q = −0.5

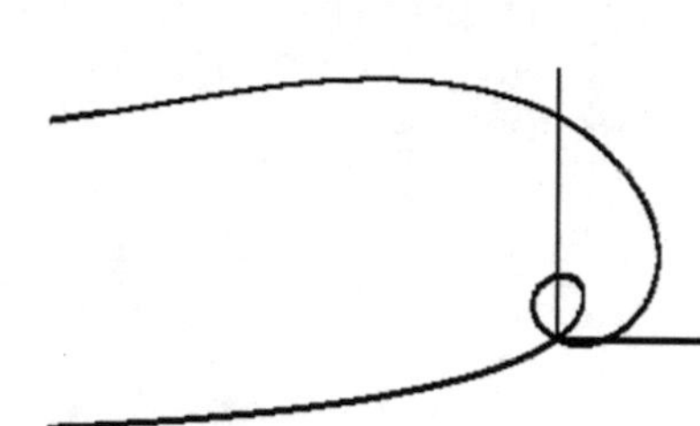

Fig. 9.33 q = −0.6

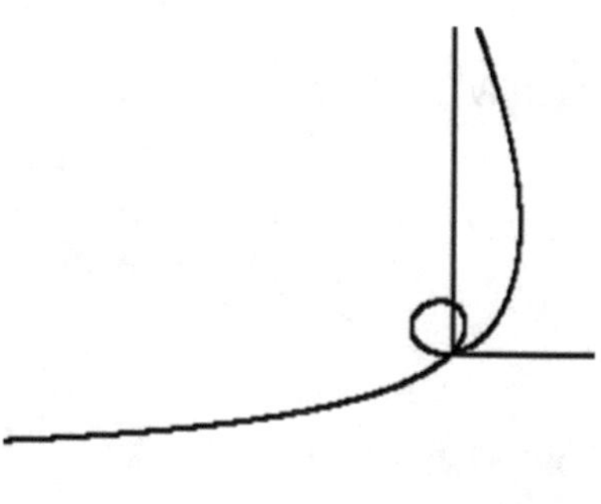

Fig. 9.34 q = −0.7

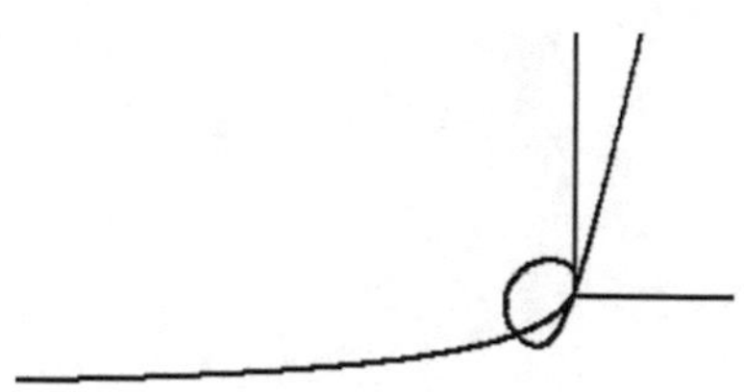

Fig. 9.35 q = −8

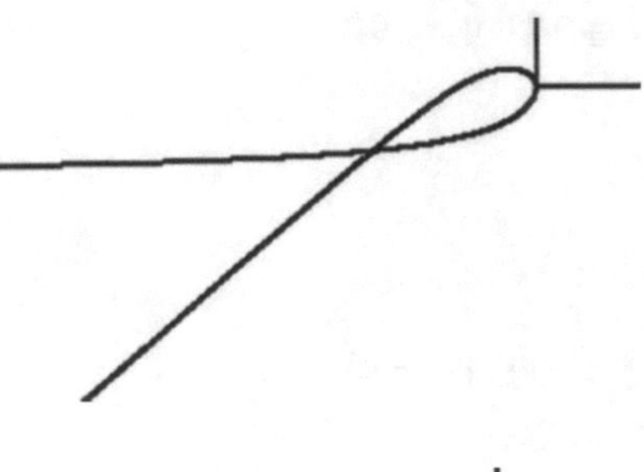

Fig. 9.36 q = −0.9

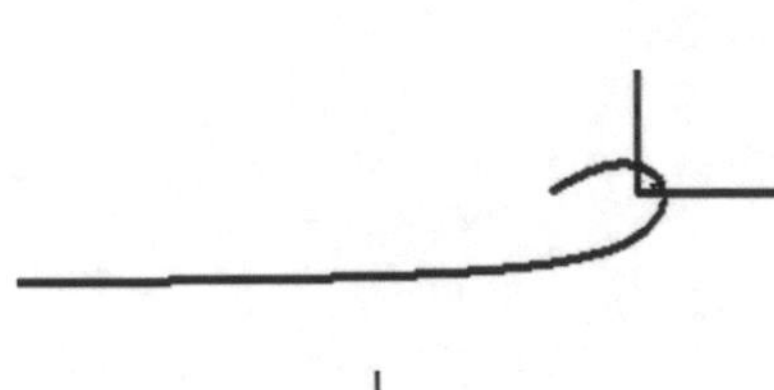

Fig. 9.37 q = −1

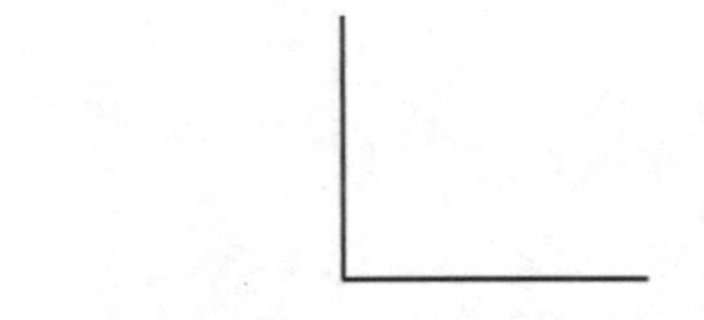

Fig. 9.38 q = −1.2

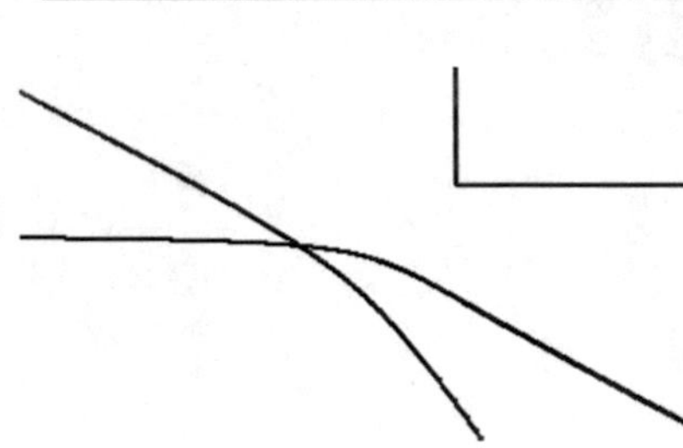

Fig. 9.39 q = −1.5

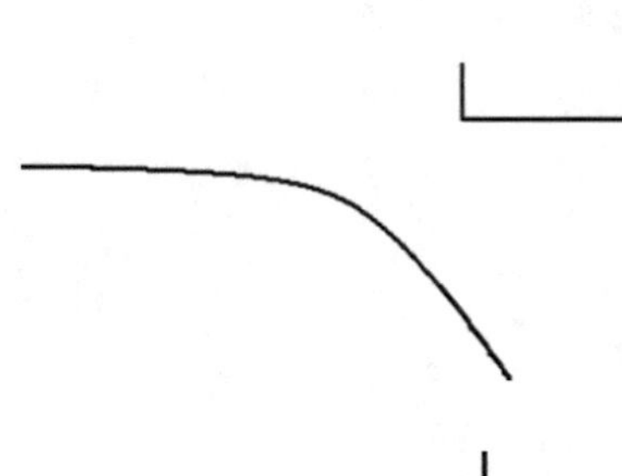

Fig. 9.40 q = −1.3

Fig. 9.41 q = −2

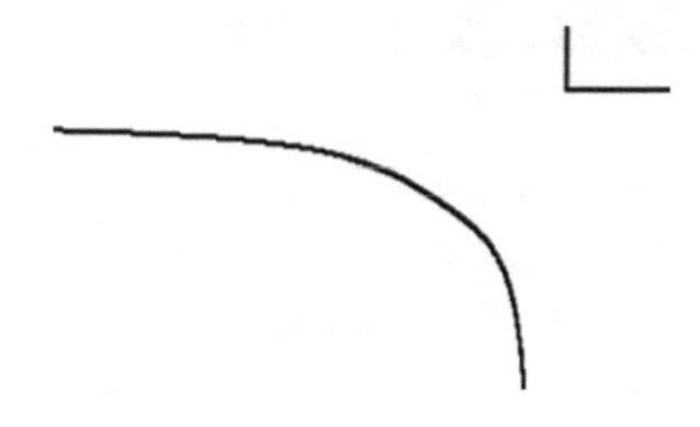

Fig. 9.42 q = −2.5

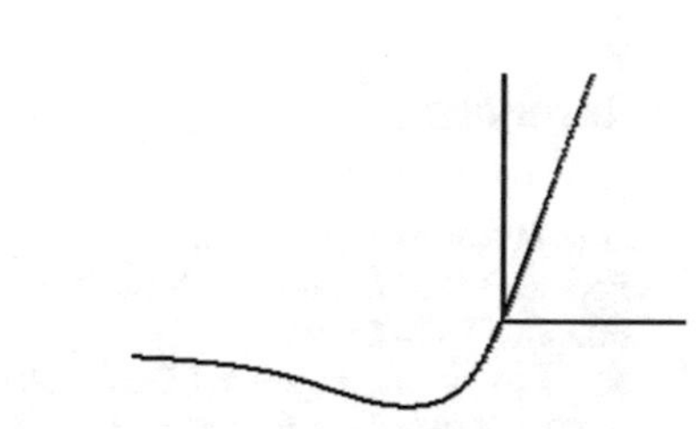

Fig. 9.43 q = −3

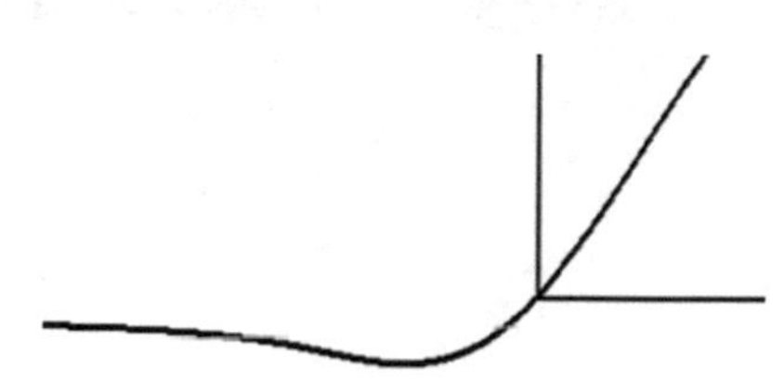

Fig. 9.44 q = −5

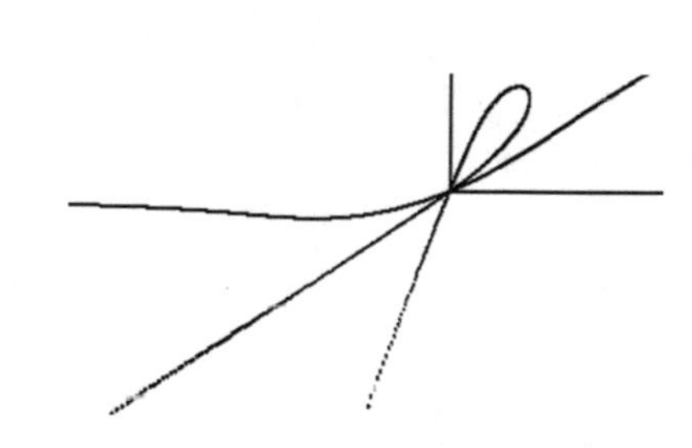

Fig. 9.45 q = −6

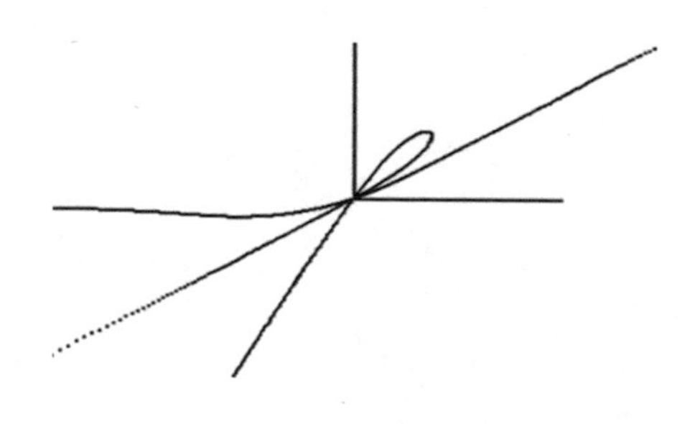

Fig. 9.46 $q = -7$

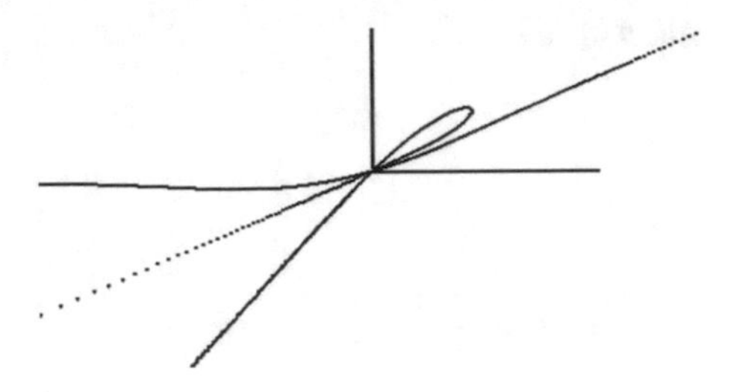

References

1. https://www.mathcurve.com
2. Popescu I (2017) Curbe de bielă în diferite plane şi locuri geometrice generate de mecanisme. Editura Sitech, Craiova
3. Sass L, Duţă A, Popescu I (2017) Nature gives us beauty, insoiring the technical creativity. Journal of industrial design and engineering graphics. Papers of the international conference on engineering graphics and design, section 3: engineering computer graphics. 12(1), 41–46

Chapter 10
Nephroida and Rhodonea as Loci

Abstract On two intersected lines in C, slide the slides 2 and 5, having the line BG = constant. The trajectories of points E and F on these slides are required. The trajectory of E is a woody, and of F a half branch of nephroid. Taking the symmetrical curves, the complete nephroid results, on which the woodcuts of E are also given. By modifying the distance GF, other curves are obtained. Another nephroid generating mechanism is given. Two intersected straight lines in A fixed, rotate around A, both being conductive elements with movements correlated by the coefficient q, and the line BC = const. slides with point C on AC. Point C will draw a rhodonea. By changing the value of q, variants of rhodoneas with several branches are obtained. The plotted curve has the number of sides equal to q − 2. If the sides AB and BC are equal, other types of rhodoneas are obtained, in which the branches pass through the origin of the axis system.

10.1 Nephroida as Locus

Nephroida is a flat curve formed of two ovals [6]. The shape is similar to kidneys. It was studied by Huygens, Tschirnhausen, Jacques Bernoulli, Daniel Bernoulli and R. Proctor, who gave her name in 1878, from the Greek word nephros, that is kidneys [1]. It is a normal two-point epicicloid, described by a point on the r/2 radius circle that turns on the r-circle [2, 3]. Figure 10.1 shows the nephroida curve for r1 = 100 and r2 = 50.

In [4] it is presented the mechanism shown in Fig. 10.2, where D point traces a nephroid if: AO = OB = BC = CD = a and the angle BCD = angle BOA. The leading element is AB.

The equation of this nephroide is:

$$\left(x^2 + y^2\right)\left(x^2 + y^2 - a^2\right)^2 = 4a^2\left(x^2 + y^2 - ax\right)^2 \tag{10.1}$$

Starting from the study of this mechanism, we have come to a problem of locus, on which the mechanism of Fig. 10.3 was built. The GAB line turns around point A,

I. Popescu et al., *Problems of Locus Solved by Mechanisms Theory*,
Springer Tracts in Mechanical Engineering,
https://doi.org/10.1007/978-3-030-63079-9_10

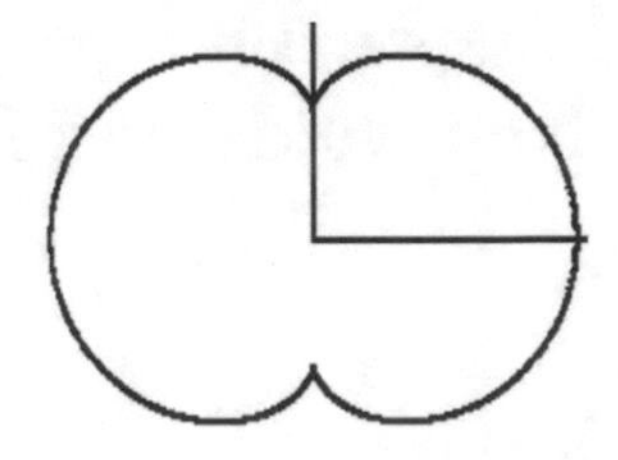

Fig. 10.1 The nephroida

and points G and B slide on two lines concurrent in point C, around which they can turn.

The following expressions are given:

$$x_B = \mathrm{AB}\cos\varphi = x_C + \mathrm{BC}\cos\psi \tag{10.2}$$

$$y_B = \mathrm{AB}\sin\varphi = y_C + \mathrm{BC}\sin\psi \tag{10.3}$$

$$x_G = \mathrm{AG}\cos(\varphi + \pi) \tag{10.4}$$

$$y_G = \mathrm{AG}\sin(\varphi + \pi) \tag{10.5}$$

$$\mathrm{CG} = \mathrm{sqr}\left[(x_G - x_C)^2 + (y_G - y_C)^2\right] \tag{10.6}$$

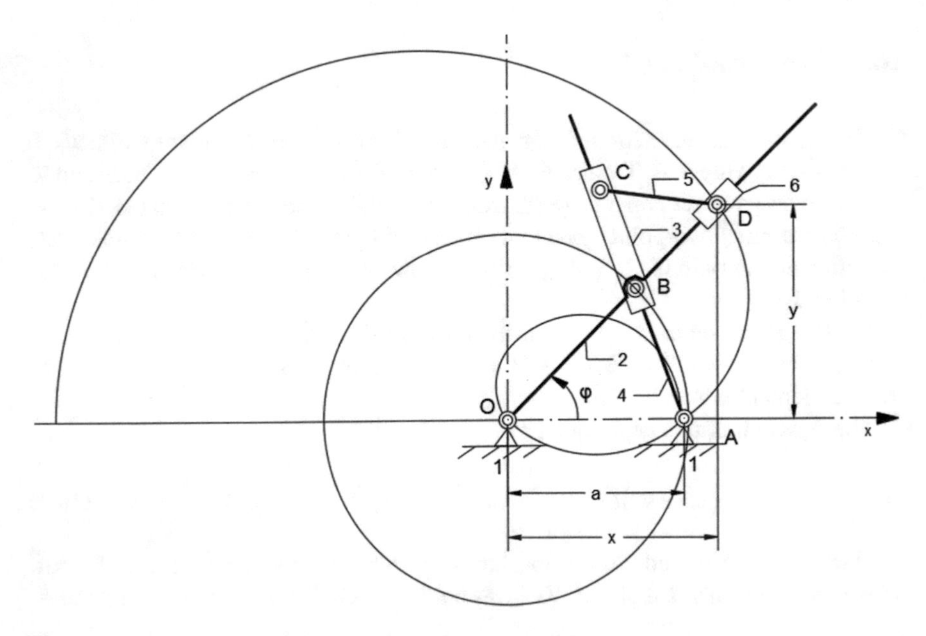

Fig. 10.2 A generating mechanism of the nephroid curve [4]

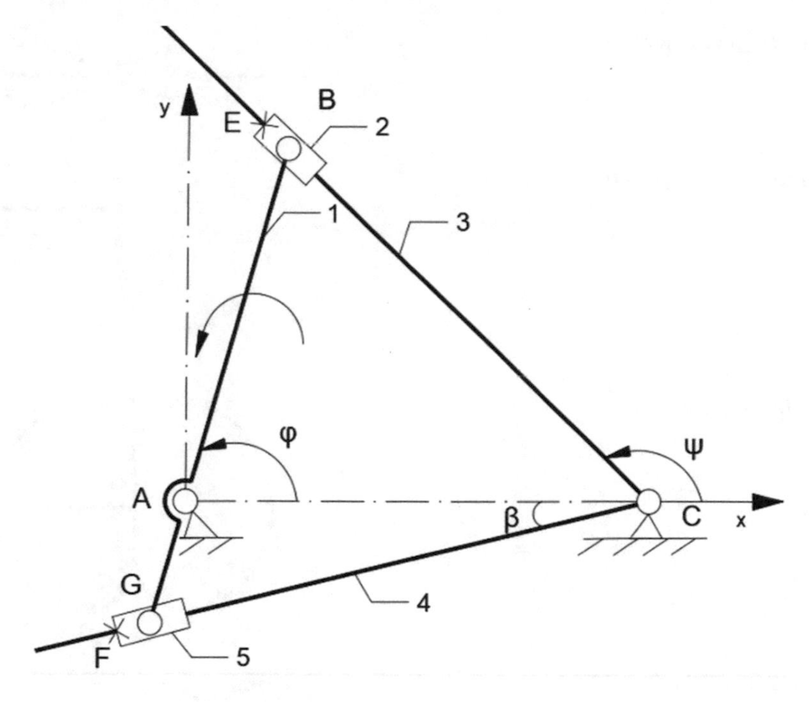

Fig. 10.3 The synthesized mechanism

$$AG^2 = x_C^2 + CG^2 - 2x_C \cdot CG \cos \beta \tag{10.7}$$

$$x_F = x_G + GF \cos(\pi + \beta) \tag{10.8}$$

$$y_F = y_G + GF \sin(\pi + \beta) \tag{10.9}$$

$$x_E = x_B + BE \cos \psi \tag{10.10}$$

$$y_E = y_B + BE \sin \psi \tag{10.11}$$

In Fig. 10.4 it is shown the trajectory of point E (a point on the slide 2), and in Fig. 10.5 it is presented the trajectory of point F (point on slide 5), for: XC = a = 30; GF = a; AB = a; AG = a; BE = 2a.

It is noticed that F point traces one part of the nephroid, and E point describes a lemniscate curve. Considering that the curves are symmetrical, F point follows a nephroid (Fig. 10.6), and both points plot the curves shown in Fig. 10.7.

Next, the above data was maintained, only a was changed, the obtained curves being presented in Figs. 10.7 and 10.8.

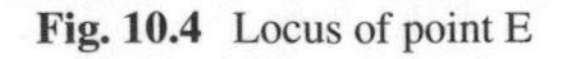
Fig. 10.4 Locus of point E

Fig. 10.5 Locus of point F

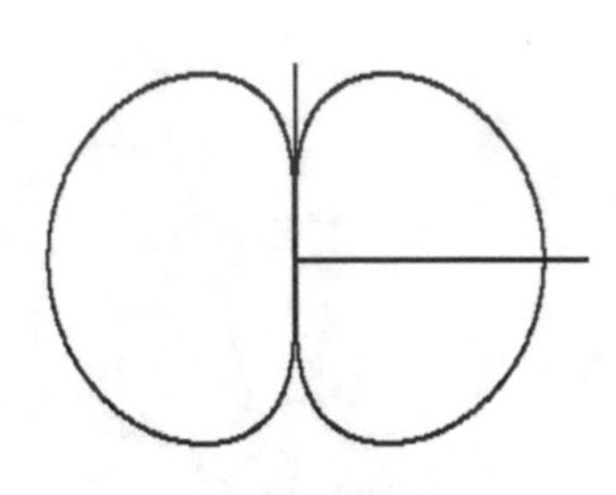

Fig. 10.6 The nephroid

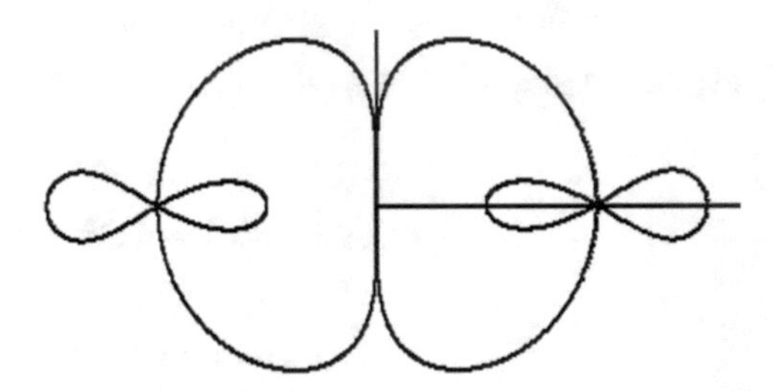

Fig. 10.7 Locus for E and F points

Furthermore, we considered again XC = 2a and GF was changed, the obtained curves being presented (Figs. 10.9, 10.10, 10.11 and 10.12).

In conclusion, this mechanism also plots nephroids, and by modification, plots other loci as well.

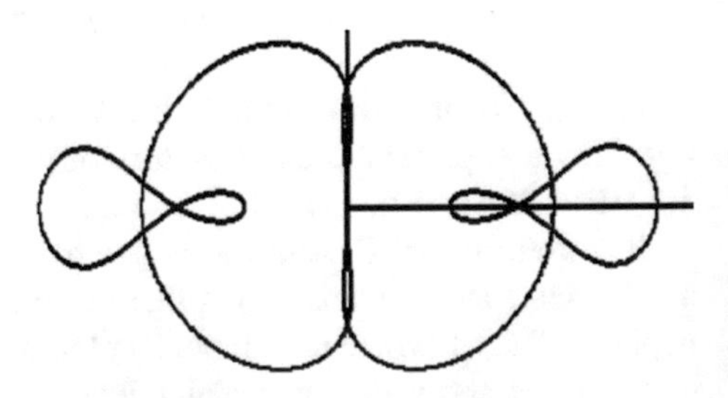

Fig. 10.8 Locus for E and F points for XC = 50

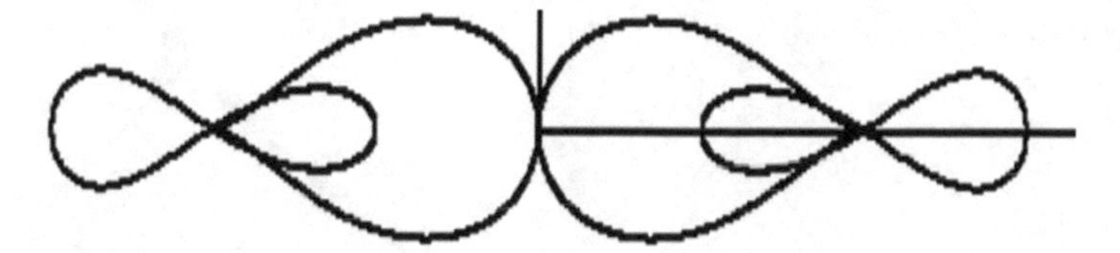

Fig. 10.9 Locus for E and F points for GF = −30

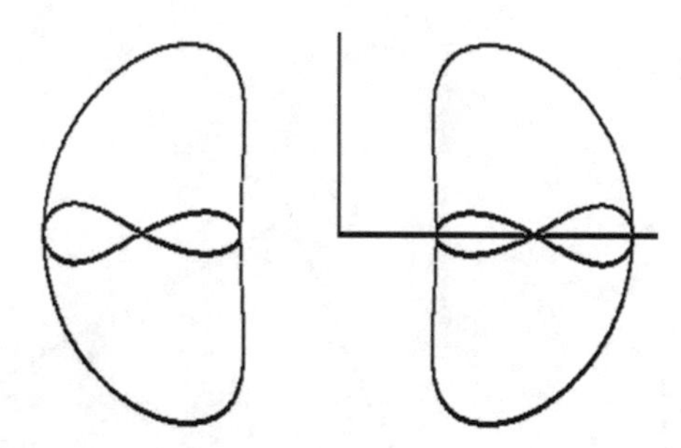

Fig. 10.10 Locus for E and F points for GF = 60

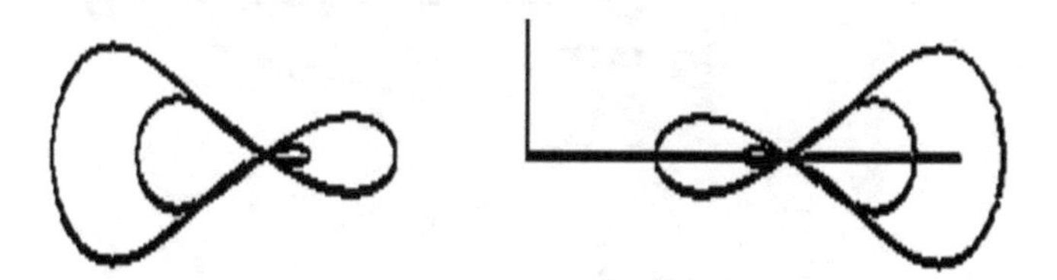

Fig. 10.11 Locus for E and F points for GF = −80

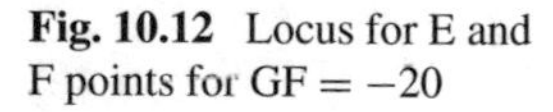

Fig. 10.12 Locus for E and F points for GF = −20

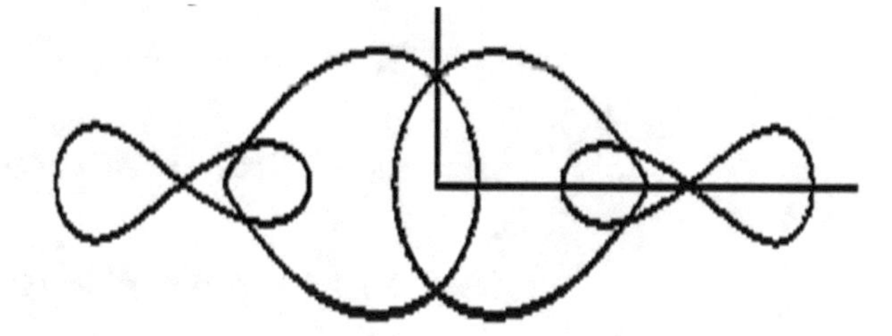

10.2 The Rhodonea as an Aesthetic Locus

The AB and AC lines were turning on around A point, and the C point of the BC line slides on the AC line. The locus of C point was required (Fig. 10.13) [5].

The mechanism presented in Fig. 10.14 was built, where the AC segment was variable, so a slider was placed in C point.

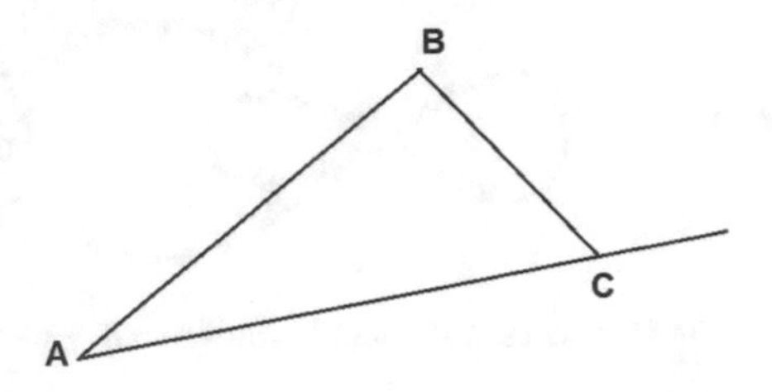

Fig. 10.13 Geometric conditions

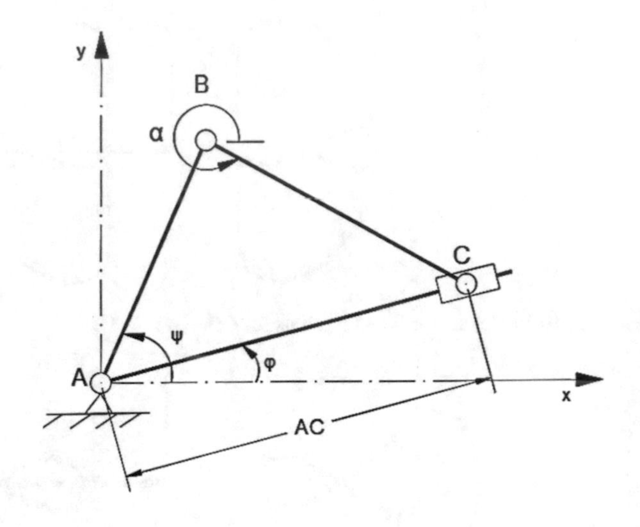

Fig. 10.14 The mechanism

The following expressions were given:

$$x_B = AB \cos \psi + x_A \tag{10.12}$$

$$y_B = AB \sin \psi + y_A \tag{10.13}$$

$$x_C = x_B + BC \cos \alpha = AC \cos \varphi \tag{10.14}$$

$$y_C = y_B + BC \sin \alpha = AC \sin \varphi \tag{10.15}$$

$$\psi = c \cdot \phi \tag{10.16}$$

The AB and AC bars had rotation motions that were liniarly correlated by the q coefficient. We searched loci obtained for different values of q. For AB = 50 mm and BC = 90 mm, the results were the curves called rhodonea shown in Figs. 10.15, 10.16, 10.17, 10.18, 10.19, 10.20, 10.21, 10.22, 10.23, 10.24 and 10.25.

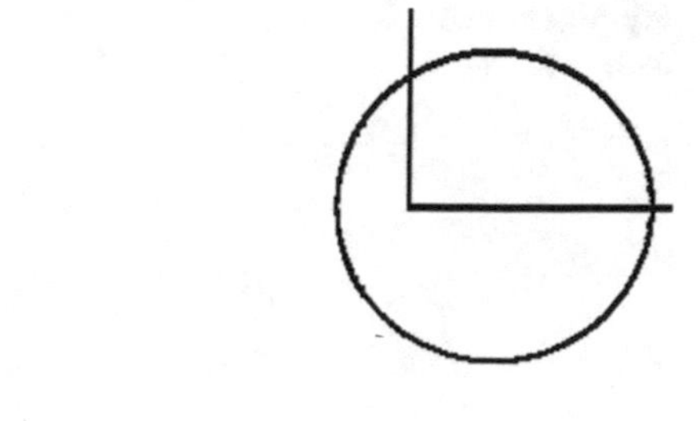

Fig. 10.15 Rhodonea curve for $q = 2$

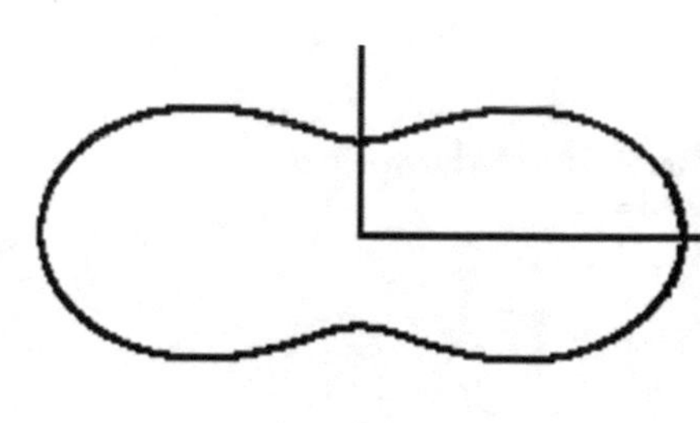

Fig. 10.16 Rhodonea curve for $q = 3$

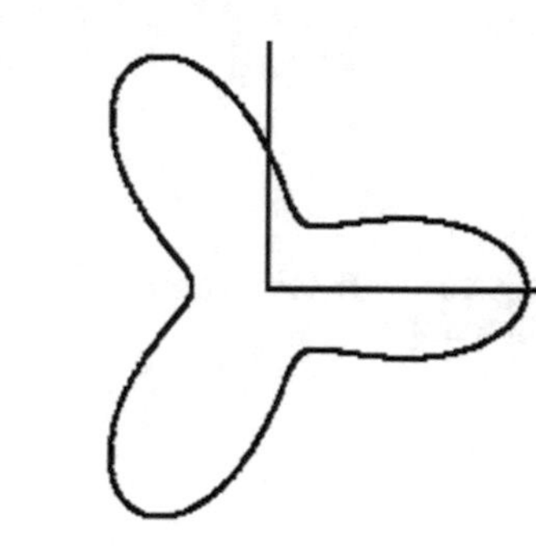

Fig. 10.17 Rhodonea curve for $q = 4$

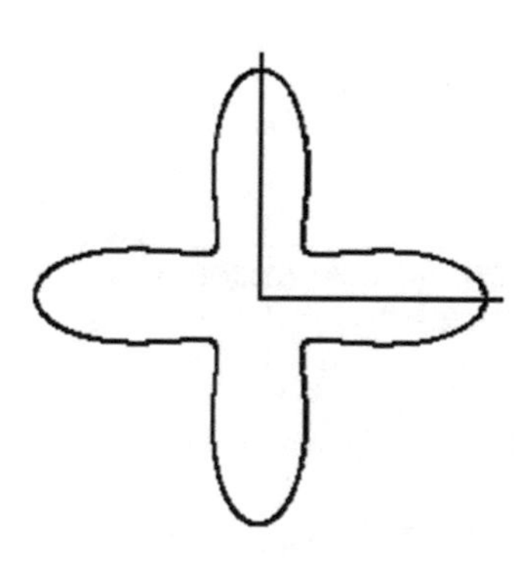

Fig. 10.18 Rhodonea curve for $q = 5$

We found that the number of loops is equal to $c - 1$. The generated rhodoneas were aesthetic loci.

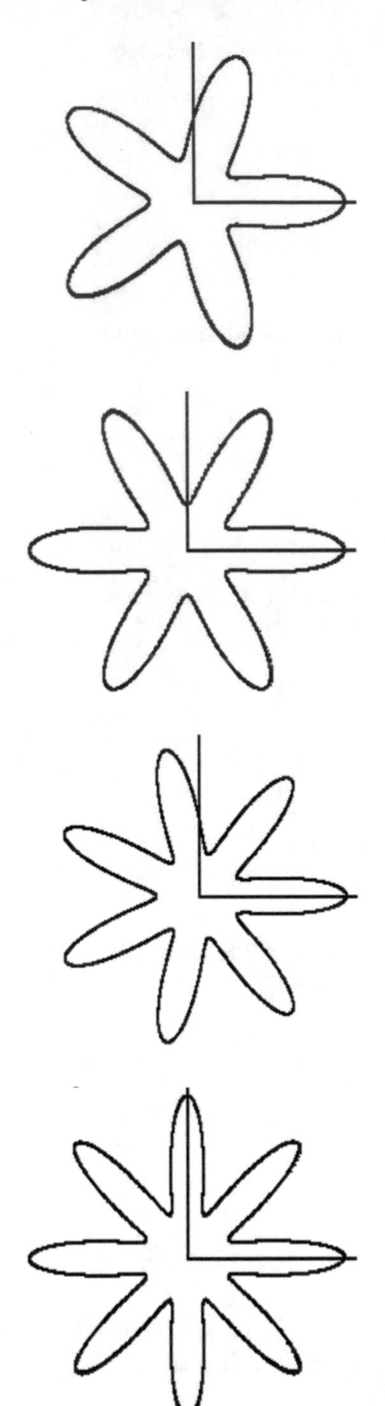

Fig. 10.19 Rhodonea curve for q = 6

Fig. 10.20 Rhodonea curve for q = 7

Fig. 10.21 Rhodonea curve for q = 8

Fig. 10.22 Rhodonea curve for q = 9

Fig. 10.23 Rhodonea curve for q = 10

Fig. 10.24 Rhodonea curve for q = 20

Fig. 10.25 Rhodonea curve for q = 50

If it was considered AB = BC = 50 mm, other rhodoneas were obtained with the centers in the origin of the coordinate system (Figs. 10.26, 10.27, 10.28, 10.29 and 10.30).

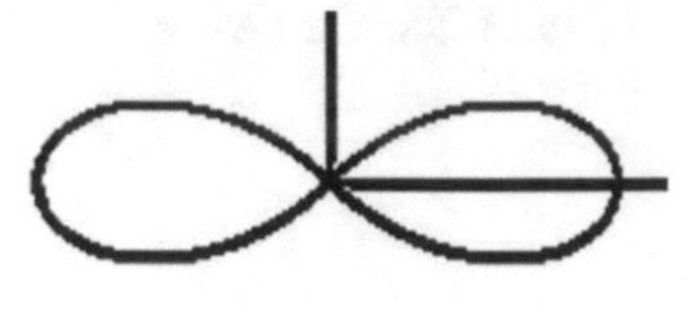

Fig. 10.26 Rhodonea curve for q = 3

Fig. 10.27 Rhodonea curve for q = 4

Fig. 10.28 Rhodonea curve for q = 5

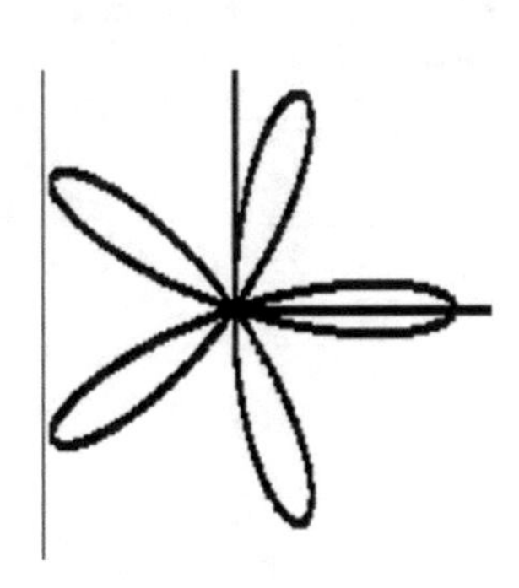

Fig. 10.29 Rhodonea curve for q = 6

Fig. 10.30 Rhodonea curve for q = 8

References

1. Cherciu M, Popescu I (2018) Problem of locus resolved by the theory of mechanisms. In: Symposium "durability and reliability of mechanical systems"—SYMECH 2018, Târgu-Jiu, 11–12 May 2018. Issue 1(21):133–138
2. Luca L, Popescu I, Ghimiş Ş (2012) Studies regarding of aesthetics surfaces with mechnanisms. In: Mathematical methods for information science & economics. Proceedings of the 17th WSEAS international conference on applied mathematics (AMATH'12), Montreux, Dec 2012, pp 249–254
3. Teodorescu ID, Teodorescu SD (1975) Culegere de probleme de geometrie superioară. Editura Didactică şi Pedagogică, Bucureşti
4. Artobolevskii II (1959) Teoria mehanizmov dlia vosproizvedenia ploskih crivâh. Izd. Academii Nauk CCCP, Moskva
5. Luca L, Popescu I (2014) Aesthetic effects of epicycloids generator mechanisms. In: Analele Universităţii « Constatin Brâncuşi » din Târgu Jiu, Seria Inginerie, Nr. 4/2014, pp 30–38
6. https://www.mathcurve.com/courbes2d/nephroid/nephroid.shtml

Chapter 11
Successions of Aesthetic Rhodonea

Abstract We start from the problem in the previous chapter but point A is moving vertically while AB and AC rotate. Sequences of rhodoneas are formed which form aesthetic surfaces, which differ according to the coefficient "c", where "c" is the correlation coefficient between the angles φ and ψ ($\psi = c \cdot \varphi$). Numerous surfaces with special aesthetics have been obtained. Decimal values for c were also used. Negative values were also taken for c, i.e. the conductive elements rotate in opposite directions. Next, other aesthetic surfaces were obtained considering that point A moves on the x-axis.

Below, we start from the idea of obtaining aesthetic images generated by a mechanism built on an issue of geometrical place [1, 3]. The mechanism can be used to generate aesthetic images on the screen, to build toys and in kinetic art.

A movable point A and two lines AB and AC (Fig. 11.1) are given on the y axis, which rotate around A. From B, with the length BC the line AC intersects, thus finding the position of point C. The locus of C is required.

The generating mechanism is given in Fig. 11.2, whose synthesis was done as in the previous chapter, but here A is mobile on the ordinate.

It is noticed that the mechanism consists of the conductive elements AA (having the stroke S_1), AC and AB, being of type PRR-RRP.

Based on Fig. 11.1 the following equations are written:

$$x_A = 0; \ \ y_A = S_1 \tag{11.1}$$

$$x_B = x_A + a\cos\psi \tag{11.2}$$

$$y_B = y_A + a\sin\psi \tag{11.3}$$

$$x_C = x_B + b\cos\alpha = S_2\cos\varphi \tag{11.4}$$

I. Popescu et al., *Problems of Locus Solved by Mechanisms Theory*,
Springer Tracts in Mechanical Engineering,
https://doi.org/10.1007/978-3-030-63079-9_11

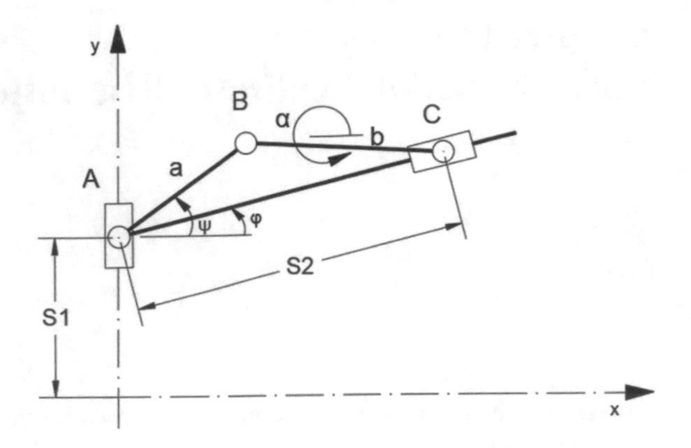

Fig. 11.1 The equivalent mechanism

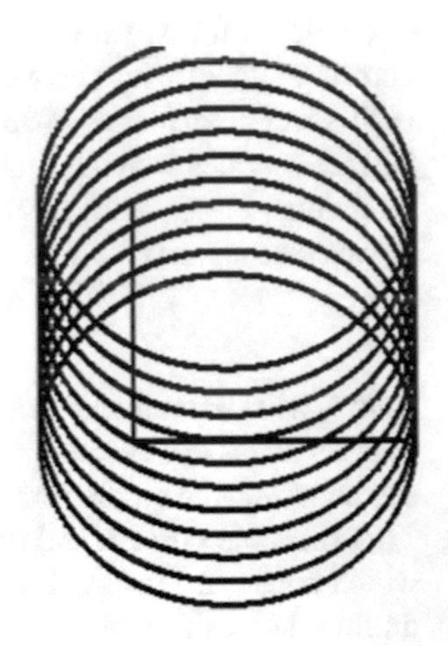

Fig. 11.2 c = 0.001

$$y_C = y_B + b\sin\alpha = S_2\sin\varphi + y_A \tag{11.5}$$

The movements for the slideway of A and for the AC element are known, and the movement of AB is considered correlated with the movement of AC through the relation $\psi = c \cdot \varphi$, and the value of the "c" factor will be changed for different cases. The coordinates of B are obtained from Eqs. (11.2) and (11.3), and from Eqs. (11.4) and (11.5) the angle α is eliminated, resulting in S_2, x_C, y_C.

We looked for aesthetic forms of the sequence of generated curves [2]. Different values for "c" were taken, hanging S_1 in a loop from 0 to 100 mm with a 10 mm pace, and φ from 0 to 360°. The dimensions: a = 35 mm, b = 70 mm were taken. Next are the sequences of loci chosen so that the images are aesthetic (Figs. 11.3, 11.4, 11.5, 11.6, 11.7, 11.8, 11.9, 11.10, 11.11, 11.12, 11.13, 11.14, 11.15, 11.16, 11.17, 11.18, 11.19, 11.20, 11.21, 11.22, 11.23, 11.24 and 11.25).

If c becomes negative, then other sequences of geometric places, given, are obtained (Figs. 11.26, 11.27, 11.28, 11.29, 11.30, 11.31, 11.32 and 11.33).

Fig. 11.3 c = 0.1

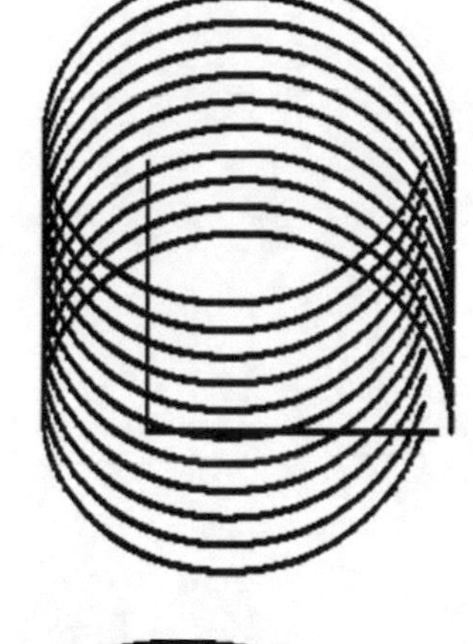

Fig. 11.4 c = 0.5

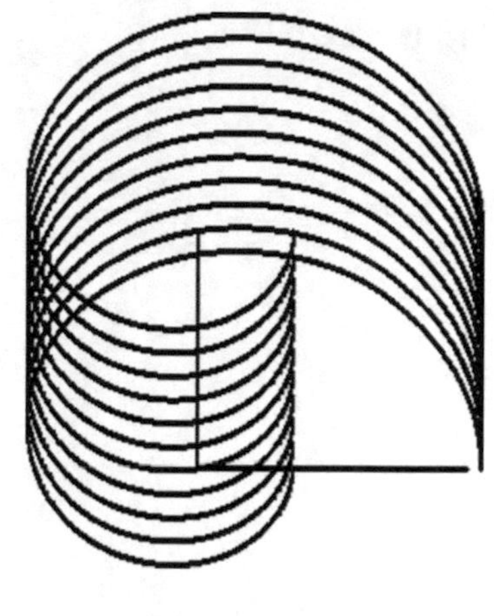

Fig. 11.5 c = 1

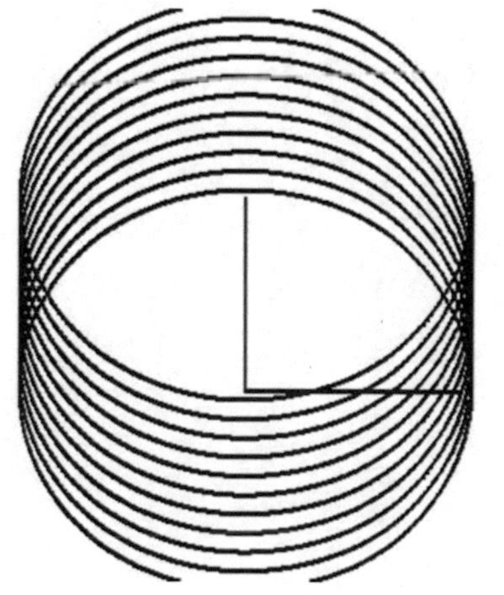

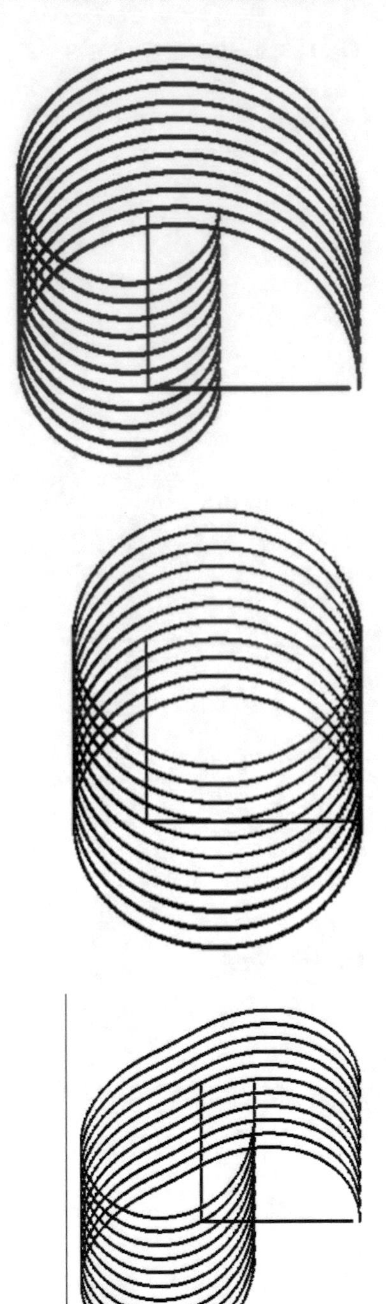

Fig. 11.6 c = 1.5

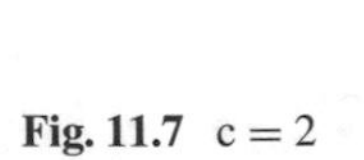

Fig. 11.7 c = 2

Fig. 11.8 c = 2.5

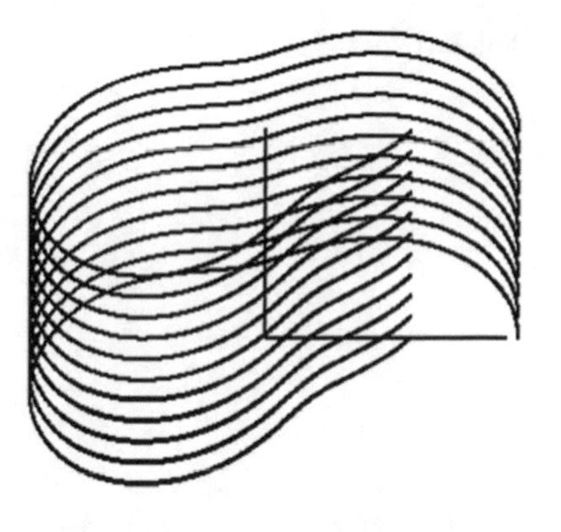

Fig. 11.9 c = 2.75

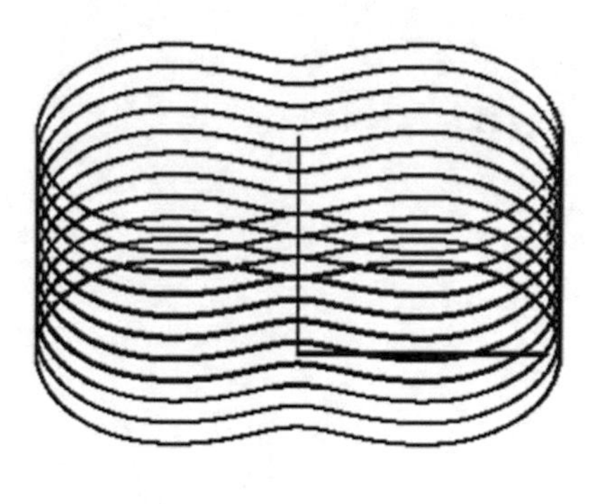

Fig. 11.10 c = 3

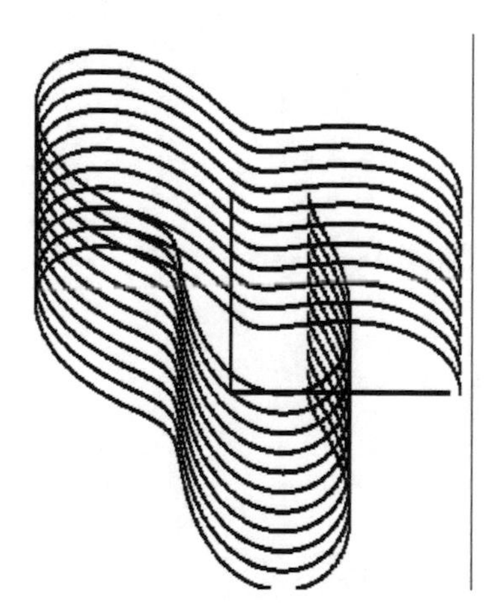

Fig. 11.11 c = 3.5

It was further considered that the point A moves along the x-axis, obtaining the sequence of curves below. They are not the sequence of curves in the cases when A moves along the y axis, rotated by 90°, but rather totally different curves (Figs. 11.34, 11.35, 11.36, 11.37, 11.38, 11.39, 11.40, 11.41, 11.42, 11.43, 11.44, 11.45, 11.46, 11.47, 11.48, 11.49, 11.50, 11.51, 11.52, 11.53, 11.54 and 11.55).

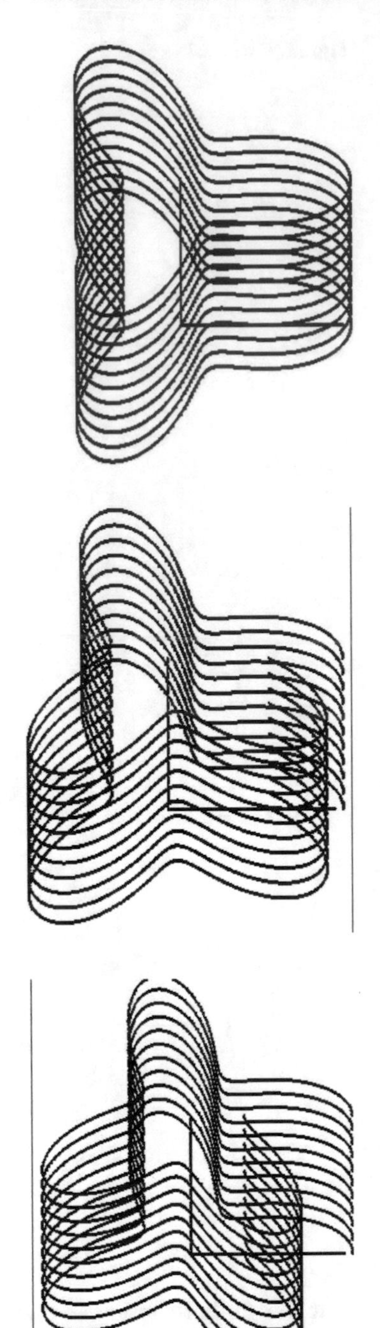

Fig. 11.12 c = 4

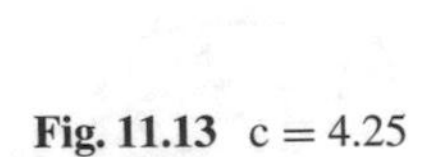

Fig. 11.13 c = 4.25

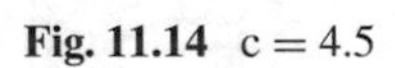

Fig. 11.14 c = 4.5

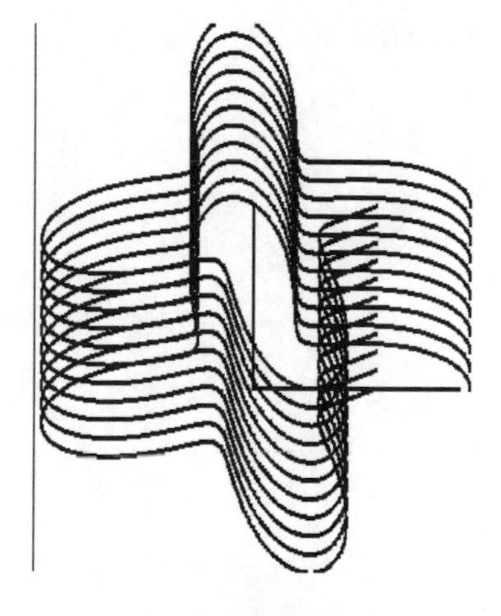

Fig. 11.15 c = 4.75

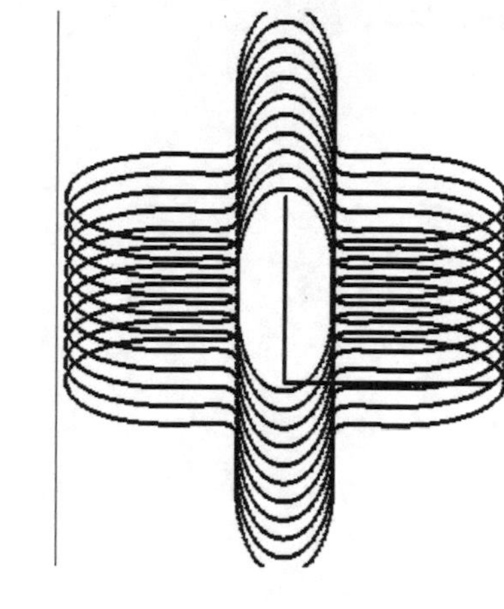

Fig. 11.16 c = 0.5

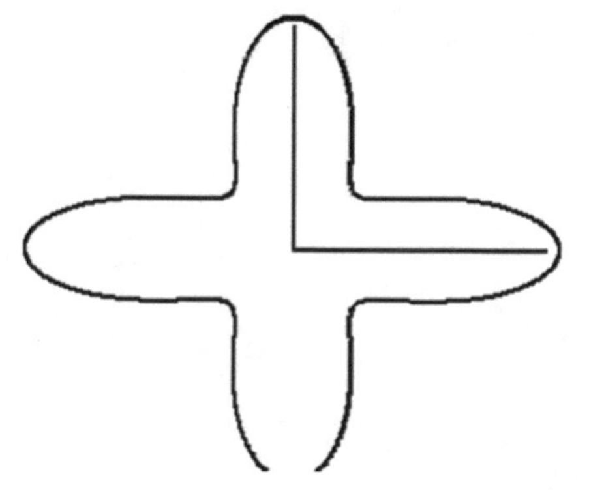

Fig. 11.17 c = 5

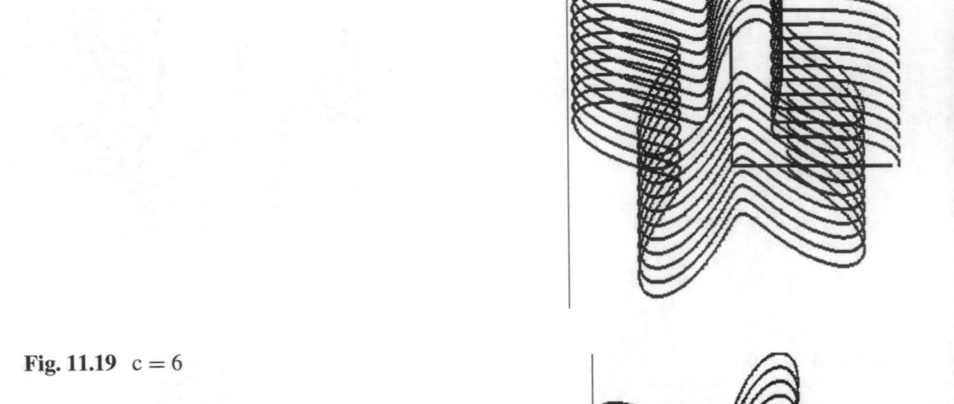

Fig. 11.18 c = 5.5

Fig. 11.19 c = 6

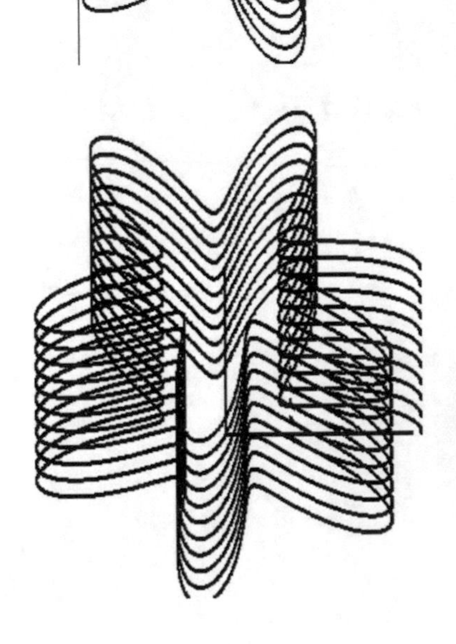

Fig. 11.20 c = 6.5

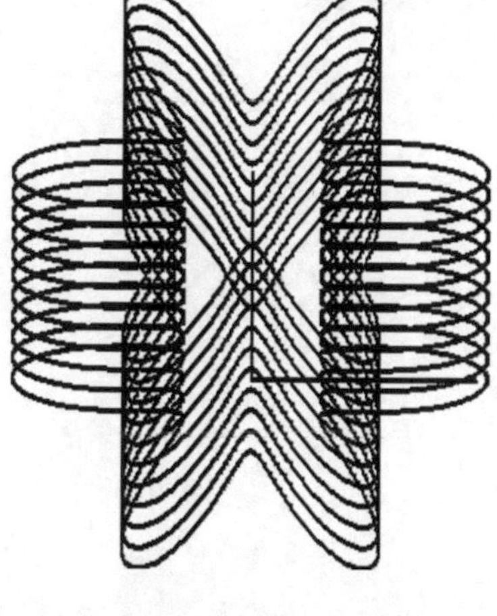

Fig. 11.21 c = 7

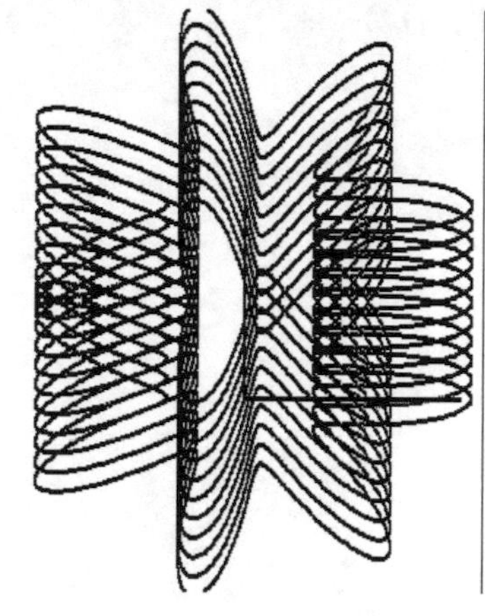

Fig. 11.22 c = 8

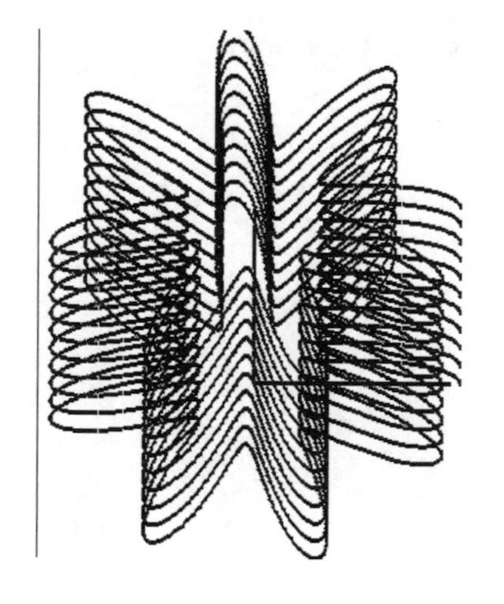

Fig. 11.23 c = 8.5

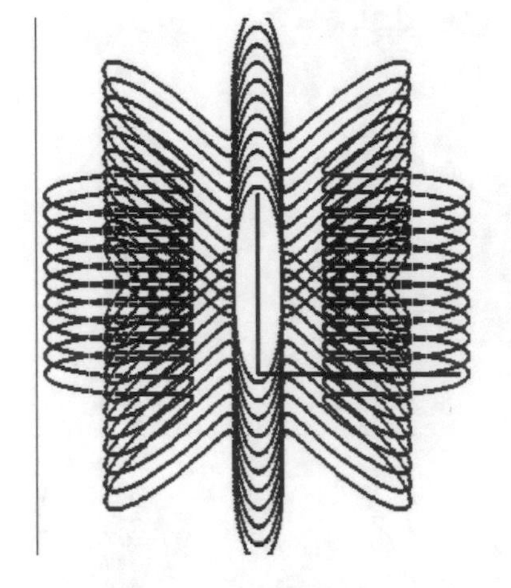

Fig. 11.24 c = 9

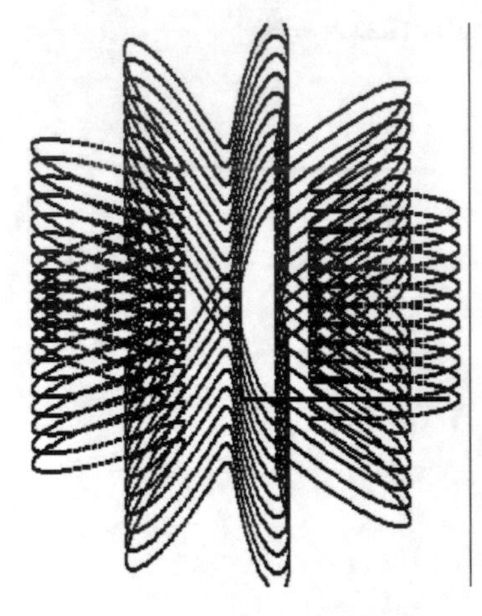

Fig. 11.25 c = 10

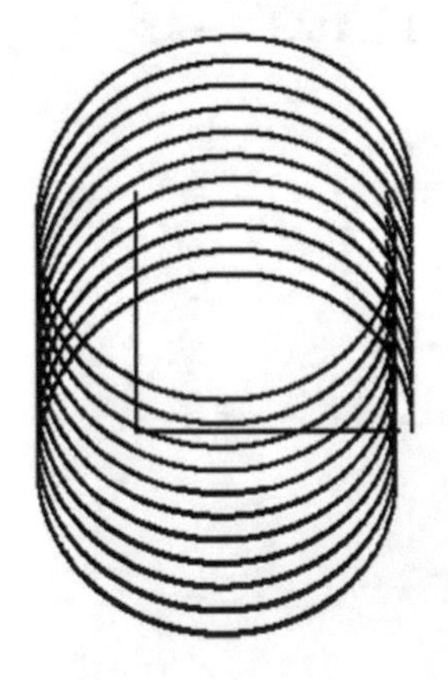

Fig. 11.26 c = −0.1

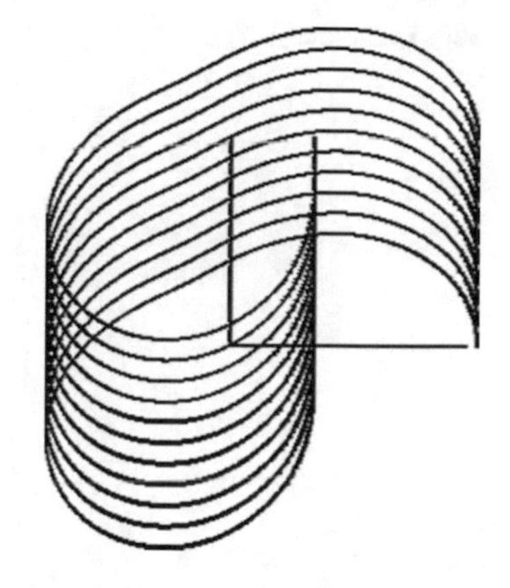

Fig. 11.27 c = −0.5

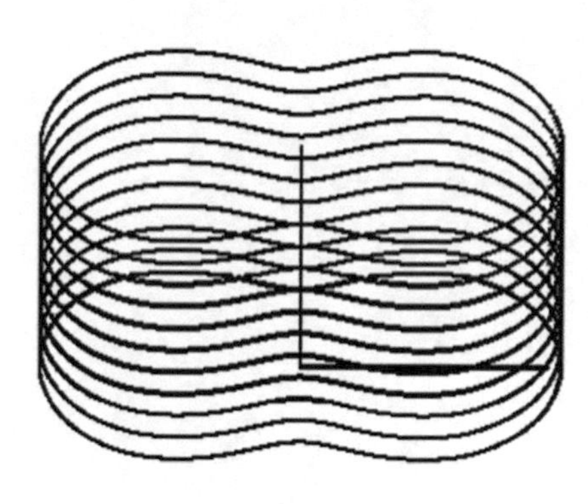

Fig. 11.28 c = −1

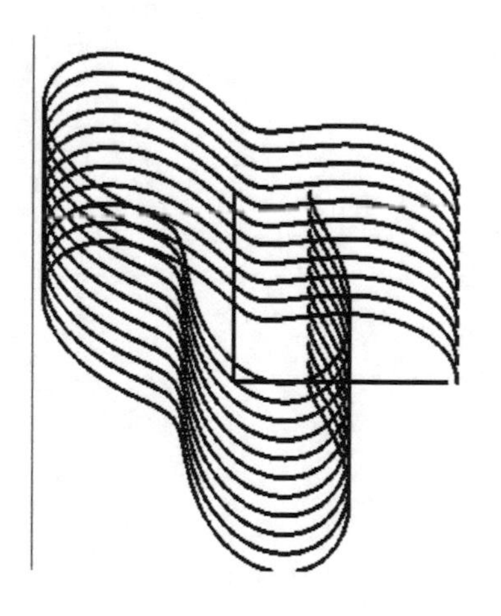

Fig. 11.29 c = −1.5

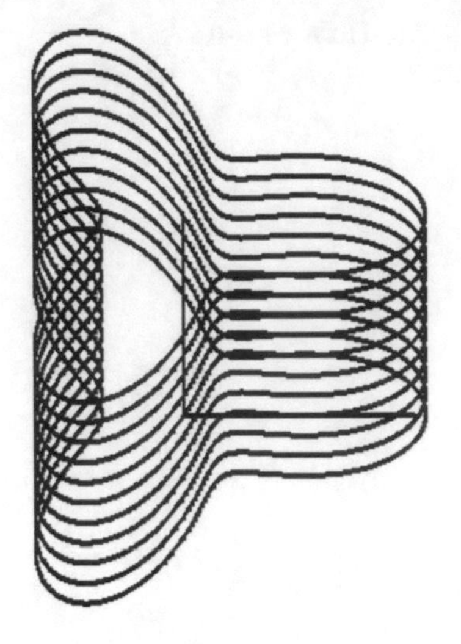

Fig. 11.30 c = −2

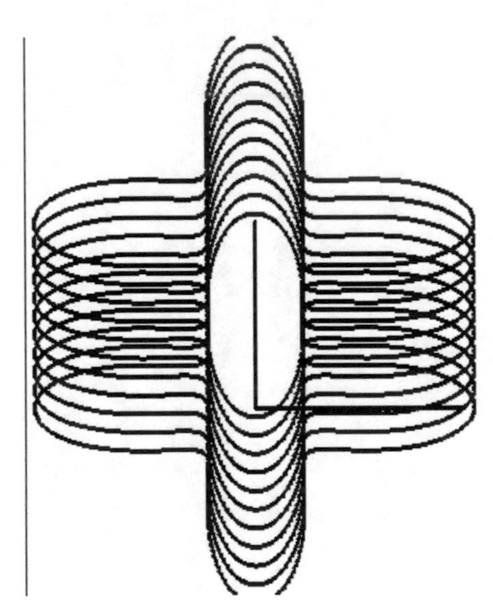

Fig. 11.31 c = −3

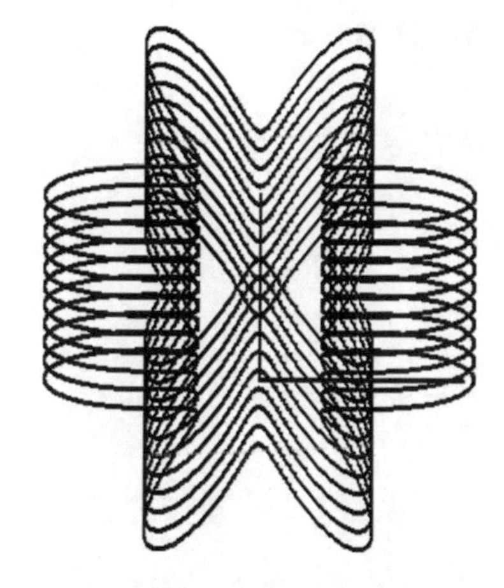

Fig. 11.32 c = −5

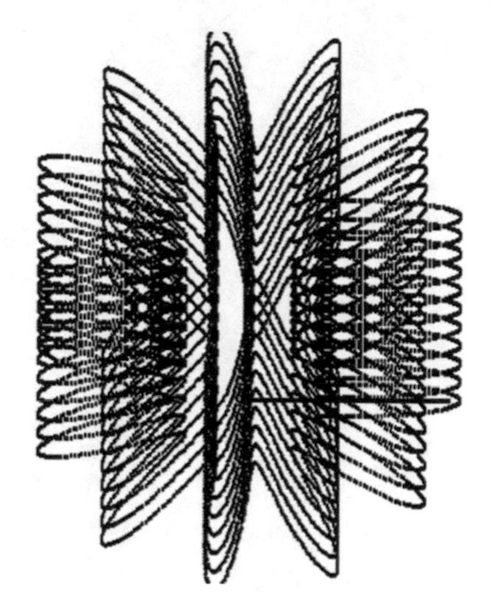

Fig. 11.33 c = −10

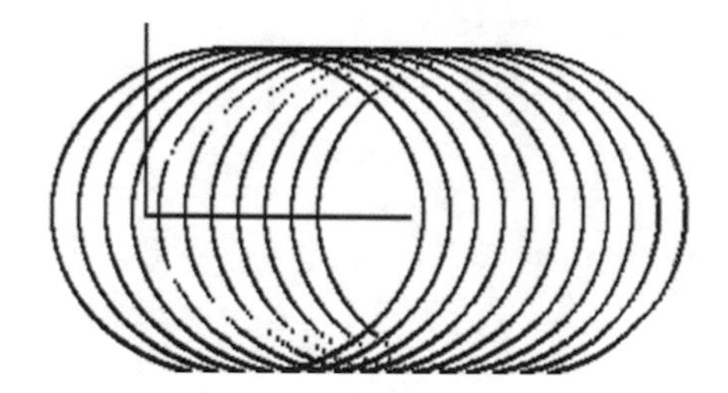

Fig. 11.34 c = 0.01

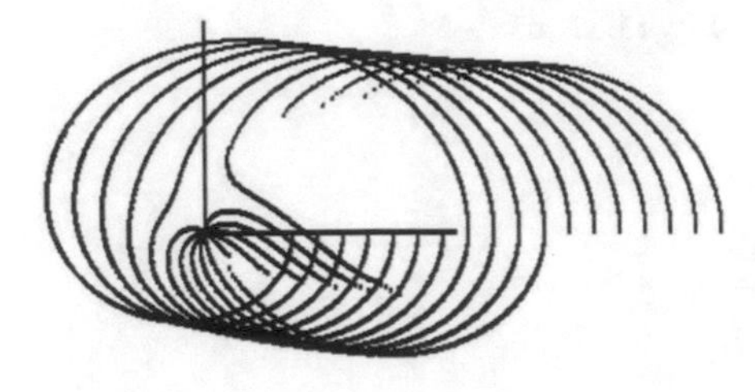

Fig. 11.35 c = 0.5

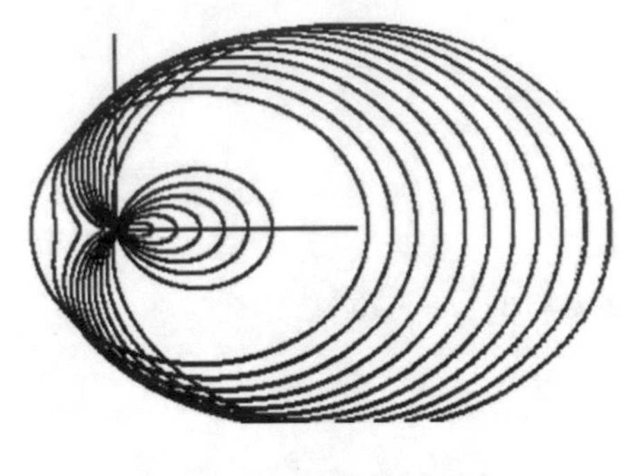

Fig. 11.36 c = 2

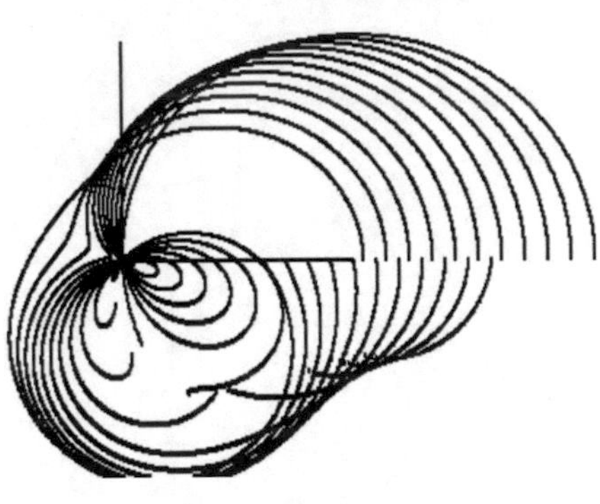

Fig. 11.37 c = 2.25

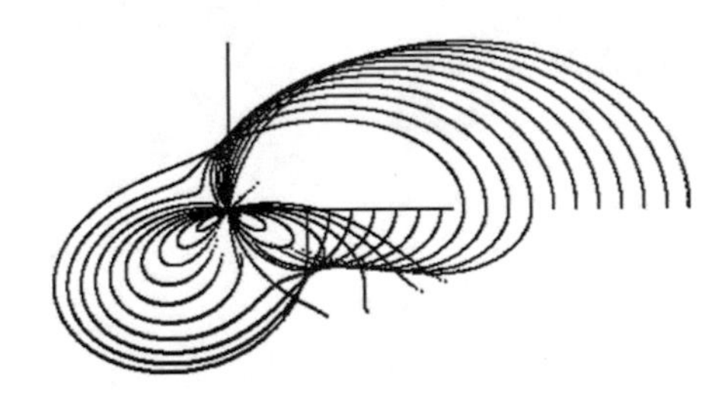

Fig. 11.38 c = 2.5

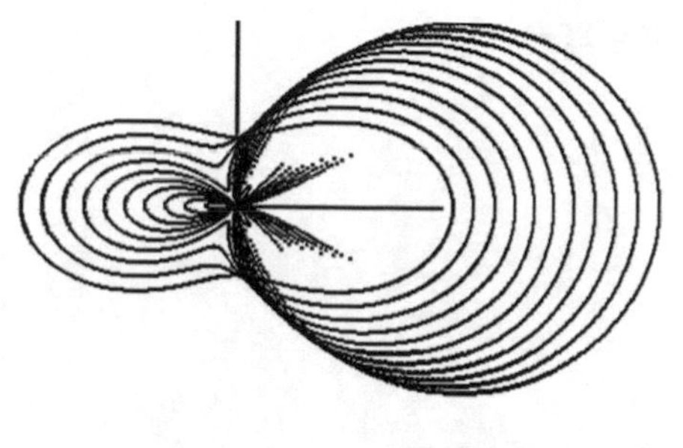

Fig. 11.39 c = 3

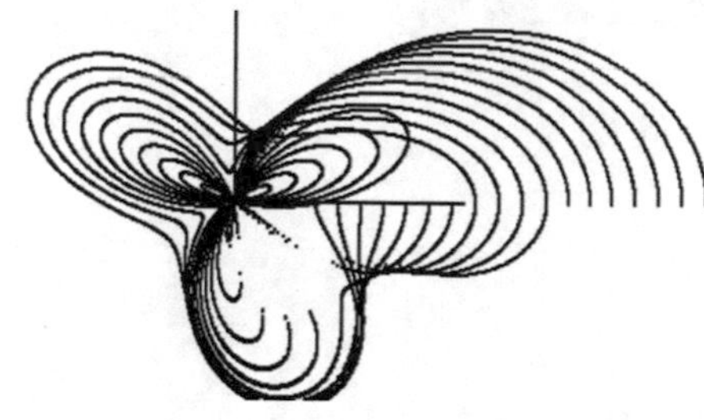

Fig. 11.40 c = 3.5

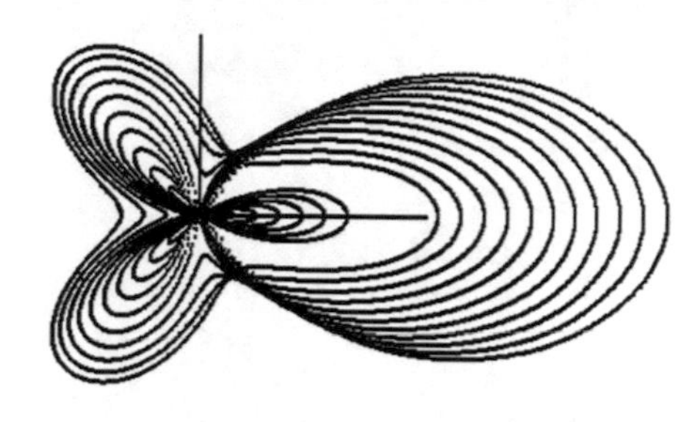

Fig. 11.41 c = 4

Fig. 11.42 c = 4.5

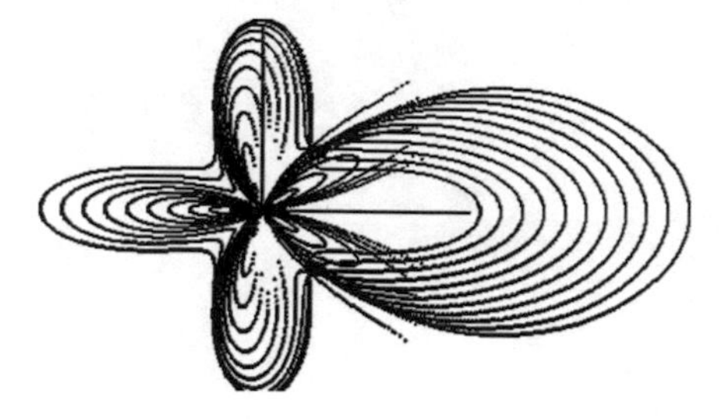

Fig. 11.43 c = 5

Fig. 11.44 c = 6

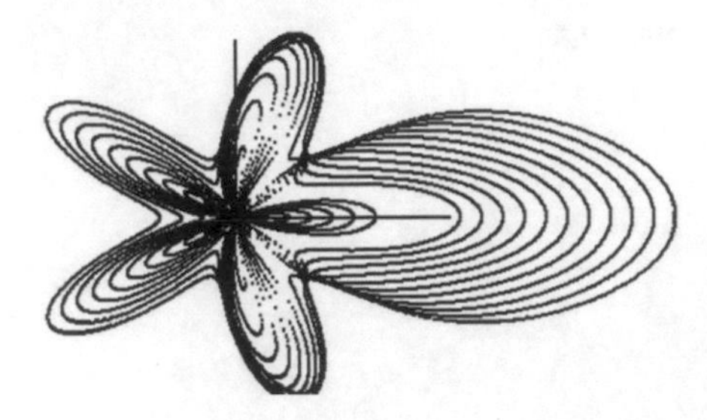

Fig. 11.45 c = 7

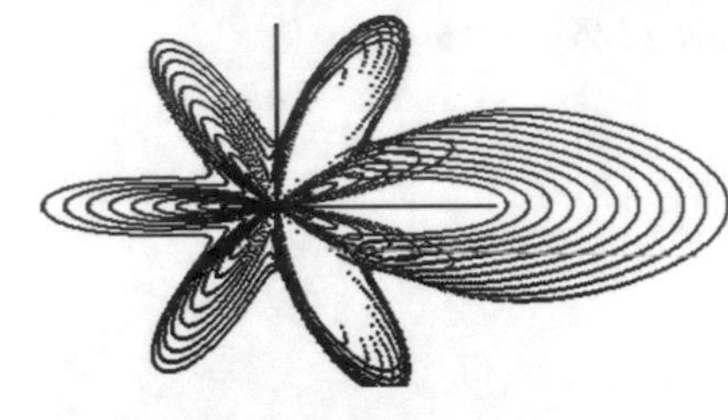

Fig. 11.46 c = 8

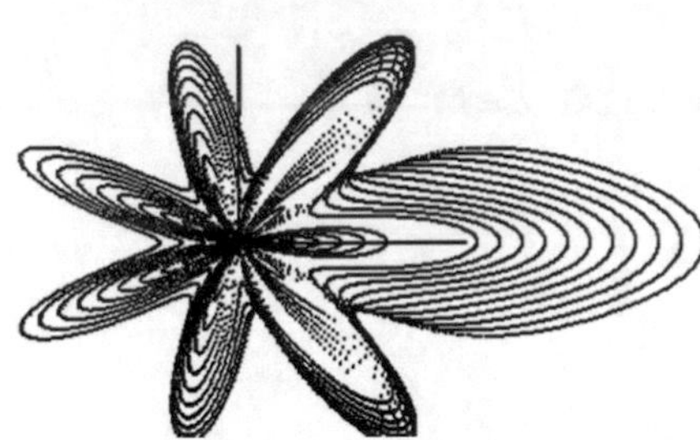

Fig. 11.47 c = 9

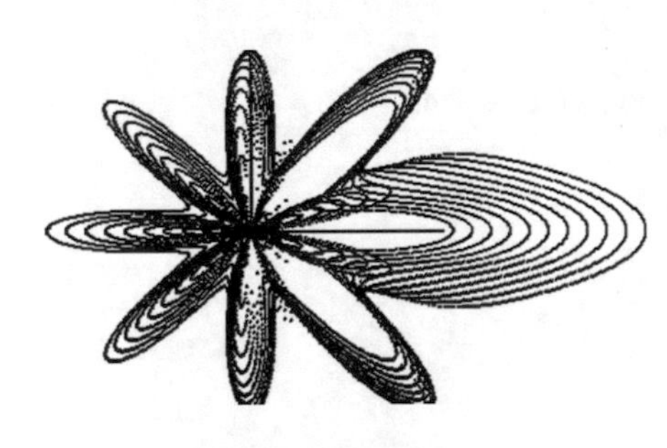

Fig. 11.48 c = 10

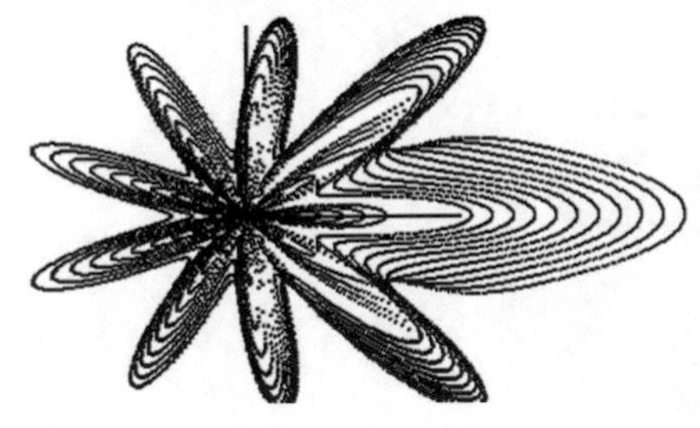

Fig. 11.49 c = 15

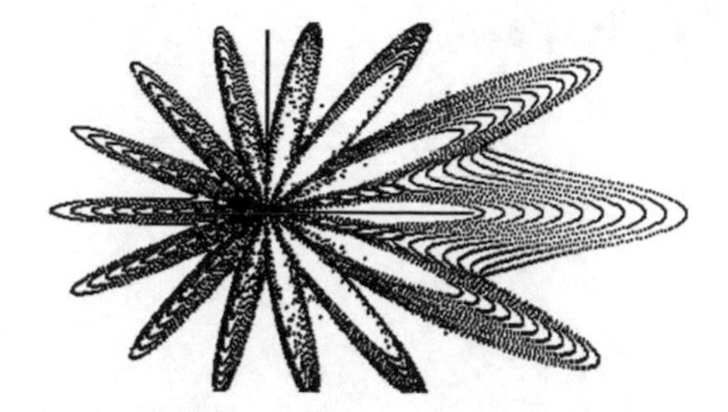

Fig. 11.50 c = −1

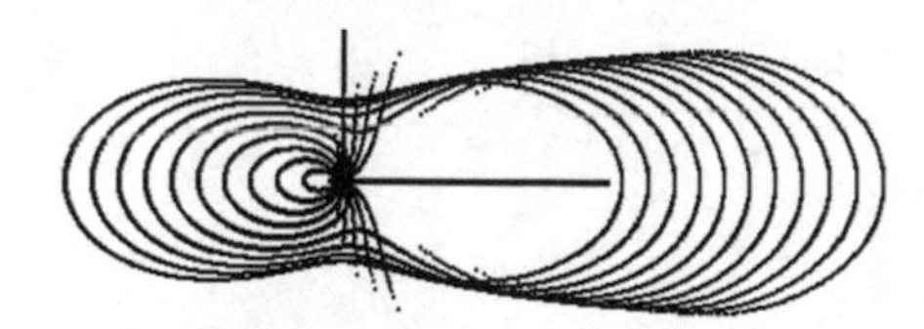

Fig. 11.51 c = −2

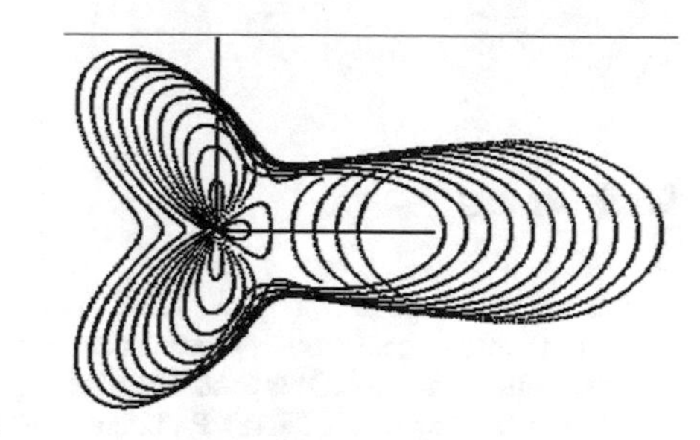

Fig. 11.52 c = −3

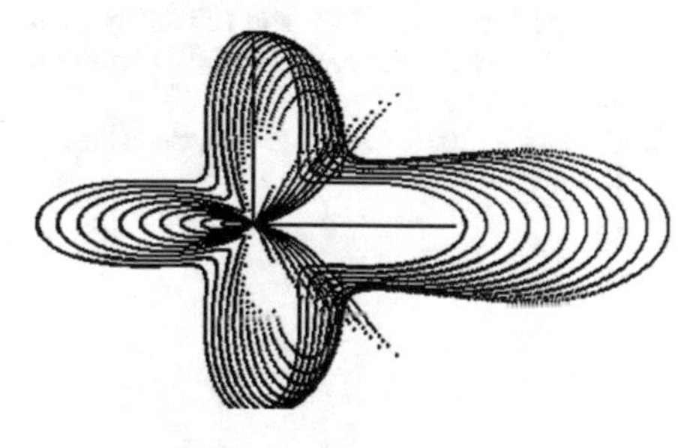

Fig. 11.53 c = −4

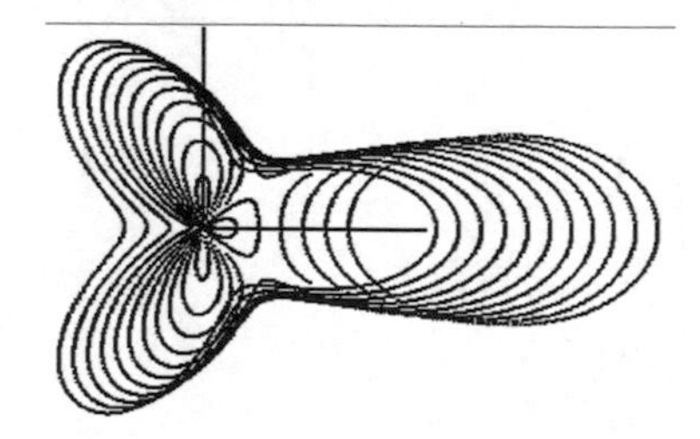

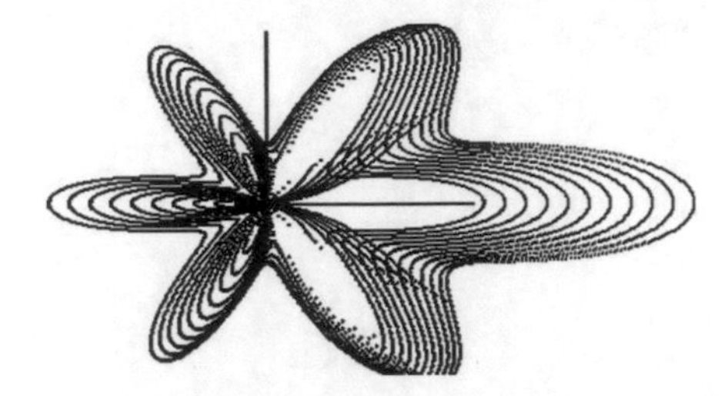

Fig. 11.54 c = −5

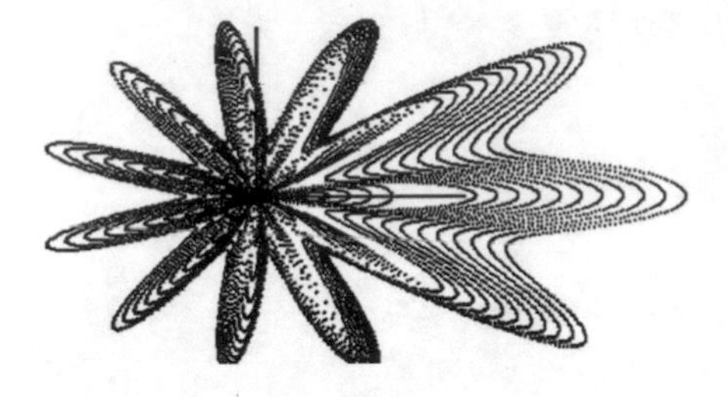

Fig. 11.55 c = −10

References

1. Cherciu M, Marghitu D, Popescu I (2018) Geometric aesthetic locus generated by a mecanisms. In: International conference of mechanical engineering ICOME 2017, Craiova, "Scientific.Net", achievements and solutions in mechanical engineering. Appl Mech Mater 880:81–86
2. Cherciu M, Popescu I (2018) Problem of locus resolved by the theory of mechanisms. In: Symposium "durability and reliability of mechanical systems"—SYMECH 2018, Târgu-Jiu, 11–12 May 2018. "Fiabilitate şi Durabilitate"—fiability & durability". Issue 1(21). Univ. "C. Brâncuşi", Târgu
3. Popescu I (2016) Locuri geometrice şi imagini estetice generate cu mecanisme. Editura Sitech, Craiova

Chapter 12
Loci in the Triangle

Abstract It is known that bisectors, medians, mediators, heights can be drawn in a triangle. If two heights are taken, for example, they intersect at a point if the triangle is static. However, if the triangle moves, having a single fixed side, several variants of equivalent mechanisms are obtained, resulting in the trajectories of the points of intersection of different shapes. The resulting curves and successive positions are given. Some mechanisms are very complicated, only that they are used only as calculation artifices, not being necessary to build them and only in real cases when it is necessary to draw these trajectories on pieces. The resulting curves are of a great diversity of shapes.

A triangle can be built if there are known [1]:

(1) the three sides;
(2) two sides and the included angle;
(3) two sides and the angle, opposite to the largest of them;
(4) two angles and a non-included side.

The mid perpendiculars in a triangle are the lines perpendicular to the sides, passing through their mid-points. They intersect at a point that is the center of the circle circumscribed to the triangle.

The heights of the triangle are the lines that descend from the peaks on opposite sides, being perpendicular to them. Their intersection point is called orthocenter.

The medians of the triangle are lines joining the triangle's peaks with the mid-points of the opposing sides, intersecting in a point called the center of gravity of the triangle. This point divides each median by the ratio 2:1, starting at its tip.

The bisectors of a triangle are lines from the peaks, making equal angles to the sides of each peak. They concur in the center of the inscribed circle; its radius being given by the segment between the bisectors point of intersection and the basis of the perpendicular line drawn from that point on either side.

In order to find the loci of the intersection points for the the triangle lines, the studied triangle must have a variable element, that is, a side or an angle.

I. Popescu et al., *Problems of Locus Solved by Mechanisms Theory*,
Springer Tracts in Mechanical Engineering,
https://doi.org/10.1007/978-3-030-63079-9_12

The following cases are distinguished:

I. one side and the opposite angle are given and kept constant, and another side is the variable element;
II. two sides are given and kept constant and the third one is the variable element;
III. an angle and an adjacent side are given and maintained constant, and the opposite side to the given angle is the variable element;
IV. one side and one angle are given, and maintained constant, and another angle is the variable element.

- **First case—one side and one angle are constant**

In Fig. 12.1 it is shown a triangle that has the AC and AB sides with fixed directions, and α = constant = 50, BC = constant = 50 (the angles in the whole book are measured in degrees and the lengths in millimeters, the values here are also kept below). It is continually extended the AC side, having the variable S3 length, and the positions of B point are determined. The equivalent mechanism [2] is given in Fig. 12.2.

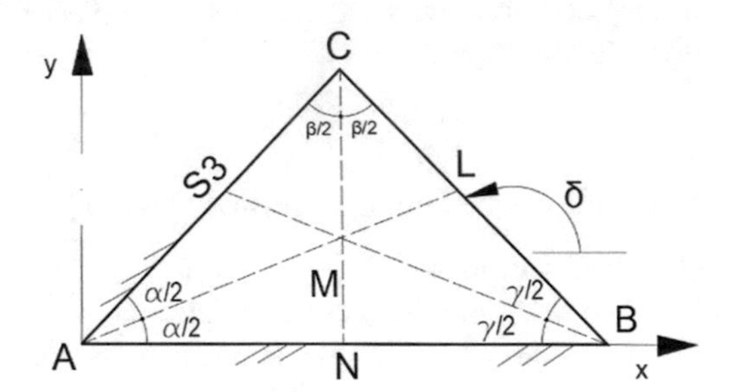

Fig. 12.1 The initial triangle

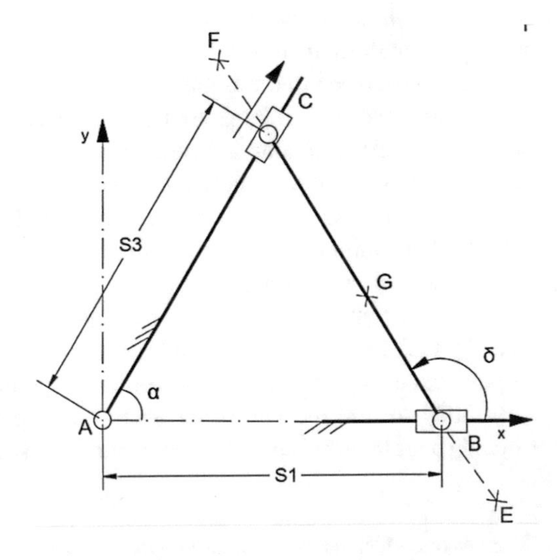

Fig. 12.2 The slider in C point is actuator

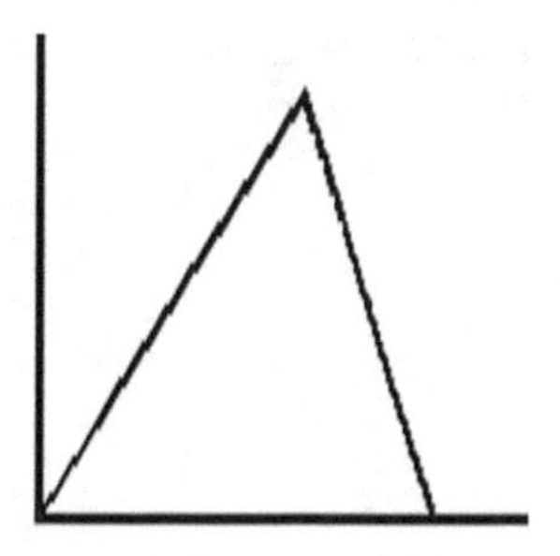

Fig. 12.3 Mechanism in one position

In C point, it was placed a slider slipping in the direction of the fixed AC line, and in B point it was placed another slider slipping in the fixed direction AB. The length of BC line was maintained constant.

For now, trajectories of G (BC mid-point), E and F points are ploted, observing the effect of the slider motion on AC line. The mechanism has the slider in C point as actuator.

The following expressions are written:

$$x_C = S_3 \cos \alpha \tag{12.1}$$

$$y_C = S_3 \sin \alpha \tag{12.2}$$

$$x_C = S_1 + BC \cos \delta \tag{12.3}$$

$$y_C = y_B + BC \sin \delta \tag{12.4}$$

having $y_B = 0$, from where S_1 şi δ are determined.

$$x_B = S_1 \tag{12.5}$$

$$x_F = x_C + CF \cos \delta \tag{12.6}$$

$$y_F = y_C + CF \sin \delta \tag{12.7}$$

$$x_E = x_B + BE \cos(\delta + \pi) \tag{12.8}$$

$$y_E = y_B + BE \sin(\delta + \pi) \tag{12.9}$$

In Fig. 12.3 a mechanism position for $\alpha = 50$, BC = 50, S3 = 60 is shown.

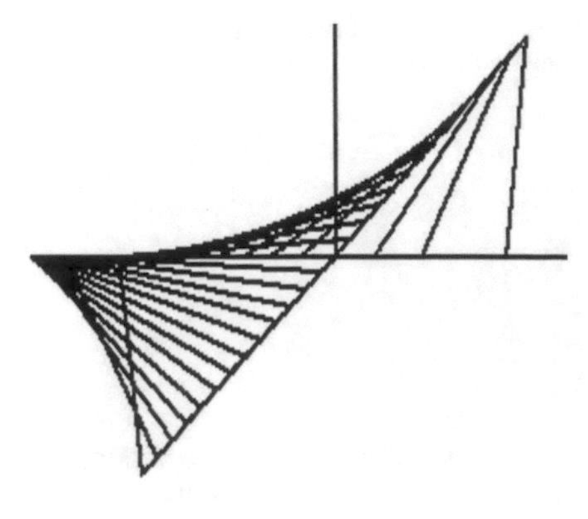

Fig. 12.4 Successive positions for $+\surd$

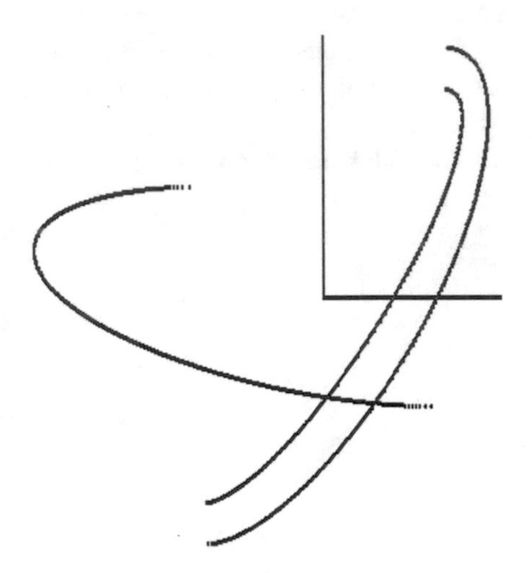

Fig. 12.5 Loci for $+\surd$

Successive positions of the mechanism for S3 variying between −200 and +200 are shown in Fig. 12.4. The mechanism exists only in the interval S3 = −65 … + 65 (the values obtained by running the computer program outside this interval the mechanism is not functional). Figures 12.4 and 12.5 were obtained for the "+" sign in front of the radical in the cos expression; for the "−" sign it is obtained the second position of the mechanism, having successive positions and trajectories presented in Figs. 12.6 and 12.7.

The lengths CG = BG = BC/2 and F = 40; BE = 40 were considered and the trajectories of points G, F, E were determined. They are presented in Fig. 12.5: G's trajectory in the middle, E's trajectory is on the left and F's trajectory is on the right. [Below the sizes of the pictures were imposed by the editing conditions].

- **Intersection of bisectors**

In Fig. 12.1, there were also drawn the bisectors of the triangle, which intersect (concur) at the M point. Having $\alpha =$ constant and the fixed AC and AB directions, it

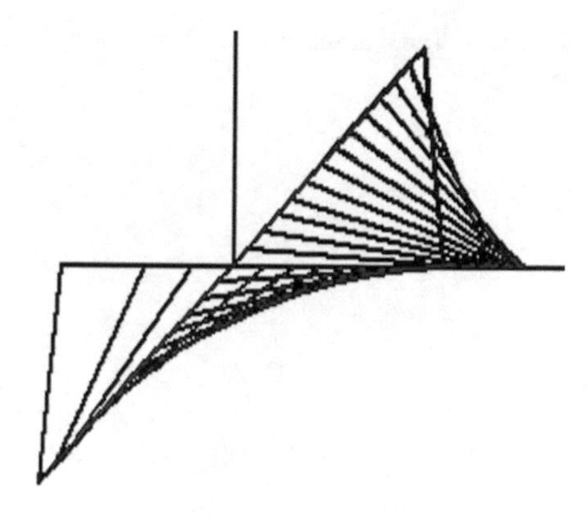

Fig. 12.6 Successive positions for $-\surd$

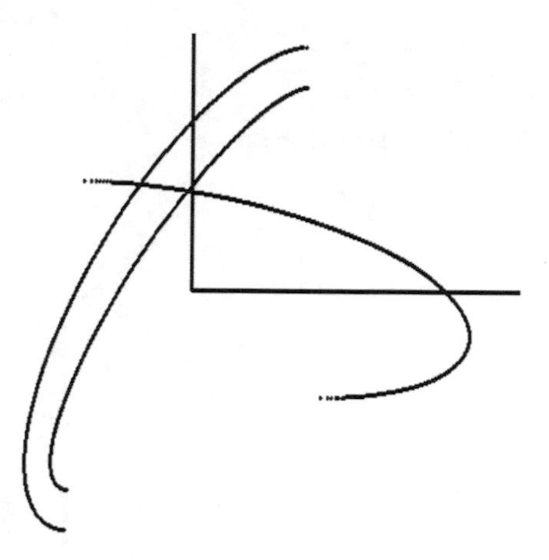

Fig. 12.7 Loci for $-\surd$

results that the bisector AM is fixed, therefore locus of the M point is on this bisector, that is, a line.

- **Intersections of heights**

In Fig. 12.8 the triangle was drawn in one position and its heights as well. We knew the data calculated above for the same triangle.

The following relationships (AB being the Y-coordinate) may be written:

$$x_M = x_N = S_3 \cos \alpha \tag{12.10}$$

$$CN = S_3 \sin \alpha \tag{12.11}$$

$$AM \cos \psi = x_N \tag{12.12}$$

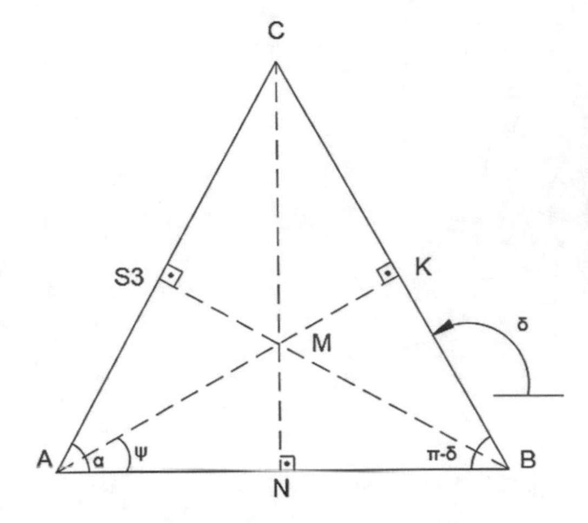

Fig. 12.8 The intersection of heights

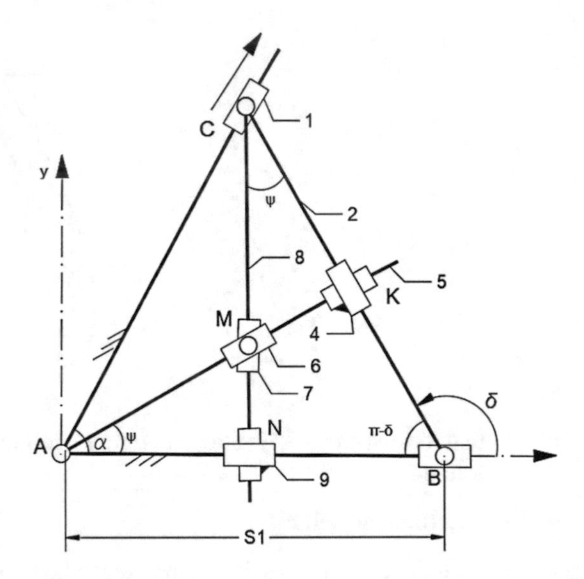

Fig. 12.9 The mechanism

$$AM \sin \psi = y_M \tag{12.13}$$

$$\psi = \delta - \pi/2 \tag{12.14}$$

The equivalent mechanism is shown in Fig. 12.9. The mechanism is complicated because it has to establish the positions of the legs of the perpendiculars on the sides. In this purpose welded perpendicular sliders are used. The point of intersection of the heights is materialized by two sliders that rotate between them, having the sides length equal to zero.

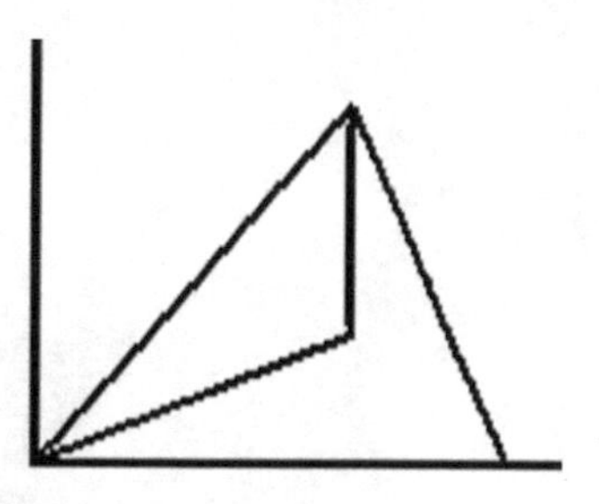

Fig. 12.10 The mechanism in one position

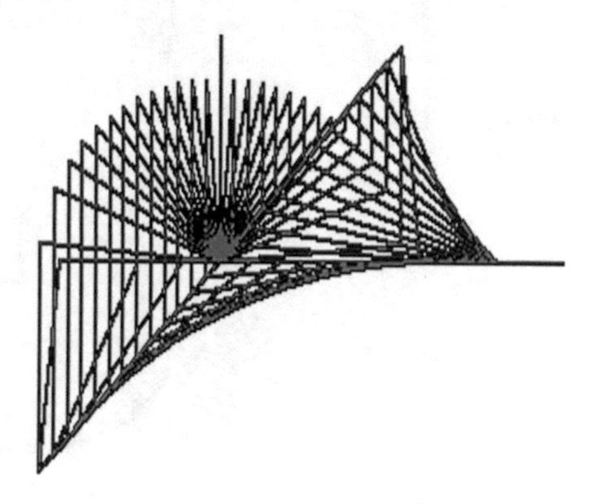

Fig. 12.11 Successive positions for $-\sqrt{}$

The above expressions were used. Figure 12.10 shows the mechanism in one position with the sign "−" in front of the radical, with only two heights (the three hights are concurring in M), and in Fig. 12.11 the successive positions of the mechanism were given.

The locus we were looking for is the one in Fig. 12.12, that is, a circle arc with the center in A.

The solution with "+" in front of the radical is presented in Figs. 12.13 and 12.14.

- **The Cross-point of mid perpendiculars**

In Fig. 12.15, there were drawn the three mid perpendiculars on the triangle sides, intersecting in M.

The equivalent mechanism is shown in Fig. 12.16.

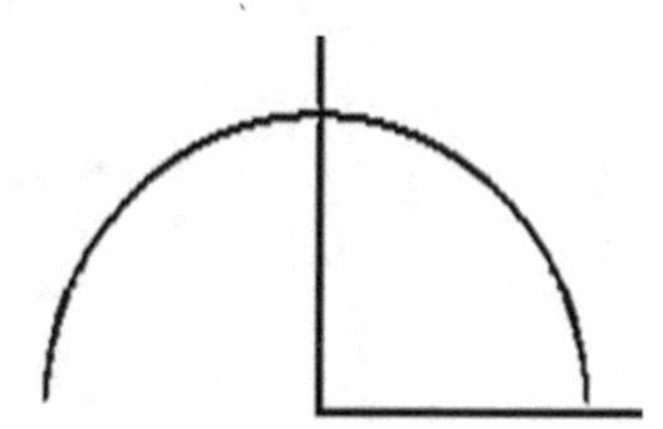

Fig. 12.12 The locus for $-\sqrt{}$

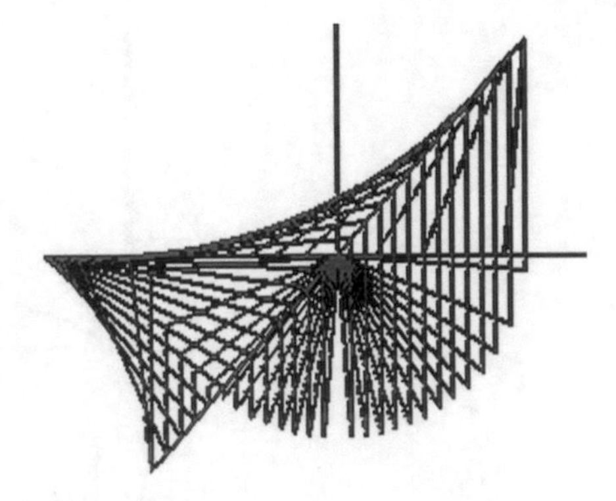

Fig. 12.13 Loci for $+\sqrt{}$

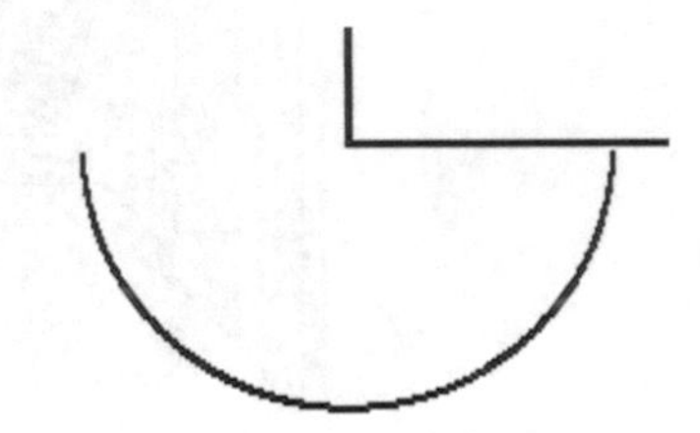

Fig. 12.14 The locus for $+\sqrt{}$

We started from the heights cazus in the ABC triangle, that is, the kinematic chain CB. Next, it was necessary for points E, F and N to be materialized. They were located at the mid-points of the sides of the ABC triangle.

For this purpose, a geometry method was used to determine the middle of a line segment. Thus, according to Fig. 12.17, C point, the mid point of the segment AB was determined as follows: there were ploted two circular arcs with equal radii,

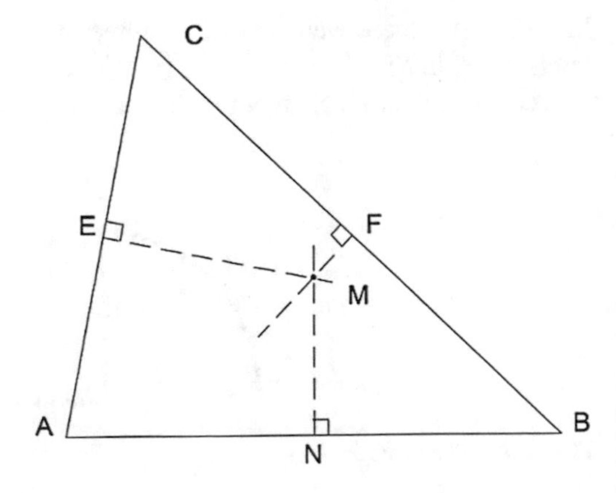

Fig. 12.15 The cross-point of mid perpendiculars

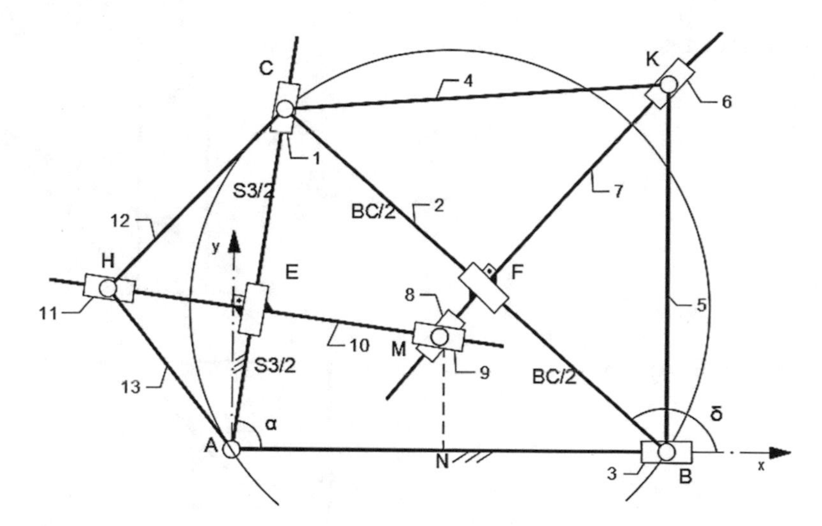

Fig. 12.16 The mechanism

Fig. 12.17 The mid-point of the line

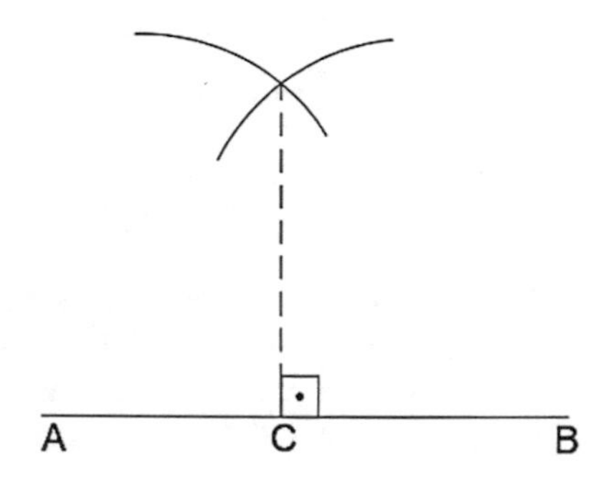

considered larger than half of the AB segment, with the needle of the compass placed in A point and then in B point. Then there was drawn a perpendicular from their cross-point on AB segment. This method was applied in Fig. 12.16 for the two sides, AC and BC. The mid-perpendiculars were concurring, so there was no need for the third mid-perpendicular to be placed by the computer program. For the AC side, two circle arcs were drawn with equal radii larger than AE. They concurred in H point, from where a perpendicular to AC was drawn, resulting into HM line. Thus, another AHCEM kinematic chain was built. The HEM element is linked to the H, E, and M pairs. Thus, the F point was obtained at the mid-point of the BC side and the EM line as mid-perpendicular. Similarly, the point F was obtained with the BCKFM kinematic chain, where F is the mid-point of the BC side, and FM is the mid-perpendicular.

Based on Fig. 12.16 the following expressions are written:

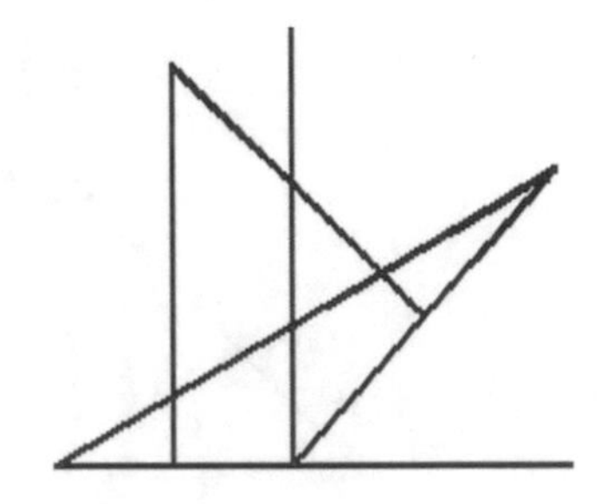

Fig. 12.18 The mechanism for $+\surd$

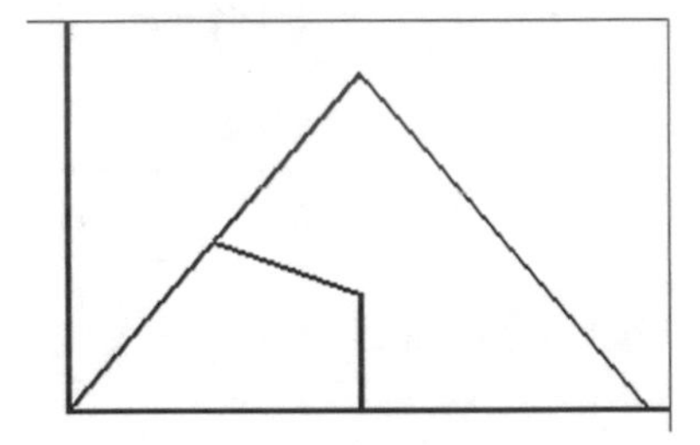

Fig. 12.19 The mechanism for $-\surd$

$$x_N = x_M = x_B/2 = FM\cos(\alpha + 270) + S_3/2\cos\alpha \tag{12.15}$$

$$y_M = MN = FM\sin(\alpha + 270) + S_3/2\sin\alpha \tag{12.16}$$

In this case also, there were two solutions of the system (including the equations above for the ABC triangle). In Fig. 12.18, the mechanism was presented for one position with only two MN and ME mid-perpendiculars for the "+" sign, and Fig. 12.19 determined the mechanism in the position with the "−" sign, both for S3 = 50.

The successive positions of the mechanism (excluding the H and K zones) were shown in Fig. 12.20 for the "+" sign and in Fig. 12.21 for the "−" sign.

The locus we wanted to determine is a planar curve, different from a circle arch, presented in Fig. 12.22 (for "+") and in Fig. 12.23 (for "−").

By modifying α and maintaining the "+" sign, different trajectories were obtained, as shown in Figs. 12.24, 12.25 and 12.26. In figures, the values of α were specified.

- **The Cross-point of medians**

The equivalent mechanism is shown in Fig. 12.27. The ABC kinematic chain was built, then the mid-points for the sides of the ABC triangle were determined, having the AKF and BLE angles of 90°. The elements 9 and 13 were connected to the sliders that move on the AC and BC sides, the sliders 8 and 12 being rotated with respect to the slides 9 and 13. The method of Fig. 12.17 was used in order to find the mid-points of the triangle sides.

Based on Fig. 12.27 the following expressions were written:

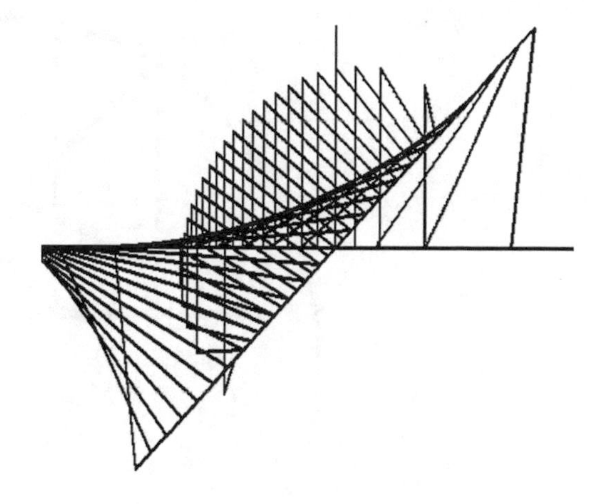

Fig. 12.20 Successive positions for $+\sqrt{}$

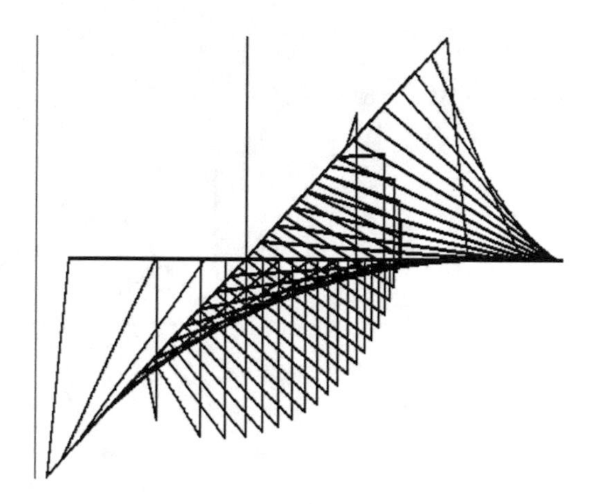

Fig. 12.21 Successive positions for $-\sqrt{}$

$$x_K = S_3/2 \cos \alpha \tag{12.17}$$

$$y_K = S_3/2 \sin \alpha \tag{12.18}$$

getting the position of K point.

$$x_L = x_B + BC/2 \cos \delta \tag{12.19}$$

$$y_L = BC/2 \sin \delta \tag{12.20}$$

getting the position of L point.

$$AL = sqr\left[(x_L - x_A)^2 + (y_L - y_A)^2\right], \text{ sqr means radical;} \tag{12.21}$$

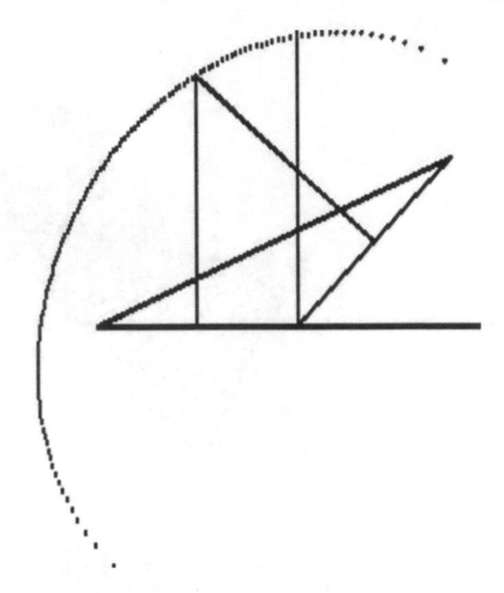

Fig. 12.22 The locus for $+\sqrt{}$

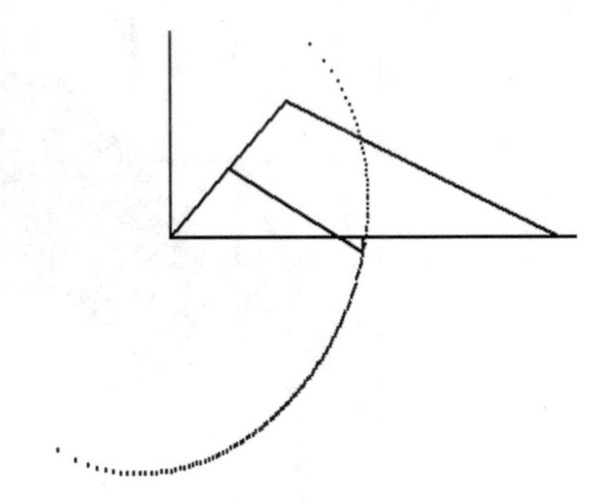

Fig. 12.23 The locus for $-\sqrt{}$

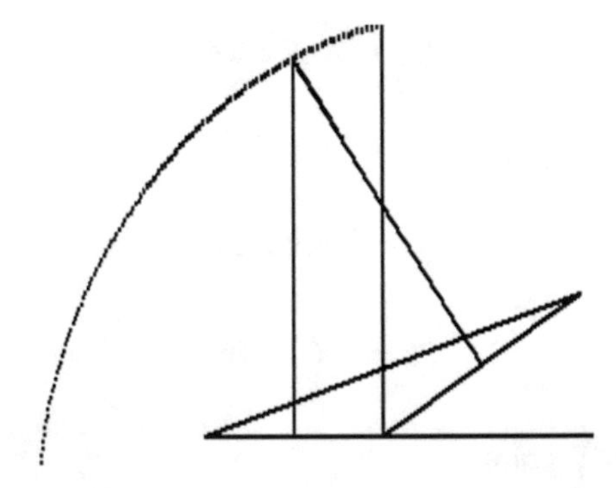

Fig. 12.24 Trajectory for $\alpha = 35$

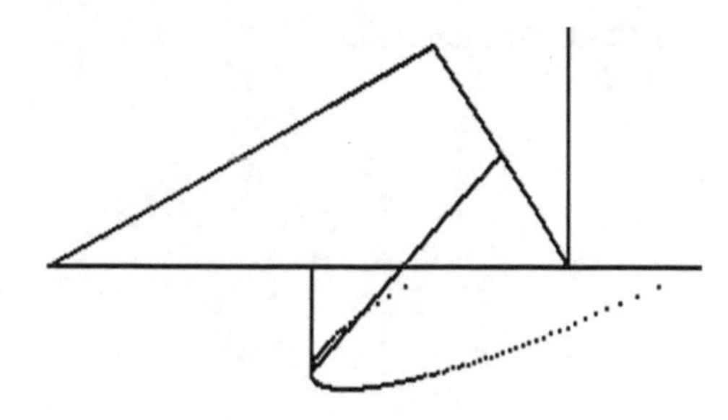

Fig. 12.25 Trajectory for $\alpha = 120$

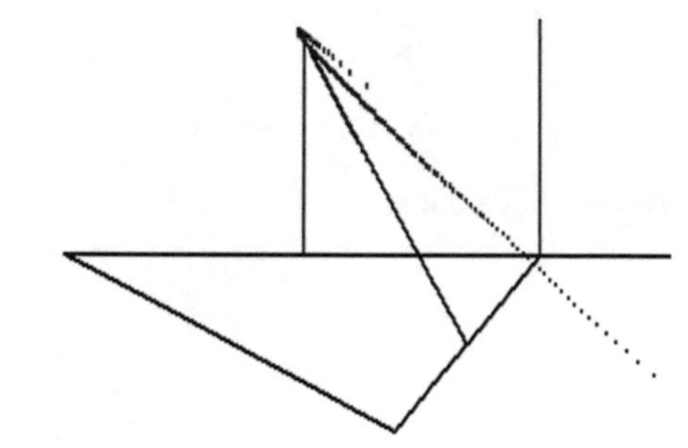

Fig. 12.26 Trajectory for $\alpha = 230$

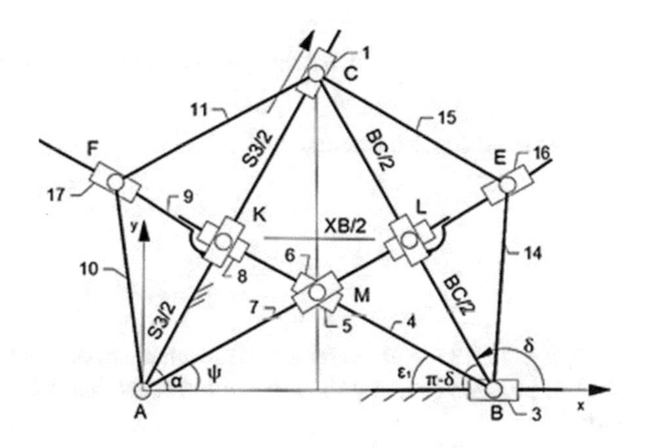

Fig. 12.27 The mechanism

$$(BC/2)^2 = AL^2 + x_B^2 - 2ALx_B \cos \psi \tag{12.22}$$

from where the angle ψ is calculated.

$$BK = sqr\left[(x_B - x_K)^2 + (y_B - y_K)^2\right], \tag{12.23}$$

$$BK/\sin \alpha = (S_3/2)/\sin \varepsilon_1 \tag{12.24}$$

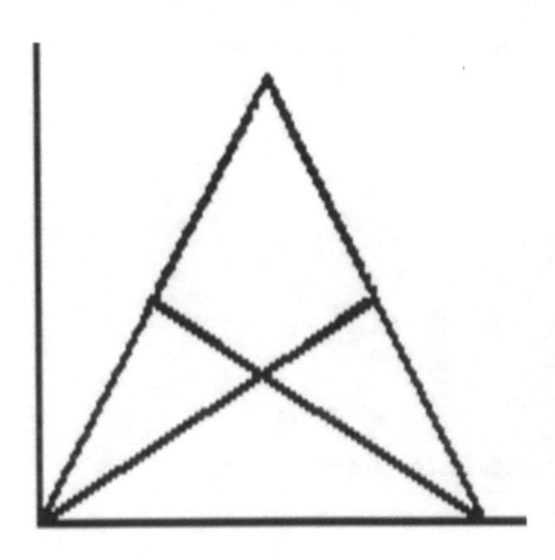

Fig. 12.28 The mechanism in one position

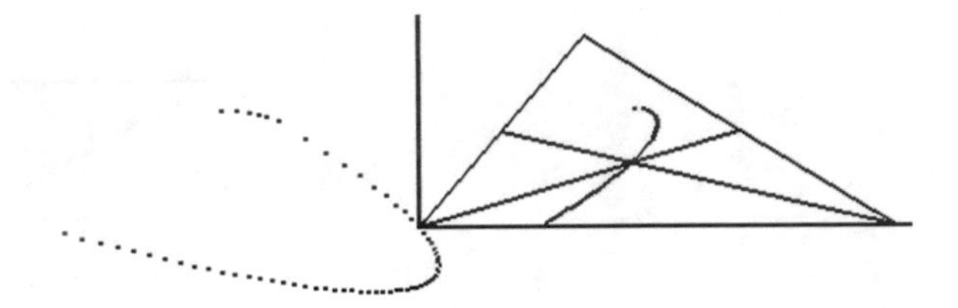

Fig. 12.29 The locus

$$AM/\sin \varepsilon_1 = x_B/\sin(\pi - \psi - \varepsilon_1) \tag{12.25}$$

from were AM is determined.

$$x_M = AM \cos \psi \tag{12.26}$$

$$y_M = AM \sin \psi \tag{12.27}$$

In Fig. 12.28 it was presented the mechanism drown by the computer programm for one position (without the kinematic chains that determine the mid-points of the sides).

In Fig. 12.29 the mechanism was shown in one position as well as the locus. We could conclude that the result is a curve with two unbound branches.

A discontinuity occurs at S3 = 0 in the mechanism operation.

The S3 variation cycle was divided into two parts. One consisted in the interval from −65 to 0, presented in Fig. 12.30 and another consisted in the interval from 0 to 65, resulting the curve presented in Fig. 12.31.

We considered the case with the sign "−" before a radical.

In Fig. 12.32, there were presented the two branches of the locus and one position of the generator mechanism.

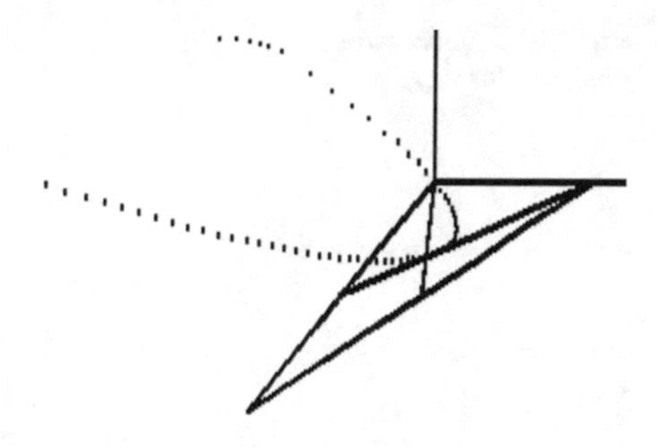

Fig. 12.30 The locus for S3 = −65 … 0

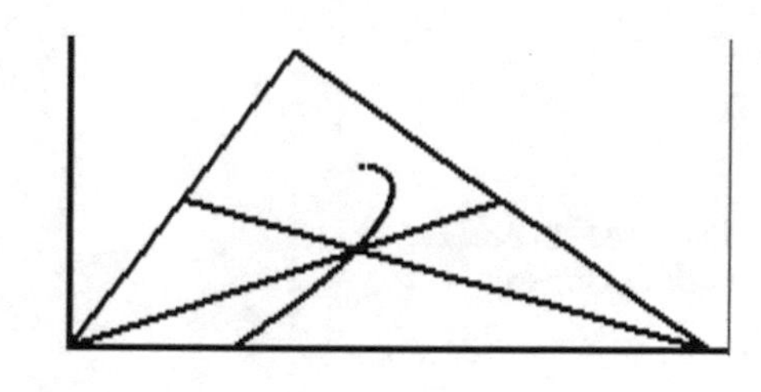

Fig. 12.31 The locus for S3 = 0 … 65

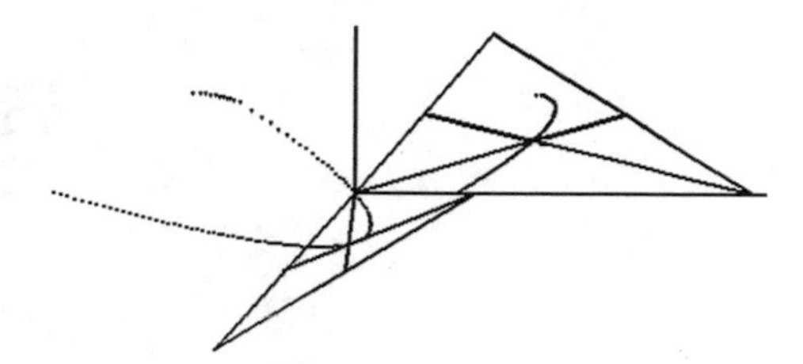

Fig. 12.32 The branches of the locus

For the case with the "+" sign in front of that radical, the curves from Fig. 12.33 were determined, the latter two pictures being symmetrical.

Figure 12.34 has shown the successive positions of the mechanism for the "−" sign in front of the radical, and Fig. 12.35 has shown the same thing, but for the "+" sign in front of the radical.

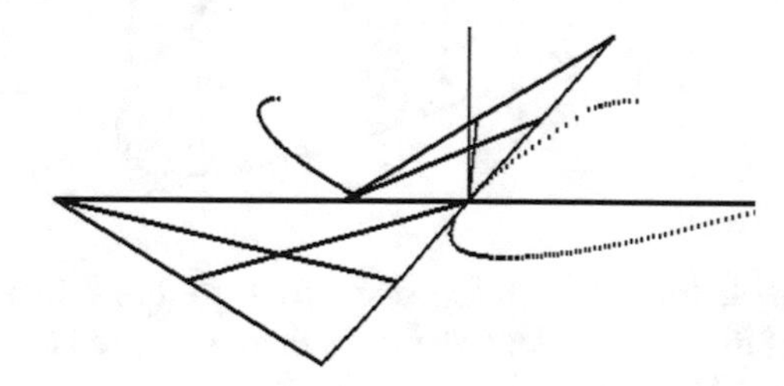

Fig. 12.33 The locus for $+\sqrt{}$

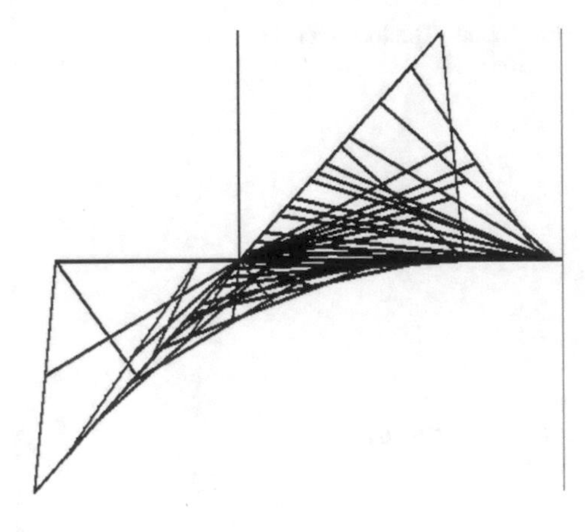

Fig. 12.34 Successive positions for $-\surd$

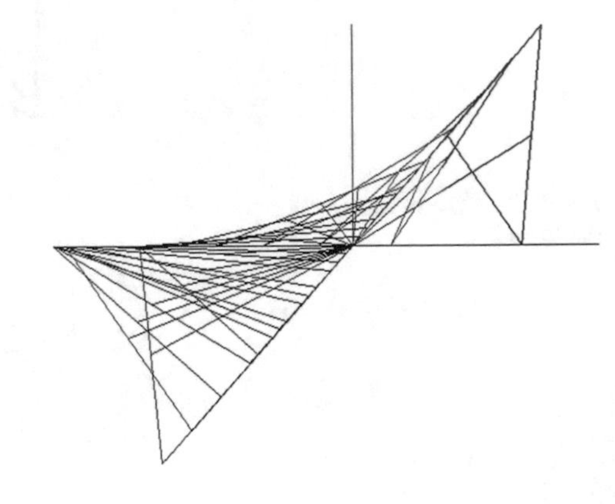

Fig. 12.35 Successive positions for $+\surd$

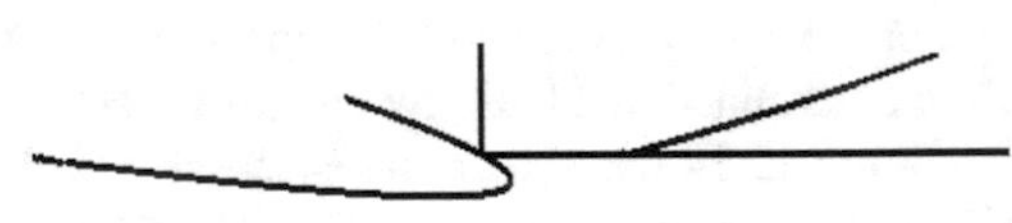

Fig. 12.36 The locus for $\alpha = 30$

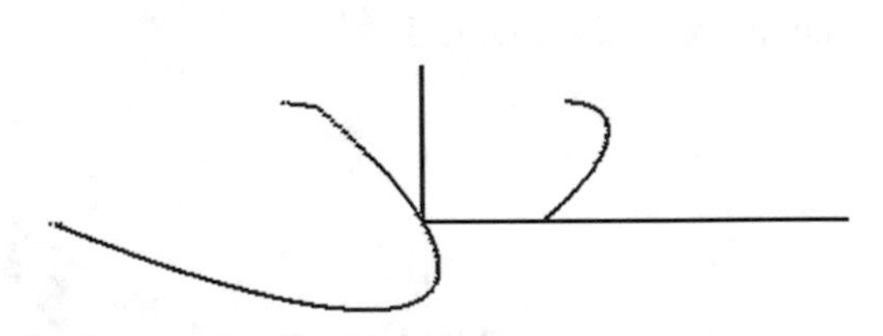

Fig. 12.37 The locus for $\alpha = 60$

Below, there are the loci for both branches, for different values of α, and for the "–" sign in front of the radical. For the above, there was used $\alpha = 50°$. The curves were shown in Figs. 12.36, 12.37, 12.38 and 12.39.

For $\alpha = 270$ there are the same branches as for $\alpha = 90$.

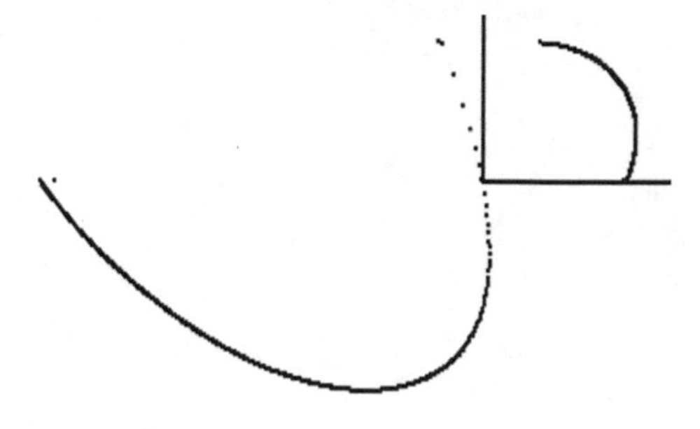

Fig. 12.38 The locus for $\alpha = 80$

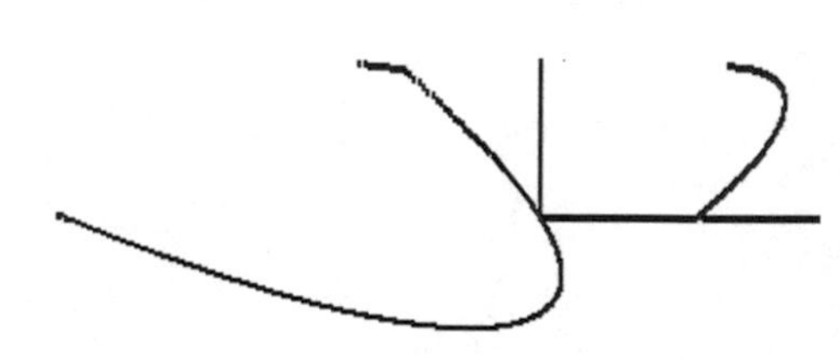

Fig. 12.39 The locus for $\alpha = 240$

References

1. Mică enciclopedie matematică (1980) Editura tehnică, Bucureşti
2. Popescu I (2016) Locuri geometrice şi imagini estetice generate cu mecanisme. Editura Sitech, Craiova

Chapter 13
Loci of Points Belonging to a Quadrilateral

Abstract If in a triangle there are few usual lines that can intersect, in a quadrilateral there are many more possibilities. Thus, heights intersecting perpendicular from each corner on the opposite sides, as well as bisectors, medians, non-curves, diagonals, can intersect here as well. For each case the synthesis of the equivalent mechanism is made and then the intersection curve of the considered lines is determined. You get lots of curves, of a great diversity of shapes. By changing some lengths of some sides, other curves are obtained. In many cases, two curves plotted by two points on the mechanism were represented. The successive positions of some generating mechanisms are also given. Particular cases of the quadrilateral are also analyzed: square, rectangle, rhombus, parallelogram. The trajectory of the point of intersection of the diagonals of a quadrilateral is searched, when one side of the quadrilateral moves, dragging them on two other sides. The trajectory of a point on the connecting rod is also given for comparison. Woody curves were obtained. Next, two lengths of the mechanism were modified, resulting in other curves, some quite interesting.

Chapter 1 has shown the loci of the points on the BC element of a quadrilateral. Below, there are studied the loci of the different points for certain specific lines of the ABCD quadrilateral shown in Fig. 13.1.

13.1 The Case of the Heights Built from A and D Points on the Connecting Rod

For the ABCD four-bar mechanism shown in Fig. 13.2 there were plotted AE and DF heights built on the connecting rod from A and D points. We needed to determine the positions for E and F points. The ABCD four-bar mechanism was initially considered and double welded sliders were mounted in E and F points in order to ensure the perpendicularity of the AE and DF lines on the BC connecting rod. The leading element is AB.

I. Popescu et al., *Problems of Locus Solved by Mechanisms Theory*,
Springer Tracts in Mechanical Engineering,
https://doi.org/10.1007/978-3-030-63079-9_13

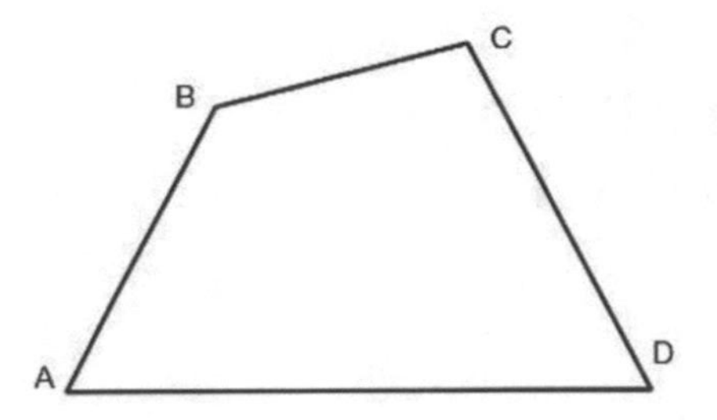

Fig. 13.1 The initial quadrilateral

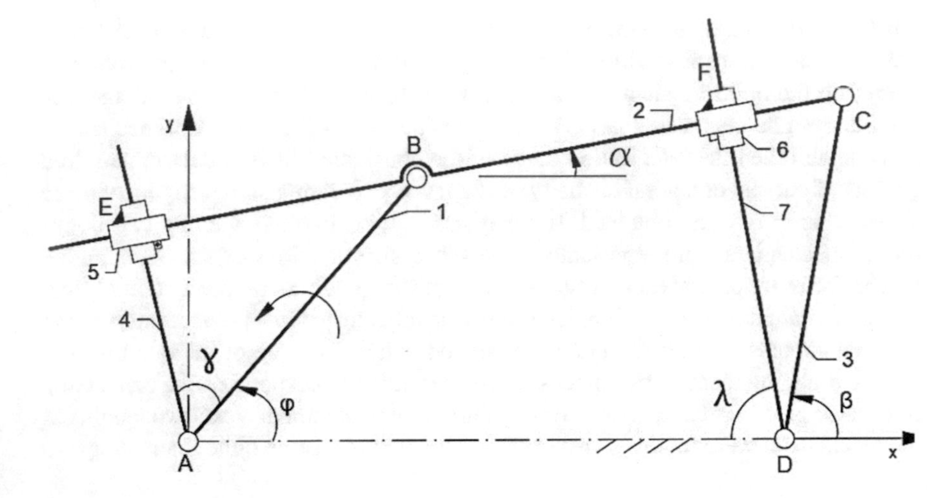

Fig. 13.2 The mechanism

The following expressions were written:

$$x_C = AB\cos\varphi + BC\cos\alpha = x_D + CD\cos\beta \quad (13.1)$$

$$y_C = AB\sin\varphi + BC\sin\alpha = y_D + CD\sin\beta \quad (13.2)$$

from where α, β, XC, YC could be determined

$$x_E = AB\cos\varphi + BE\cos(\alpha + \pi) = AE\cos(\varphi + \gamma) \quad (13.3)$$

$$y_E = AB\sin\varphi + BE\sin(\alpha + \pi) = AE\sin(\varphi + \gamma) \quad (13.4)$$

having $\gamma = 90 + \alpha - \varphi$ from where AE, BE, X_E, Y_E can be determined.

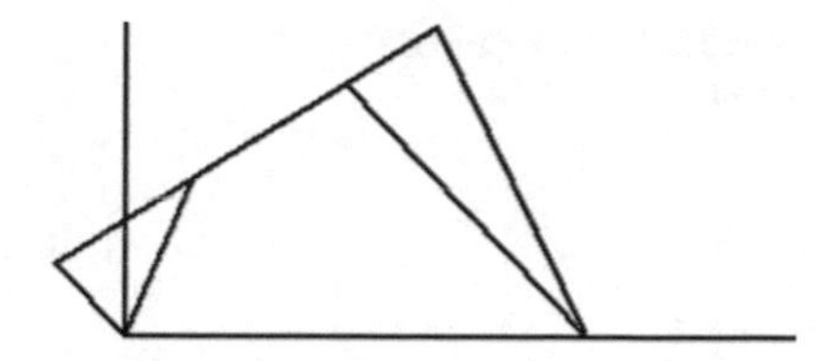

Fig. 13.3 The mechanism in one position

$$x_F = x_C + CF\cos(\alpha + \pi) = x_D + DF\cos\lambda \tag{13.5}$$

$$y_F = y_C + CF\sin(\alpha + \pi) = y_D + DF\sin\lambda \tag{13.6}$$

having $\lambda = \alpha + 90$, from where CF, DF, X_F, Y_F can be determined.

The following initial values were used: AB = 20; BC = 30; CD = 40; $X_D = 45$, so that the built-in mechanism met the Grashof conditions. In Fig. 13.3 the mechanism in one position ($\varphi = 70$) was presented. Both heights were visible.

In Fig. 13.4, the successive positions of the mechanism are drawn.

Figure 13.5 it is presented the trajectory of the E point (one locus), and in Fig. 13.6 it is given the trajectory of F point (another locus).

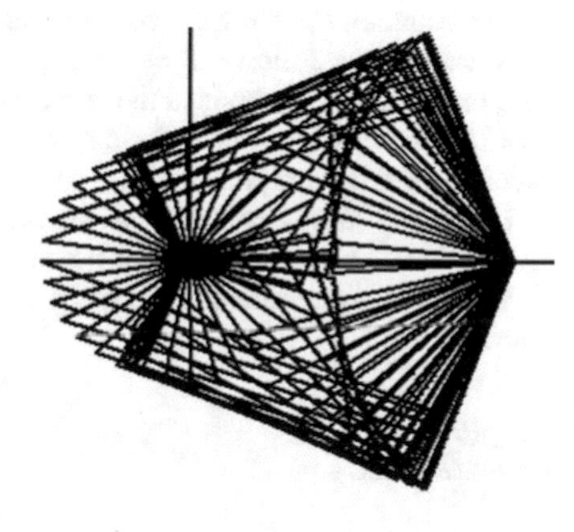

Fig. 13.4 The successive positions

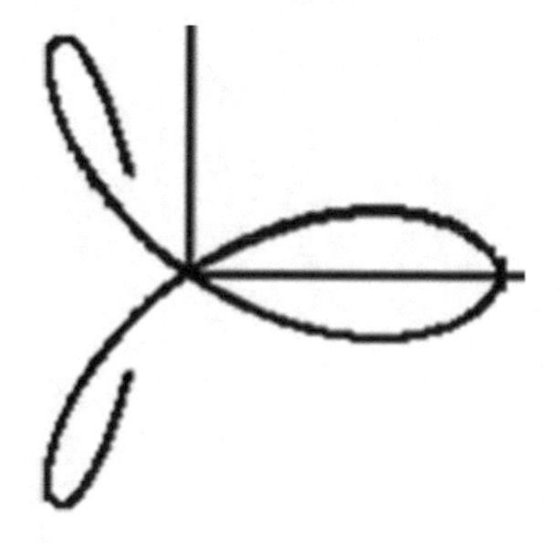

Fig. 13.5 The locus of E point

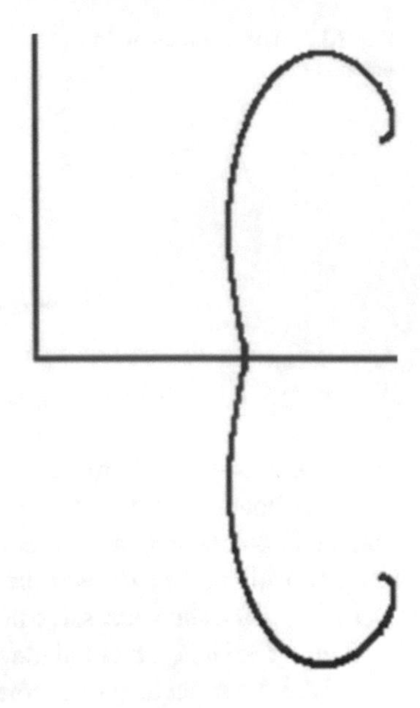

Fig. 13.6 The locus of F point

From the motion diagrams, we found out that the Y-coordinates of E and F points have leaps to $\varphi = 180$ degrees, which deformed the resulting curves.

The equation system has two solutions, so that a radical appears with the S sign to be changed as follows: S = 1 for $\varphi = 0 \ldots 179$ and S = − 1 for $\varphi = 181 \ldots 360$. In this case it is obtained the curve shown in Fig. 13.8, where there is a jump at $\varphi = 180$ degrees, but it is a passing over this critical position; in fact, the passing is accomplished through inertia.

Below, the successive positions are presented in Fig. 13.7, the locus of E point is given in Fig. 13.8 and the one for F point is shown in Fig. 13.9.

We found some interesting curves with loops, turning points and intersecting branches.

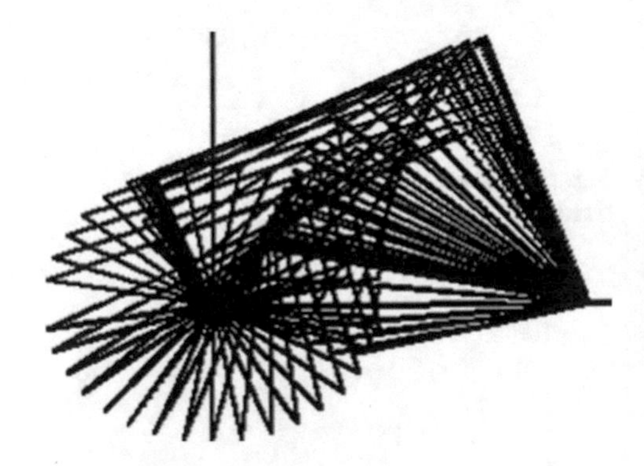

Fig. 13.7 Another successive positions

Fig. 13.8 The locus of E point

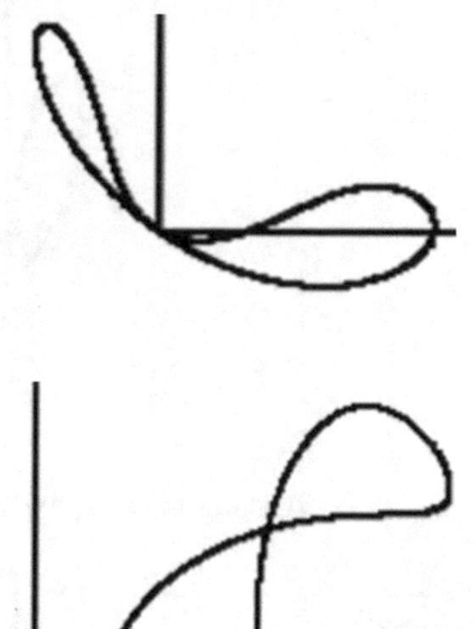

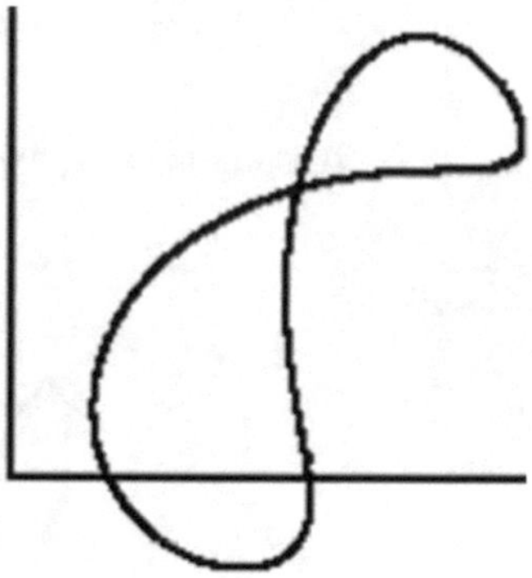

Fig. 13.9 The locus of F point

It is obvious that, *when changing the dimensions of the mechanism*, other curves are obtained as loci [1]. Below, there are examples of such curves, resulting from the changing of only some lengths of the sides. Thus, maintaining BC = 30 and CD = 40 constants, and by altering AB and XD lengths, the figures below were drawn (Fig. 13.10).

In Figs. 13.11, 13.12, 13.13, 13.14, 13.15, 13.16, 13.17, AB = BC and CD = XD were taken into consideration.

In the case shown in Fig. 13.18, 13.19 and 13.20 it was considered, AB = CD as the curves were similar.

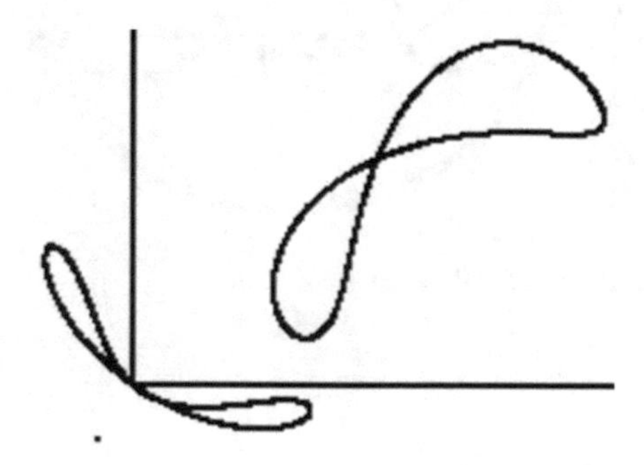

Fig. 13.10 The locus for AB = 15; XD = 50

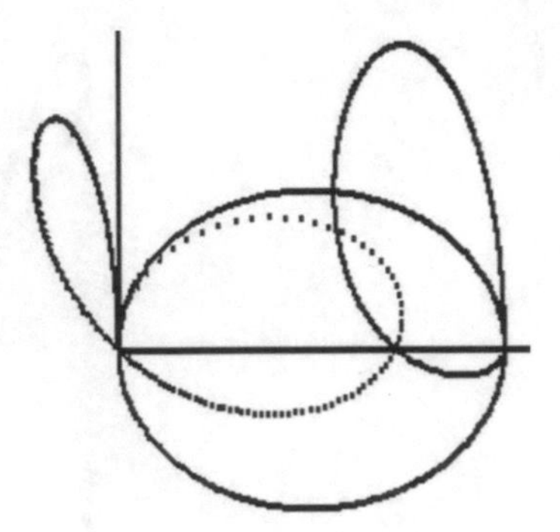

Fig. 13.11 The locus for AB = 25; XD = 45

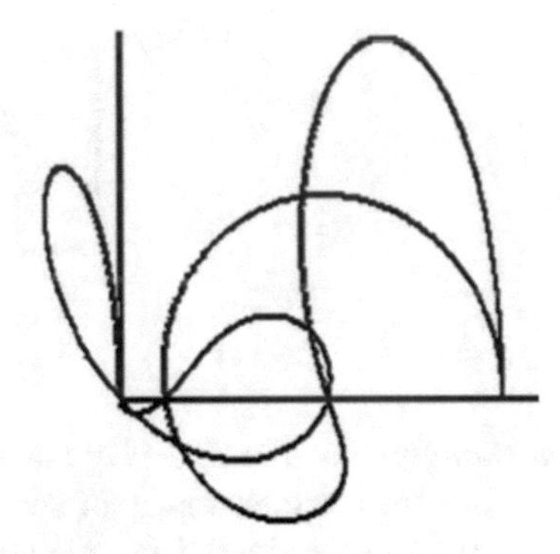

Fig. 13.12 The locus for AB = 35; XD = 40

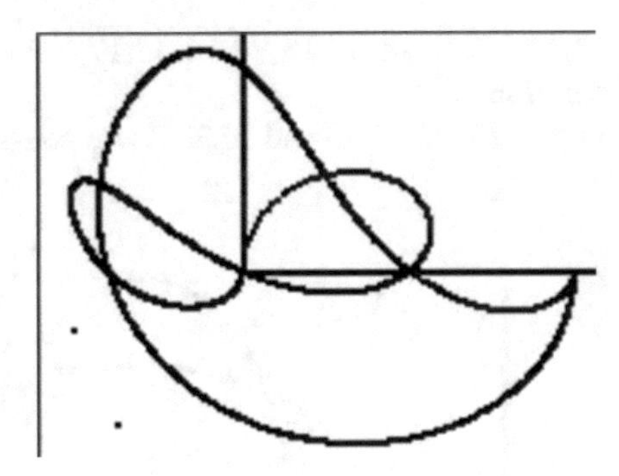

Fig. 13.13 The locus for AB = 15; XD = 25

Fig. 13.14 The locus for AB = 10; XD = 40

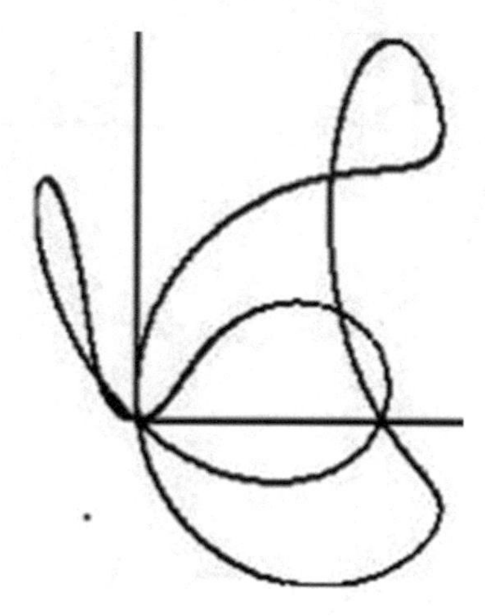

Fig. 13.15 The locus for AB = 25; XD = 40

Next, the there were kept constant AB = 20 and XD = 45 and BC and CD were changed, resulting in the following pictures (Figs. 13.21, 13.22, 13.23, 13.24, 13.25, 13.26, 13.27, 13.28, 13.29, 13.30, 13.31, 13.32, 13.33, 13.34, 13.35, 13.36, 13.37, 13.38, 13.39 and 13.40).

Furthermore, other values were used for the lenghts of the sides.

In Fig. 13.41, the case of the parallelogram was analyzed. Curves were opposite and symmetrical. The case of the rhombus is shown in Fig. 13.42.

The curves are opposite and symmetrical (Figs. 13.43 and 13.44).

Here also, curves are symmetrical, opposite, with AB = CD and BC = XD (parallelogram) (Figs. 13.45 and 13.46). Further, XD < 0 was considered, having as result the loci presented below (Figs. 13.47, 13.48, 13.49, 13.50 and 13.51).

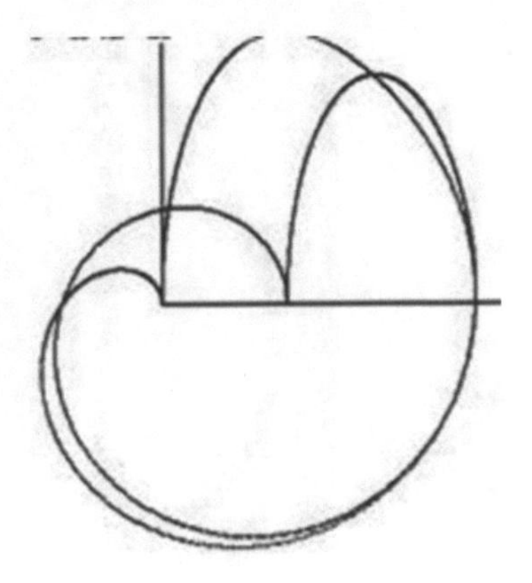

Fig. 13.16 The locus for AB = 50; XD = 20

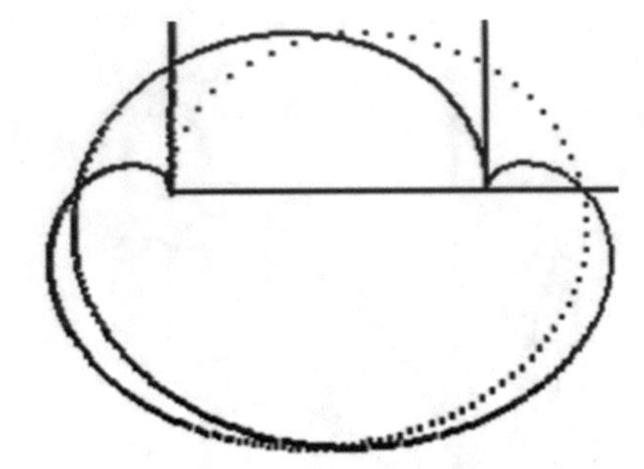

Fig. 13.17 The locus for AB = 40; XD = 30

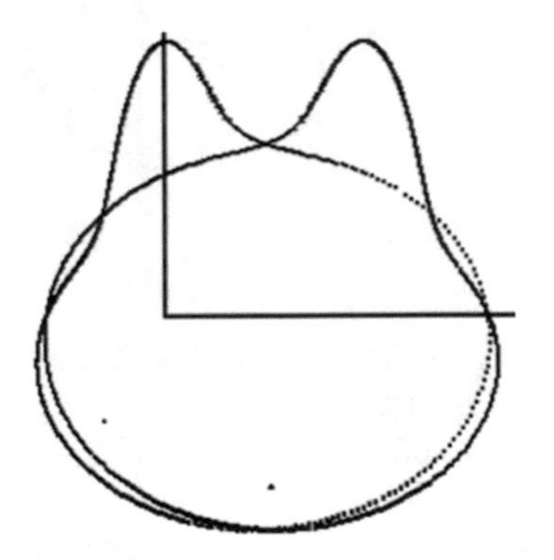

Fig. 13.18 The locus for AB = 40; XD = 25

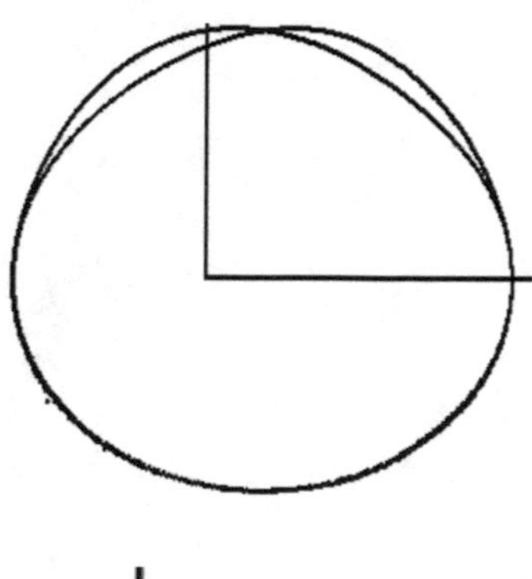

Fig. 13.19 The locus for AB = 40; XD = 15

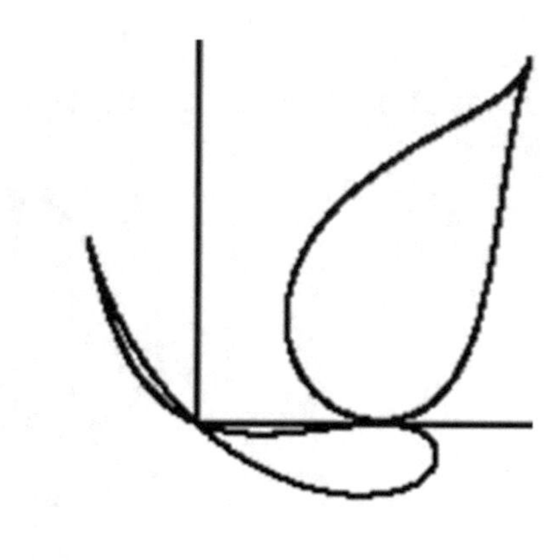

Fig. 13.20 The locus for BC = 40; CD = 40

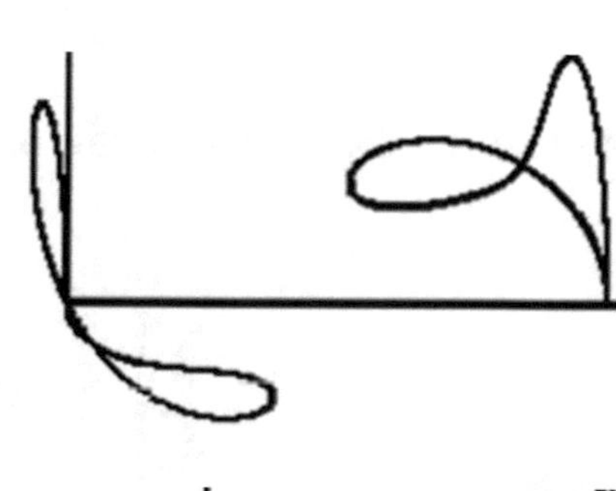

Fig. 13.21 The locus for BC = 40; CD = 25

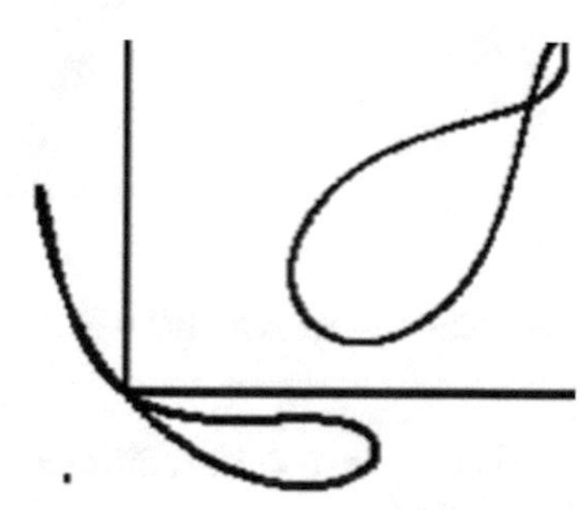

Fig. 13.22 The locus for BC = 40; CD = 35

Fig. 13.23 The locus for BC = 40; CD = 50

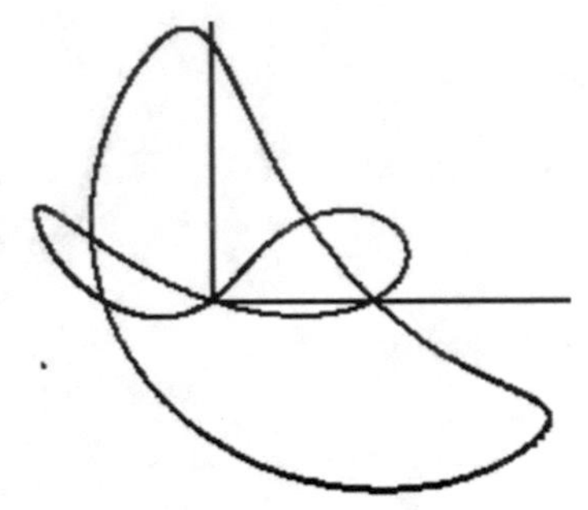

Fig. 13.24 The locus for BC = 40; CD = 60

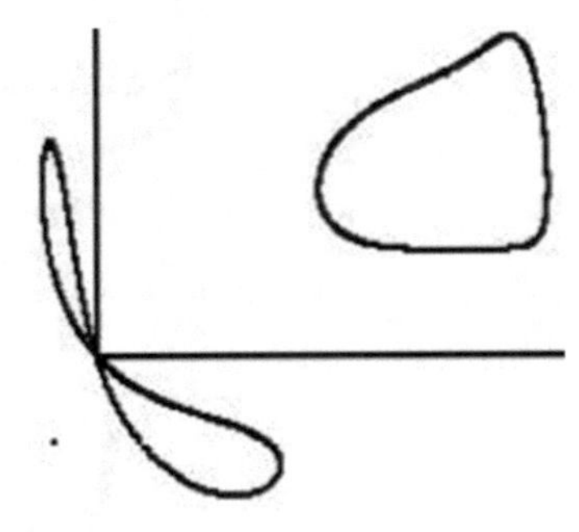

Fig. 13.25 The locus for BC = 50; CD = 30

There were obtained different curves, depending on the lengths of the elements. In conclusion, *loci are variable according to the lengths of the elements.* Other conclusions can be drawn by comparing the loci found above.

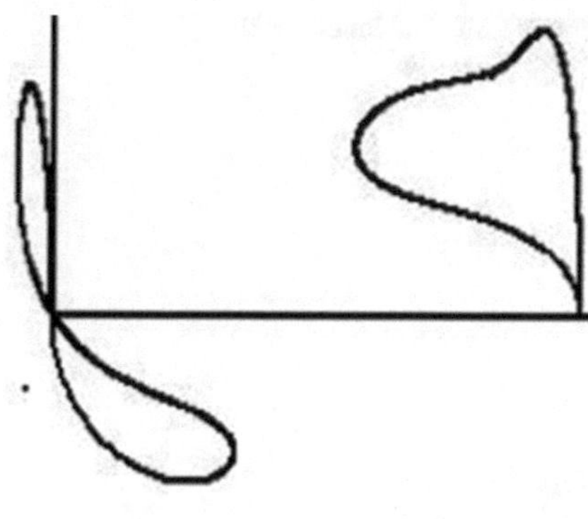

Fig. 13.26 The locus for BC = 50; CD = 25

Fig. 13.27 The locus for BC = 50; CD = 40

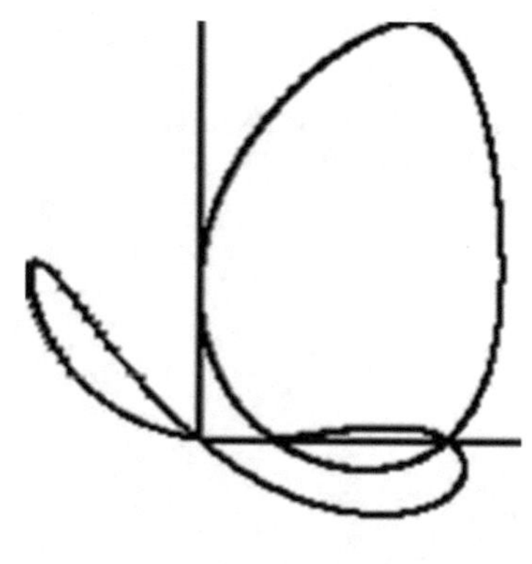

Fig. 13.28 The locus for BC = 50; CD = 50

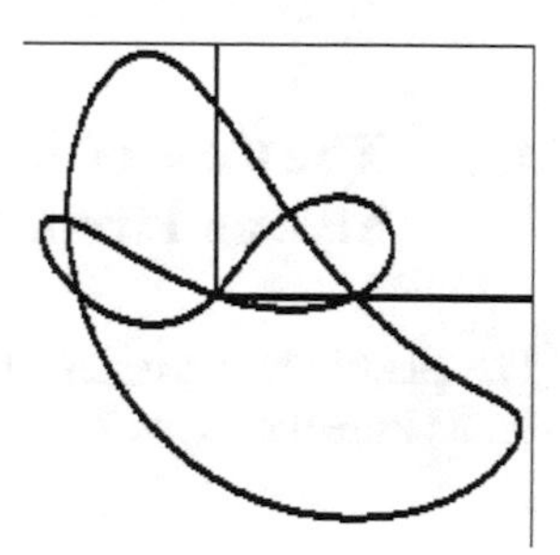

Fig. 13.29 The locus for BC = 50; CD = 70

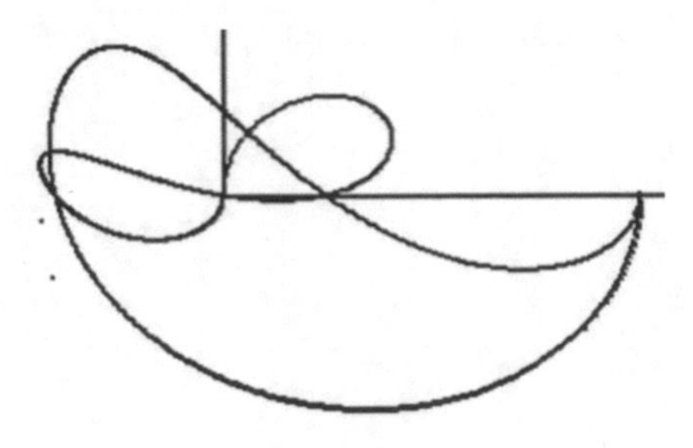

Fig. 13.30 The locus for BC = 50; CD = 75

Fig. 13.31 The locus for BC = 35; CD = 45

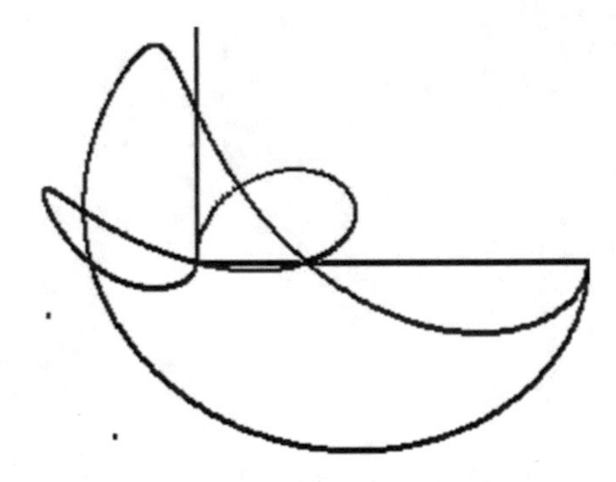

Fig. 13.32 The locus for BC = 35; CD = 60

13.2 The Case of Heights Drawn from B Point on the CD Side and from C Point on AB Side

The quadrilateral presented in the previous chapter was used, including the relations for its kinematics, so XC, YC, α and β are considered as know from Fig. 13.52.

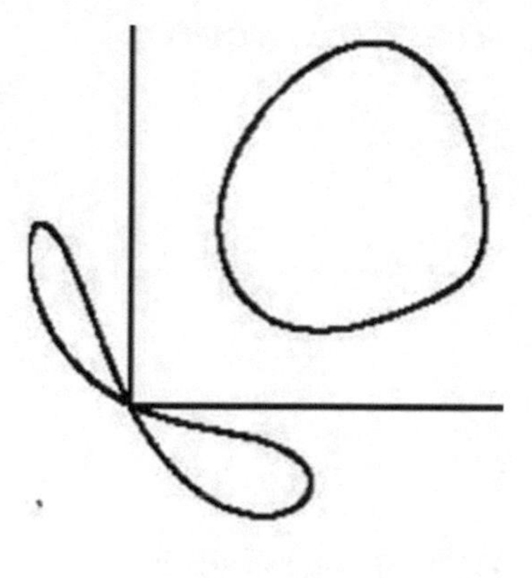

Fig. 13.33 The locus for BC = 60; CD = 45

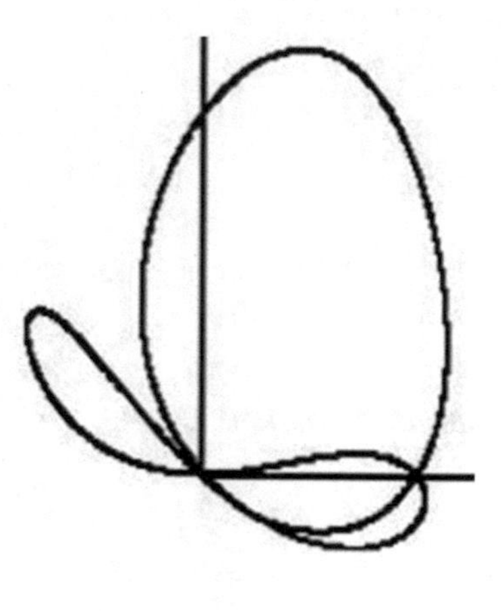

Fig. 13.34 The locus for BC = 60; CD = 60

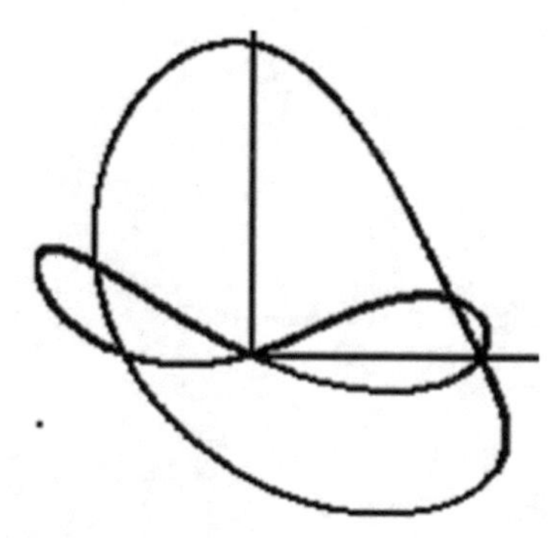

Fig. 13.35 The locus for BC = 60; CD = 70

The CH and BG heights were drawn and we needed to determine the loci of the feet of perpendiculars, i.e. H and G points. In the H and G points, welded double sliders were built in order to ensure the perpendicularity.

The following expressions were written:

Fig. 13.36 The locus for BC = 70; CD = 45

Fig. 13.37 The locus for BC = 70; CD = 55

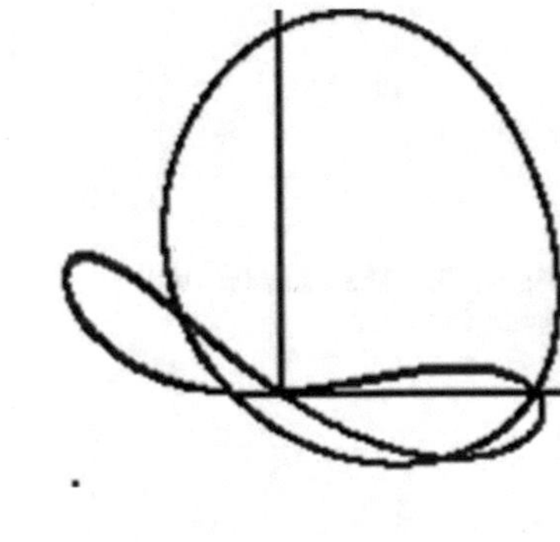

Fig. 13.38 The locus for BC = 70; CD = 70

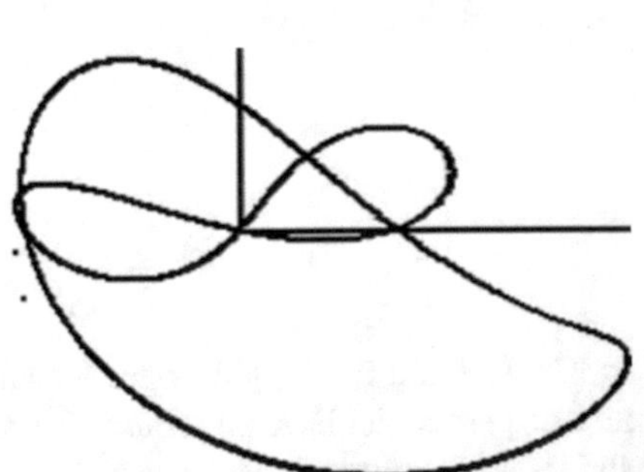

Fig. 13.39 The locus for BC = 70; CD = 80

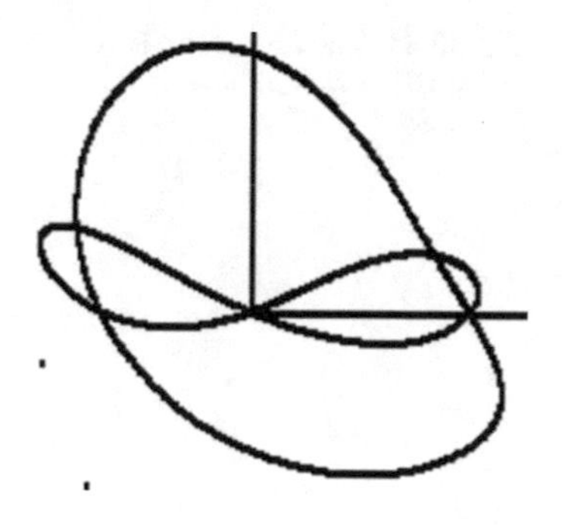

Fig. 13.40 The locus for BC = 70; CD = 90

Fig. 13.41 The locus for AB = 20; BC = 40; CD = 20; XD = 40

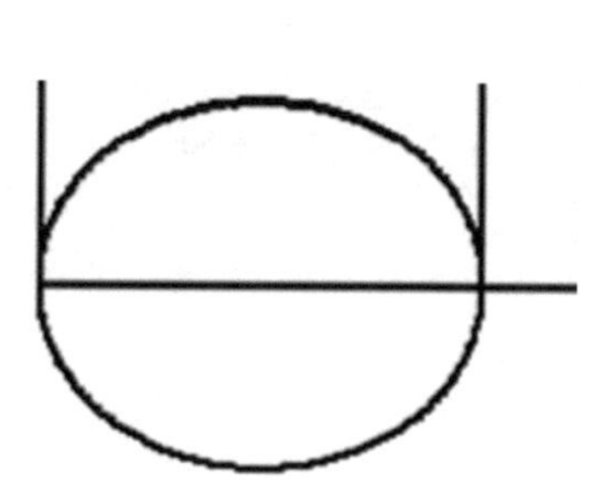

Fig. 13.42 The locus for AB = 30; BC = 30; CD = 30; XD = 30

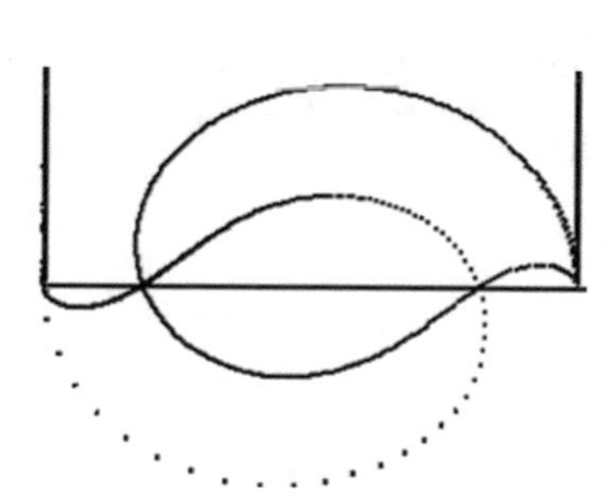

Fig. 13.43 The locus for AB = 50; BC = 60; CD = 50; XD = 60

$$x_C = AH\cos\varphi + HC\cos(\varphi + 270) \quad (13.7)$$

$$y_C = AH\sin\varphi + HC\sin(\varphi + 270) \quad (13.8)$$

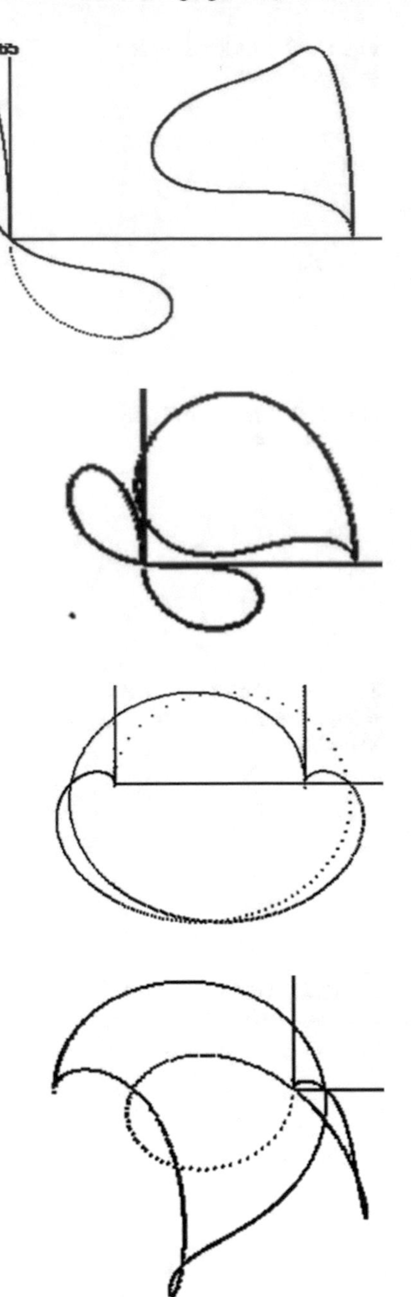

Fig. 13.44 The locus for AB = 35: BC = 70; CD = 45; XD = 65

Fig. 13.45 The locus for AB = 15; BC = 45; CD = 35; XD = 25

Fig. 13.46 The locus for AB = 50; BC = 40; CD = 50; XD = 40

Fig. 13.47 The locus for AB = 25; BC = 43; CD = 40; XD = –35

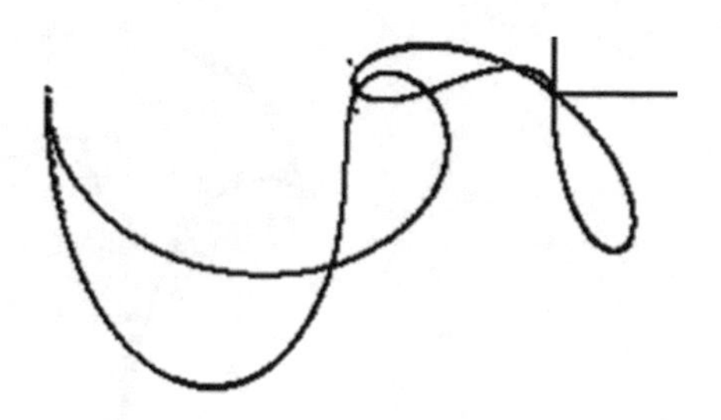

Fig. 13.48 The locus for AB = 20; BC = 30; CD = 40; XD = –50

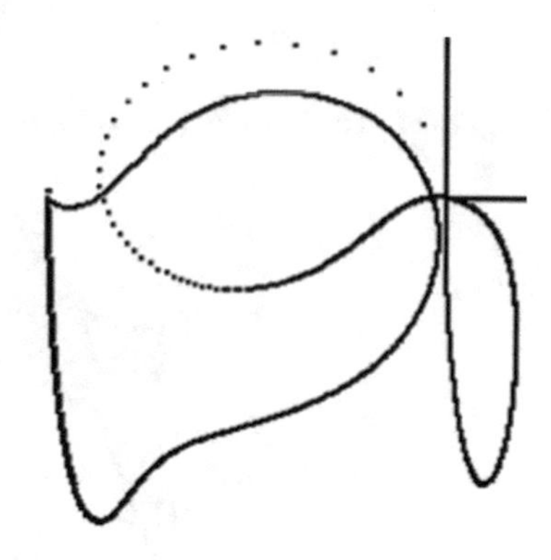

Fig. 13.49 The locus for AB = 20; BC = 30; CD = 50; XD = –50

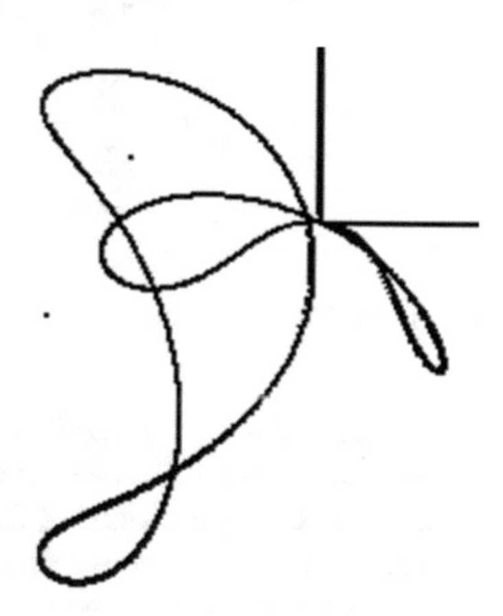

Fig. 13.50 The locus for AB = 35; BC = 45; CD = 40; XD = –40

$$x_G = x_D + DG\cos\beta = x_B + BG\cos\psi \tag{13.9}$$

$$y_G = y_D + DG\sin\beta = y_B + BG\sin\psi \tag{13.10}$$

$$\psi = \beta + 270. \tag{13.11}$$

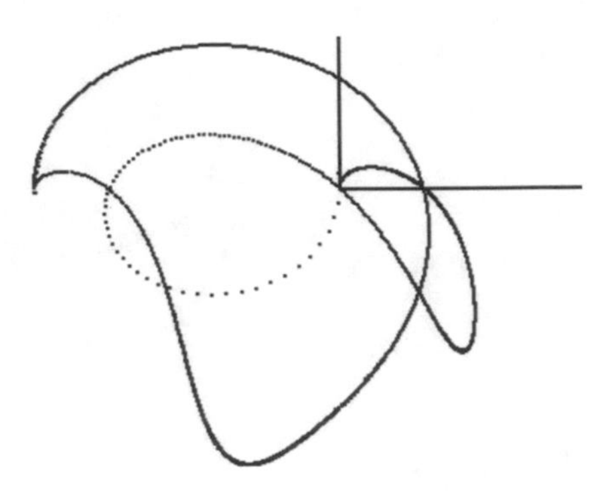

Fig. 13.51 The locus for (AB = 35; BC = 50; CD = 60; XD = –45)

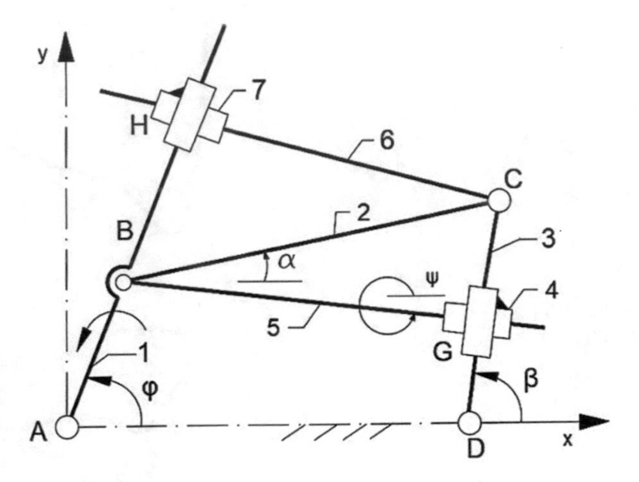

Fig. 13.52 The mechanism

In Fig. 13.53 the mechanism could be seen in one position, as the heights were positioned correctly. Initial data were: AB = 20, BC = 30, CD = 40, XD = 45.

The successive positions of the mechanism were shown in Fig. 13.54.

The locus of the H point was shown in Fig. 13.55, as a closed curve with inner loop, and the locus of G point could be seen in Fig. 13.56, i.e. a closed curve, like a

Fig. 13.53 The mechanism in one position

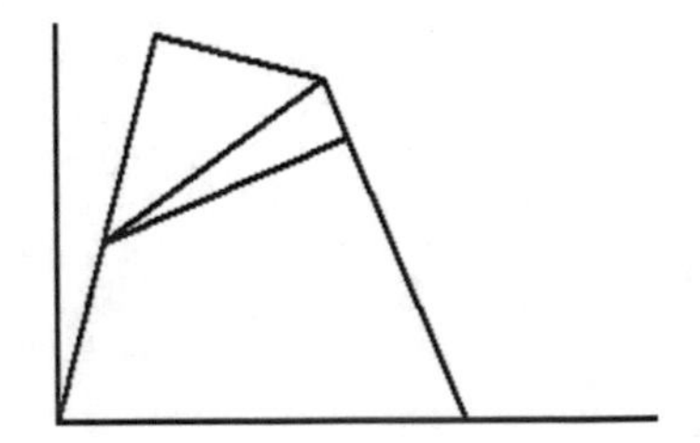

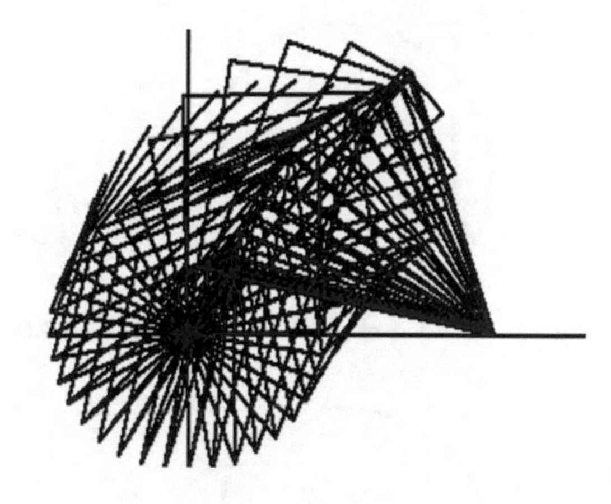

Fig. 13.54 The successive positions

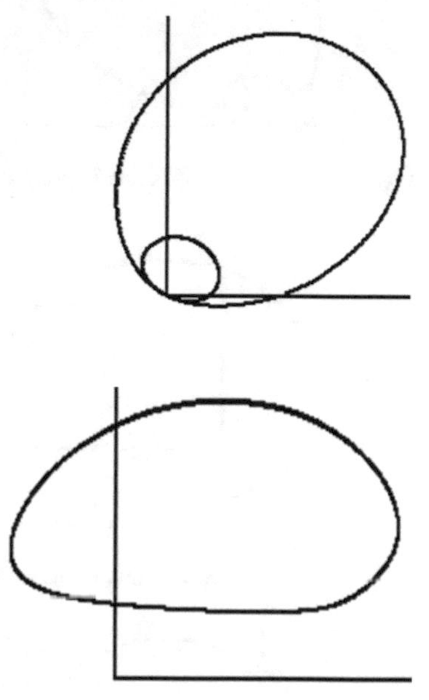

Fig. 13.55 The locus for H point

Fig. 13.56 The locus for G point

"banana". T It was obvious that the shapes of curves that resulted as loci depended on the dimensions of the sides of the quadrilateral. In order to find the shapes of these curves, the hus, BC and CD were kept constant (at baseline values) and different values of AB and XD were selected, and only the closed curves were hold. We could not represent both loci on the same picture because they intersected, so they were separated. (H—the left curve, and G—the right curve) (Figs. 13.57, 13.58, 13.59, 13.60, 13.61, 13.62, 13.63, 13.64, 13.65, 13.66 and 13.67).

Next, the initial values of the AB and XD were kept the same and the BC and CD values were changed (Figs. 13.68 and 13.69 and 13.70).

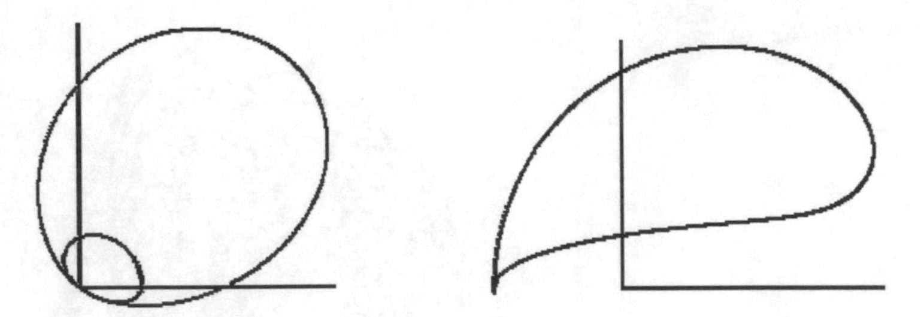

Fig. 13.57 The loci of H and G points for AB = 20; XD = 50

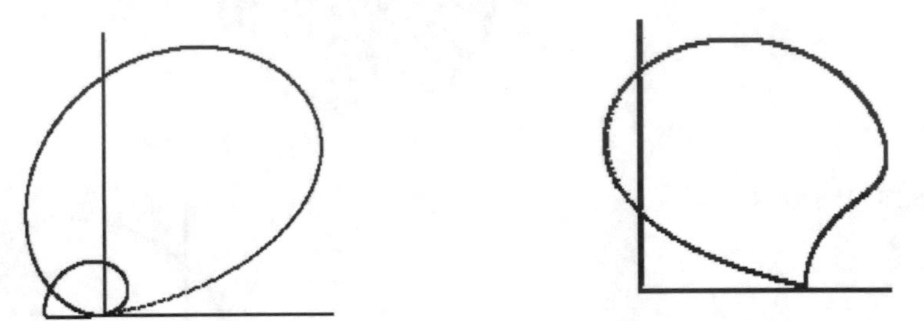

Fig. 13.58 The loci of H and G points for AB = 20; XD = 30

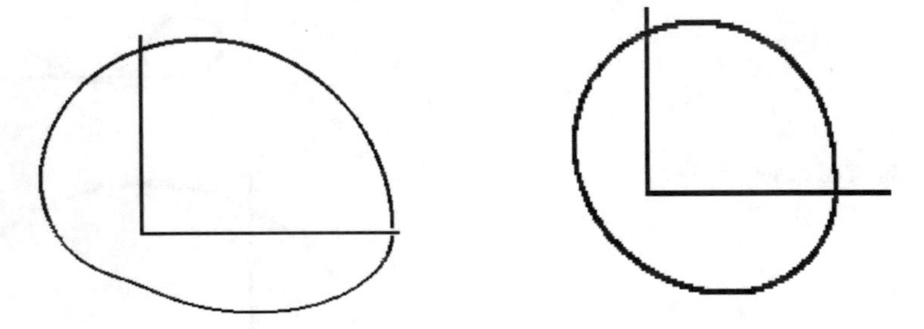

Fig. 13.59 The loci of H and G points for AB = 20; XD = 10

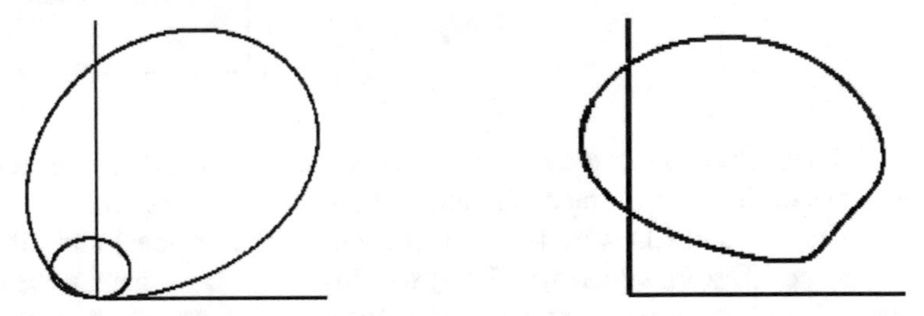

Fig. 13.60 The loci of H and G points for AB = 20; XD = 35

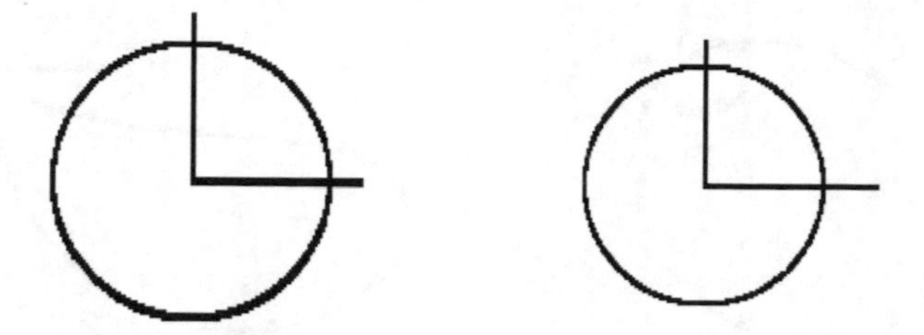

Fig. 13.61 The loci of H and G points for AB = 20; XD = 0

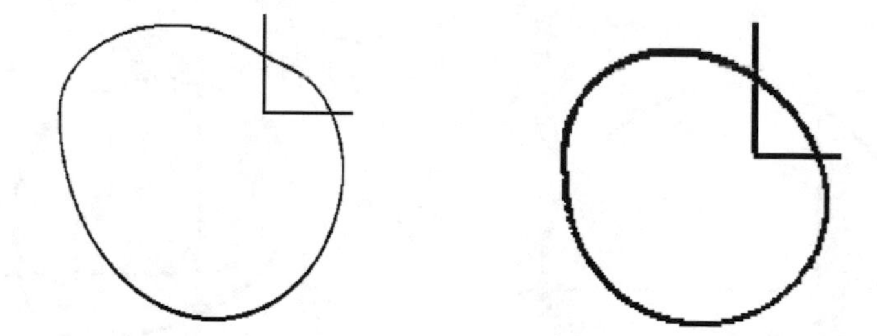

Fig. 13.62 The loci of H and G points for AB = 20; XD = –10

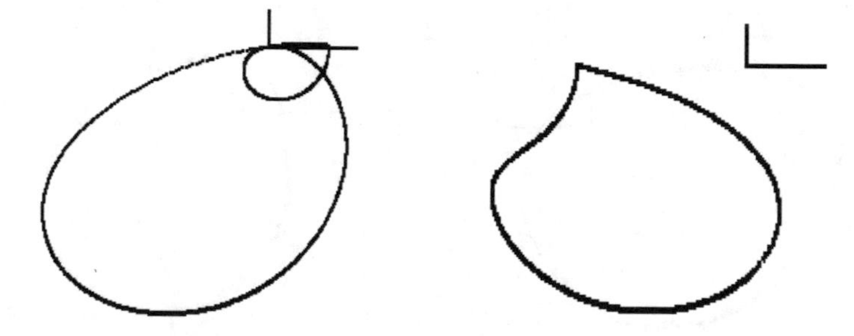

Fig. 13.63 The loci of H and G points for AB = 20; XD = –30

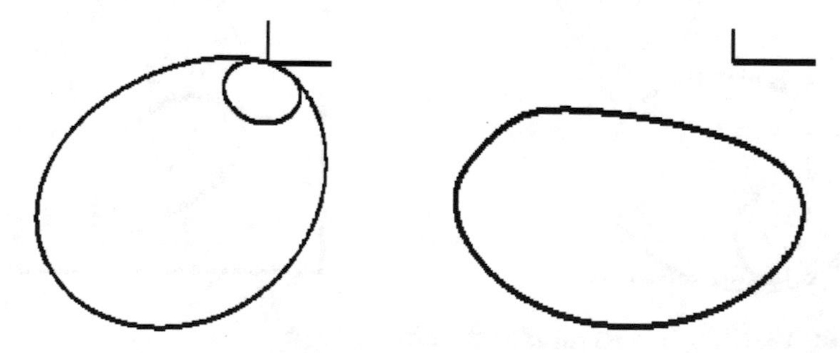

Fig. 13.64 The loci of H and G points for AB = 20; XD = −40

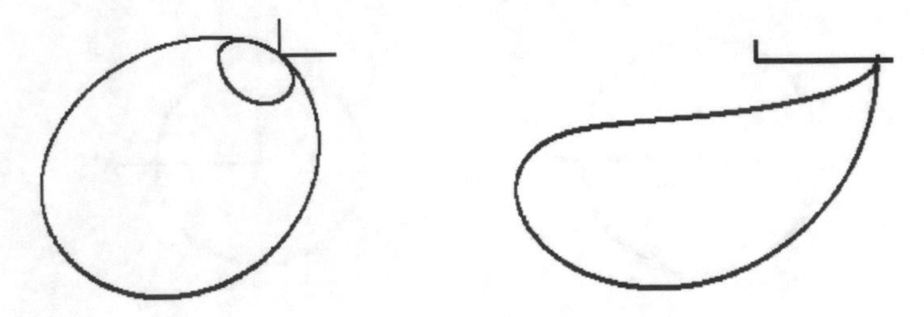

Fig. 13.65 The loci of H and G points for AB = 20; XD = –50

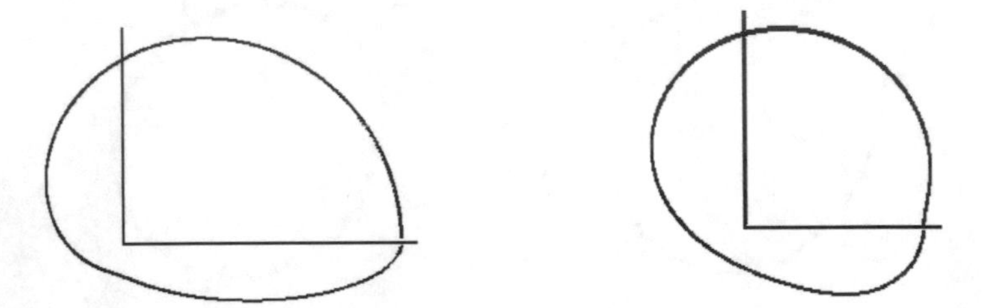

Fig. 13.66 The loci of H and G points for AB = 30; XD = 20

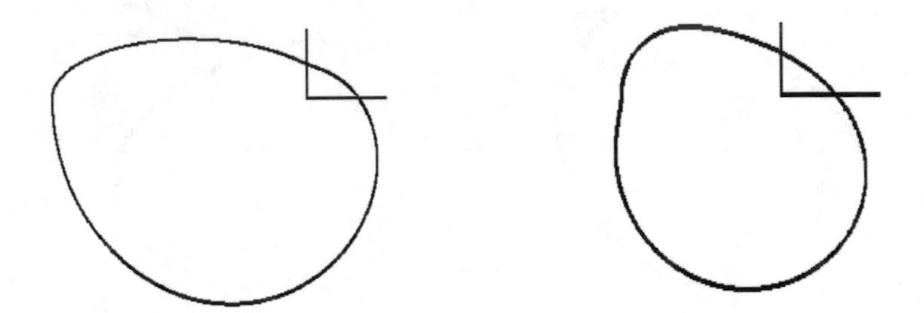

Fig. 13.67 The loci of H and G points for AB = 30; XD = –20

Fig. 13.68 The loci of H and G points for BC = 50, CD = 50

Next, only the locus of G point was represented, the curves being different from the ones above (Figs. 13.71, 13.72 and 13.73).

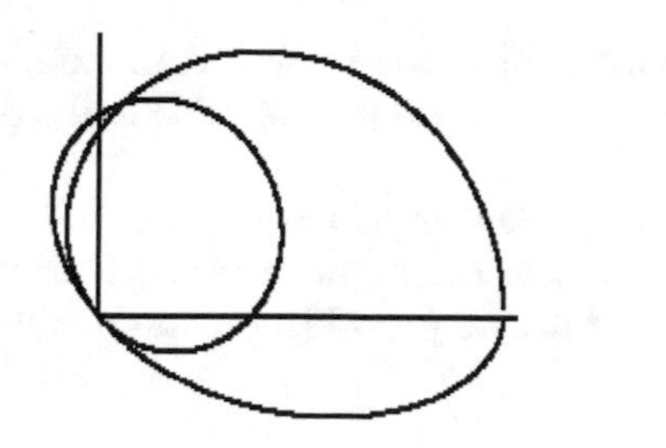
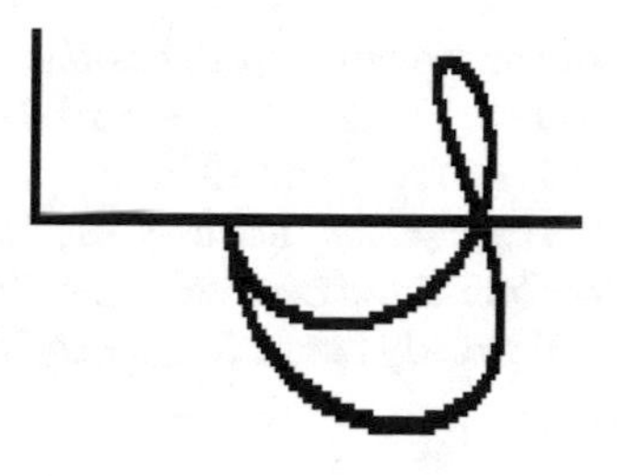

Fig. 13.69 The loci of H and G points for BC = 70, CD = 45

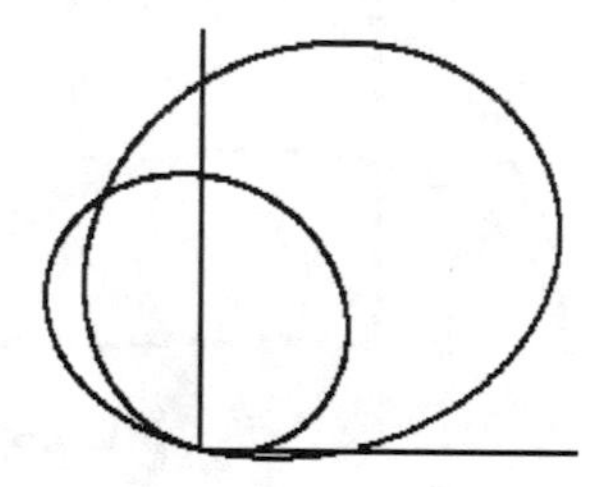

Fig. 13.70 The loci of H and G points for BC = 80, CD = 80

Fig. 13.71 The locus of G for BC = 30, CD = 50

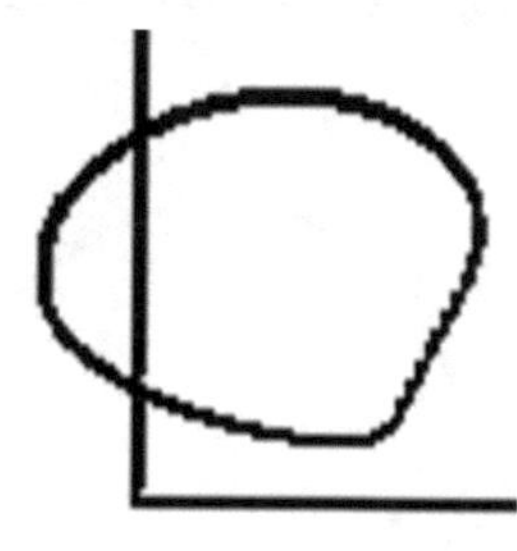

Fig. 13.72 The locus of G for BC = 60, CD = 35

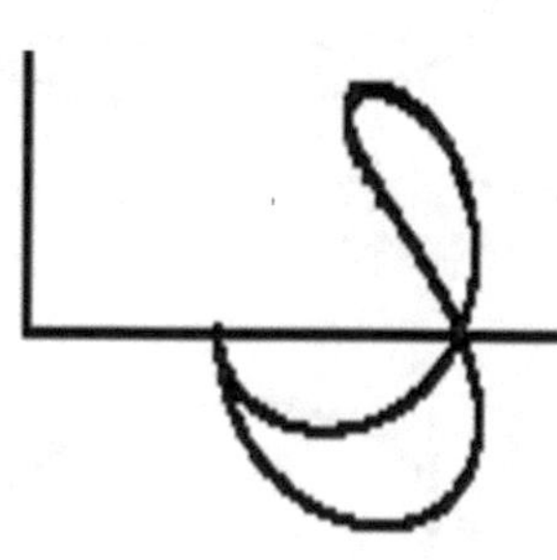

Fig. 13.73 The locus of G for BC = 30, CD = 35

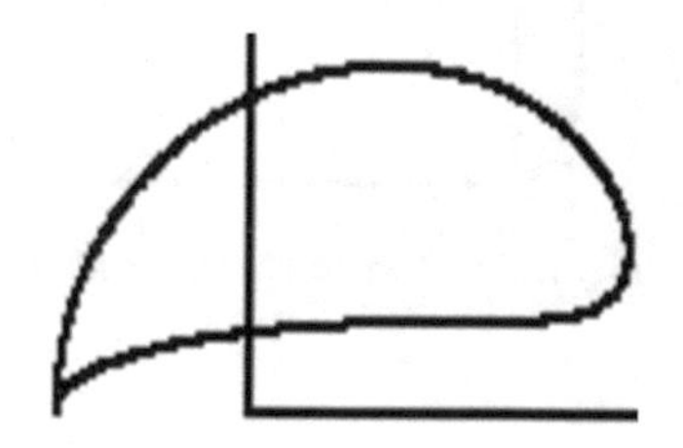

Furthermore, all the dimensions of the mechanism were changed, the following pictures being obtained (the curve of H is to the left and the one of G to the right) (Figs. 13.74, 13.75 and 13.76).

In the case of the rhombus, the curves were symmetrical and opposite.

Other loci could be determined for other dimensions of the mechanism; as many cases had already been tried, it may be deduced that they would have the same shapes as above.

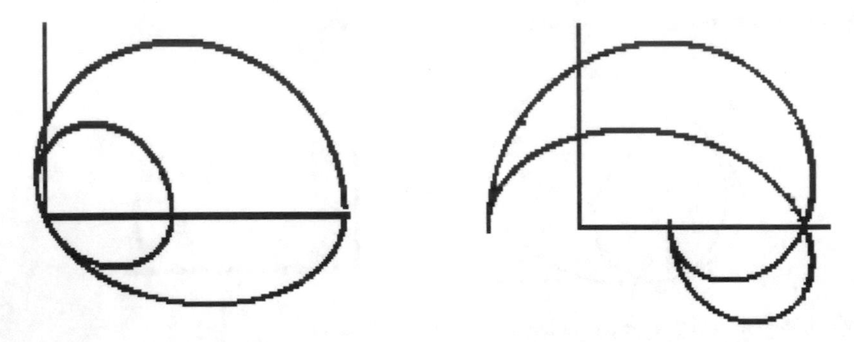

Fig. 13.74 The curves of H and G points for AB = 20; BC = 50; CD = 20; XD = 50—parallelogram

Fig. 13.75 The curves of H and G points for AB = 35; BC = 45; CD = 35; XD = 45—also parallelogram

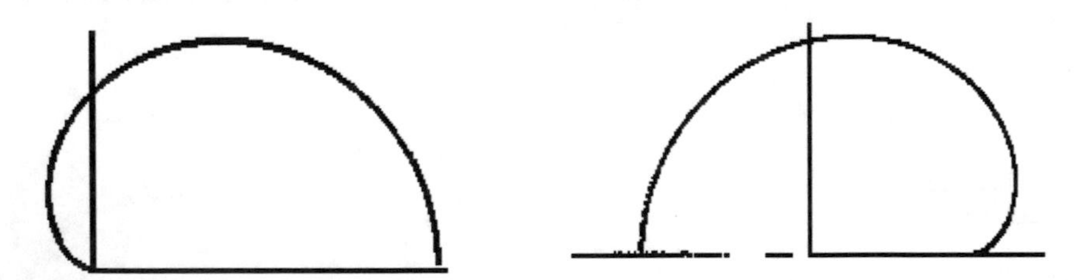

Fig. 13.76 The curves of H and G points for AB = BC = CD = XD = 40—rhombus

13.3 The Case of Heights Built from B and D Points on the AC Diagonal

It was considered the quadrilateral from the previous cases and the diagonal AC was drawn. Then, perpendiculars from B and D points to the AC diagonal were built. (Fig. 13.77)

The following expressions were written:

$$AC^2 = x_C^2 + y_C^2 \tag{13.12}$$

$$\sin\delta = y_C/AC \tag{13.13}$$

$$\cos\delta = x_C/AC \tag{13.14}$$

$$x_B = AK\cos\delta + BK\cos(\delta + 90) \tag{13.15}$$

$$y_B = AK\sin\delta + BK\sin(\delta + 90) \tag{13.16}$$

$$AL\cos\delta = x_D + DL\cos(\pi - 90 + \delta) \tag{13.17}$$

$$AL\sin\delta = y_D + DL\sin(\pi - 90 + \delta) \tag{13.18}$$

$$\beta = 90. \tag{13.19}$$

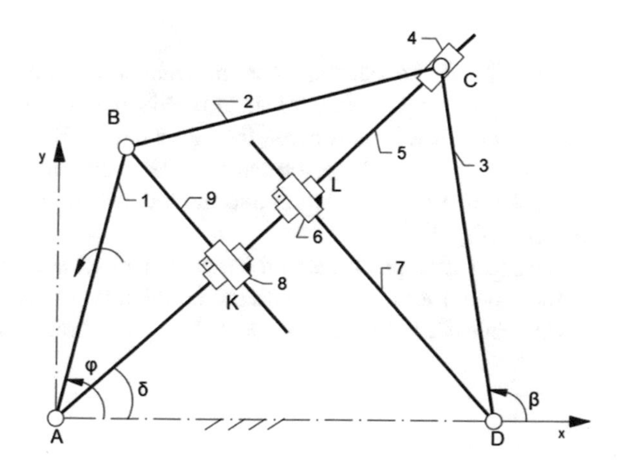

Fig. 13.77 The mechanism

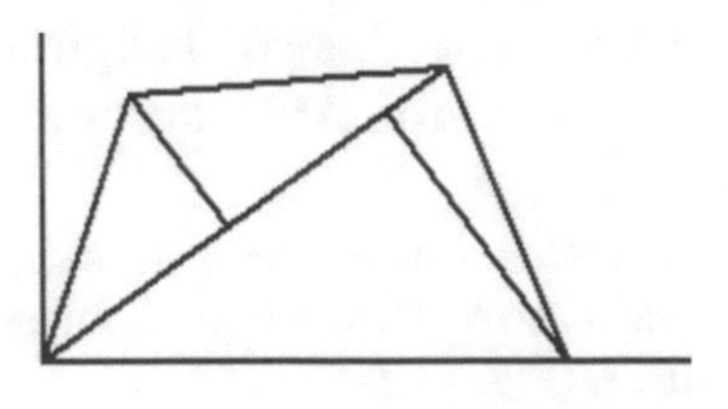

Fig. 13.78 The mechanism in one position

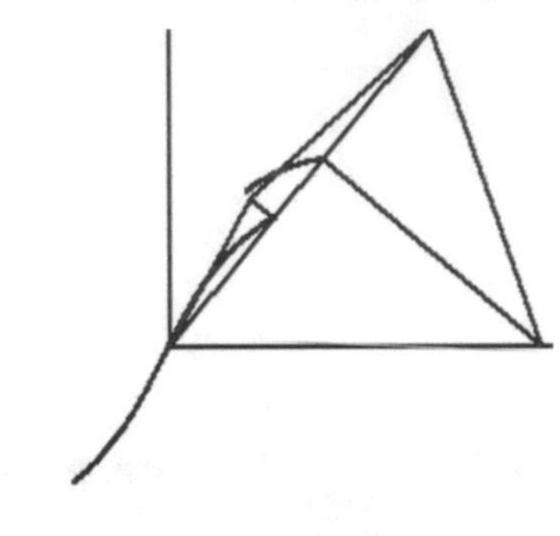

Fig. 13.79 Loci for K and L points

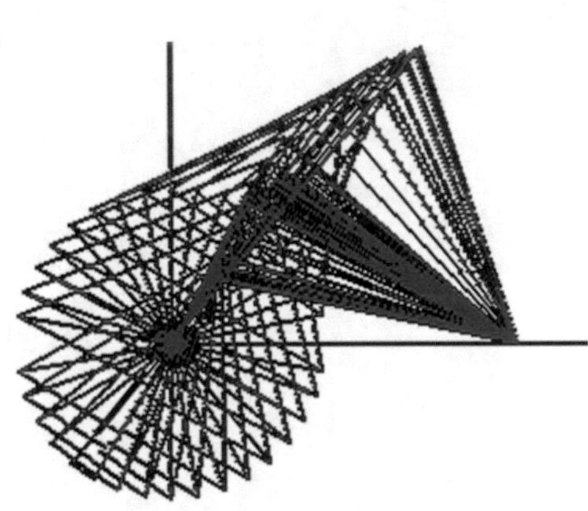

Fig. 13.80 The successive positions

In Fig. 13.78 the mechanism is shown in one position, for AB = 122; BC = 138; CD = 137; XD = 227 (data for sketch made on scale in order to verify the program).

Further on, it was considered the initial data: AB = 20; BC = 30; CD = 40; XD = 45, so that the crank performed complete rotations. In Fig. 13.78 it is presented the mechanism in one position and the loci described by points K (left) and L (on the right) (Fig. 13.79).

The successive positions of the mechanism are shown in Fig. 13.80.

The trajectories obtained with the initial data (Fig. 13.81) are not distinct.

The locus of the L point appears to be a circle arch, but it is a coupler-point curve.

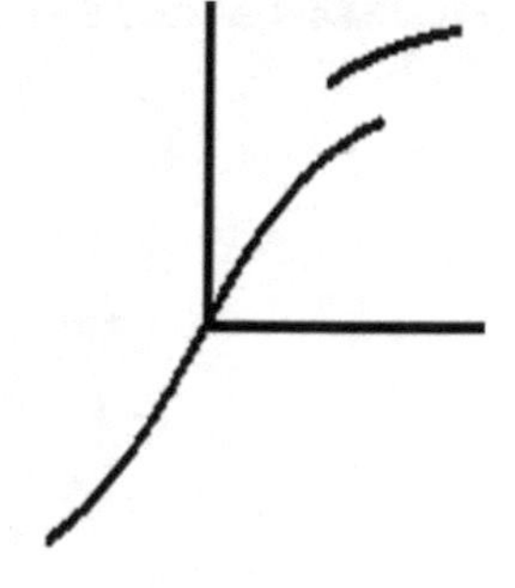

Fig. 13.81 Loci for initial data

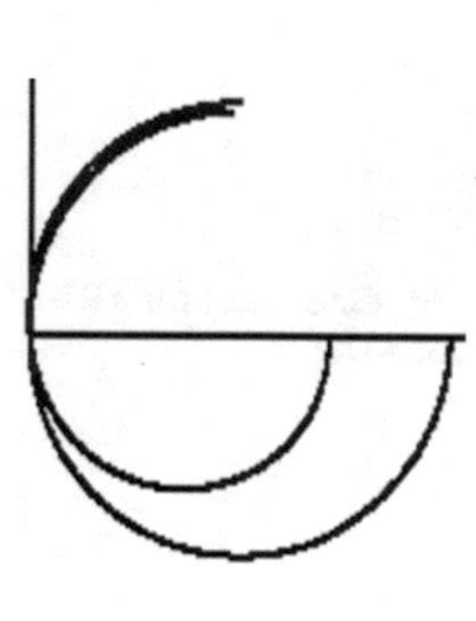

Fig. 13.82 Loci for AB = 25; XD = 35

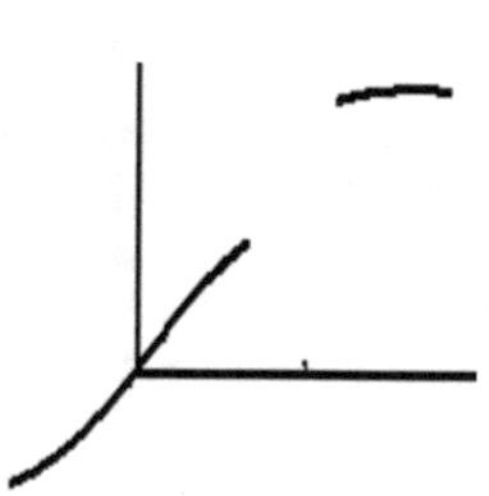

Fig. 13.83 Loci for AB = 15; XD = 50

Next, the initial values for BC and CD were kept and values for AB and XD were considered, having as result the loci presented in the following pictures (Figs. 13.82 and 13.83 and 13.84 and 13.85 and 13.86).

Further length values were chosen for different particular cases (Figs. 13.87, 13.88 and 13.89).

In the case of the rhombus, there are no solutions.

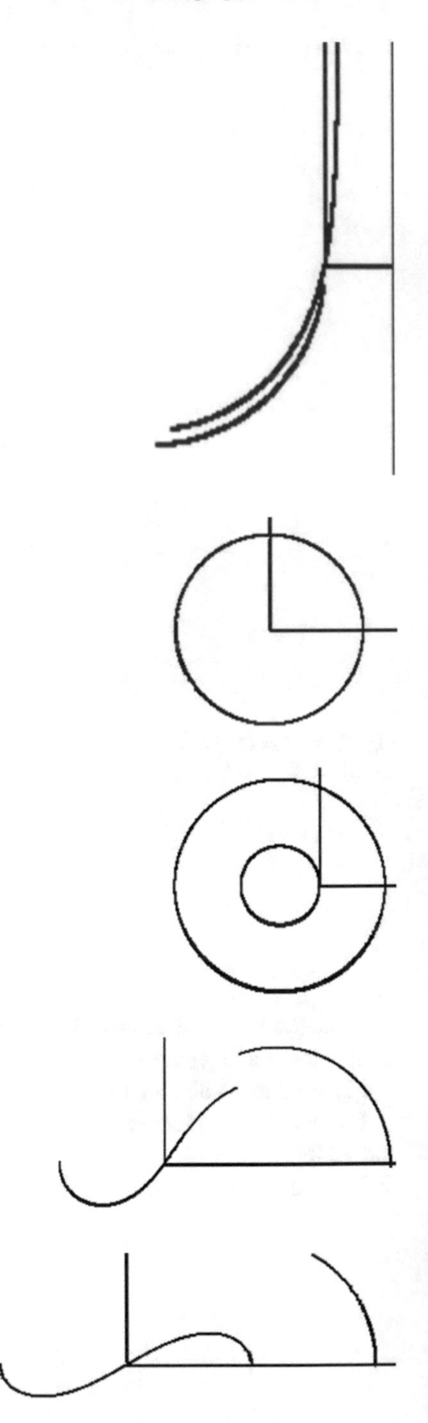

Fig. 13.84 Loci for AB = 25; XD = –40

Fig. 13.85 Loci for AB = 25; XD = 0

Fig. 13.86 Loci for AB = 30; XD = –15

Fig. 13.87 Loci for AB = 25; BC = 35; CD = 45; XD = 55

Fig. 13.88 Loci for AB = 25; BC = 50; CD = 25; XD = 50—parallelogram

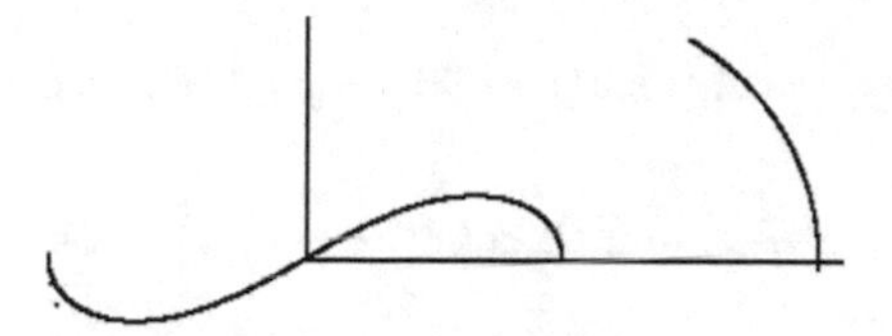

Fig. 13.89 Loci for AB = 30; BC = 60; CD = 30; XD = 60—also parallelogram

13.4 The Case of Heights Built from A and C Points on the BD Diagonal

We wanted to find the geometric locations of the M and N points, that represent the cross points of the perpendiculars built from A and C points on the BD diagonal. The generating mechanism is shown in Fig. 13.90. In M and N points, there were placed double sliders, and, a slider allowing the variation of the length of the diagonal BD was also placed in B point. Structurally, the mechanism is similar to the one in the previous case.

$$BD^2 = (x_B - x_D)^2 + (y_B - y_D)^2 \tag{13.20}$$

$$\sin\gamma = y_B/BD \tag{13.21}$$

$$\cos\gamma = (x_B - x_D)/BD \tag{13.22}$$

$$x_M = AM\cos(\gamma - 90) = x_B + BM\cos\varepsilon \tag{13.23}$$

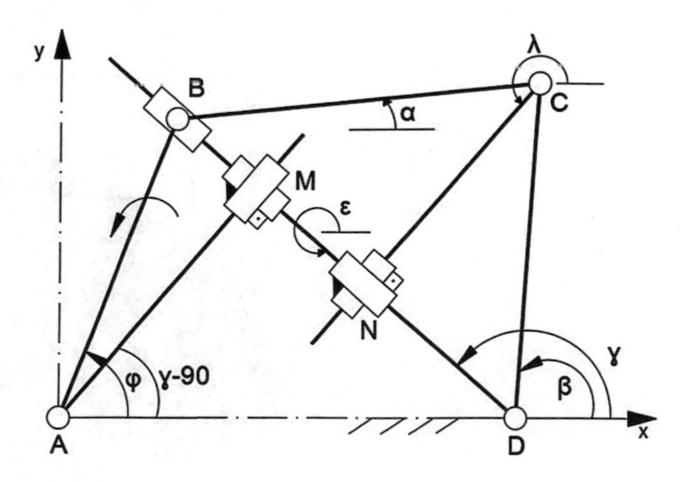

Fig. 13.90 The mechanism

$$y_M = AM\sin(\gamma - 90) = y_B + BM\sin\varepsilon \tag{13.24}$$

$$x_N = x_D + DN\cos\gamma = x_C + CN\cos\lambda \tag{13.25}$$

$$y_N = y_D + DN\sin\gamma = y_C + CN\sin\lambda \tag{13.26}$$

$$\lambda = \gamma + 90. \tag{13.27}$$

In Fig. 13.91, there was shown the mechanism in one position, and in Fig. 13.92, there were presented the successive positions of the mechanism for: AB = 20; BC = 30; CD = 40; XD = 45.

In Fig. 13.93, there are shown the mechanism in one position and the loci we were trying to determine.

The locus of the M point is a coupler-point curve, similar to a circular arc, and the one for the N point is a nephroidal closed curve.

The following are the loci for cases in which BC and CD sides were kept constant and the sides AB and XD were modified (Figs. 13.94, 13.95, 13.96, 13.97, 13.98, 13.99, 13.100, 13.101 and 13.102).

Several curve forms were determined, some of them were closed curves, when AB turned on completely, others open curves. Further, AB and XD sides were kept

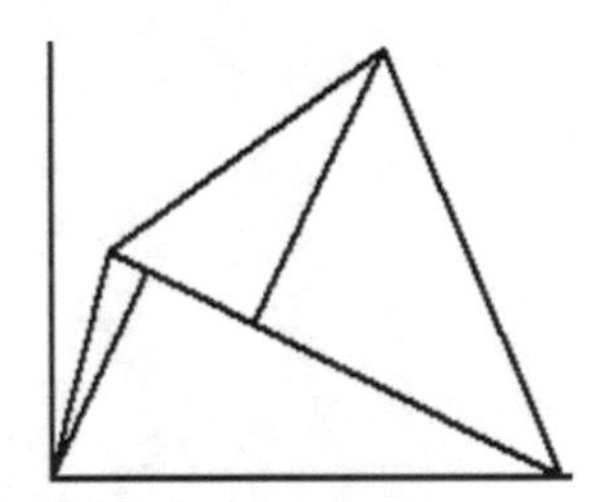

Fig. 13.91 The mechanism in one position

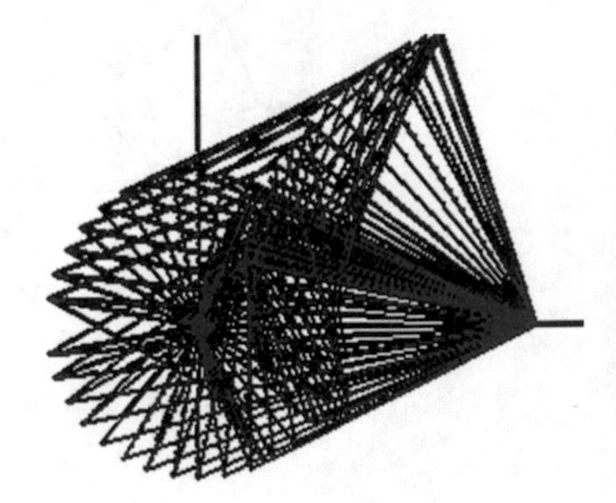

Fig. 13.92 Successive positions

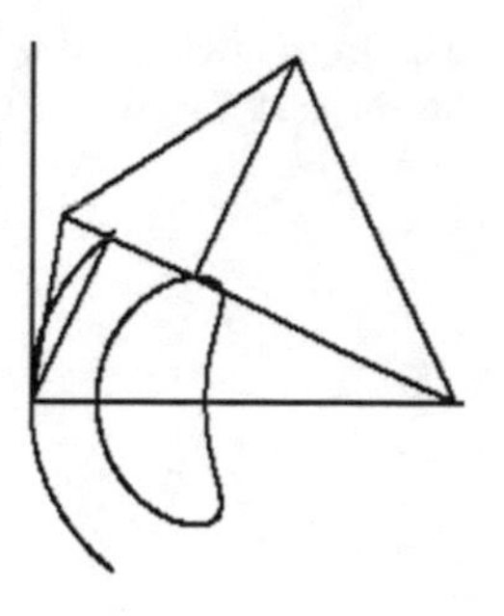

Fig. 13.93 The loci of M and N points

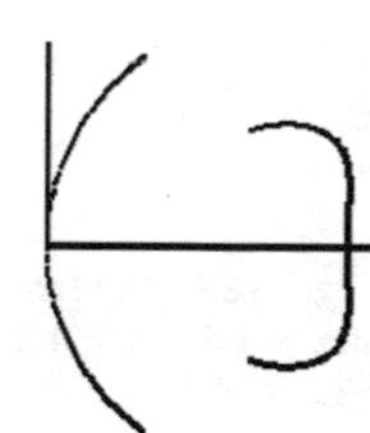

Fig. 13.94 The loci of M and N points for AB = 25; XD = 60

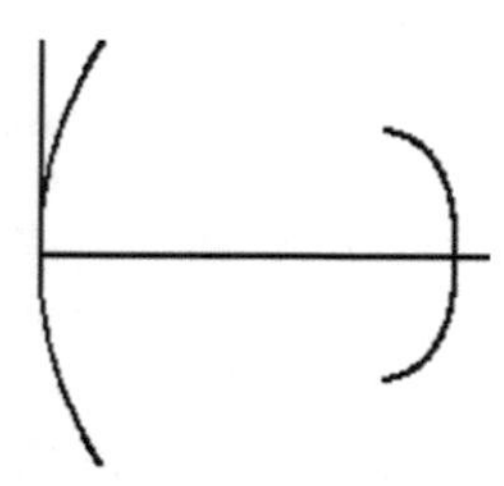

Fig. 13.95 The loci of M and N points for AB = 20; XD = 70

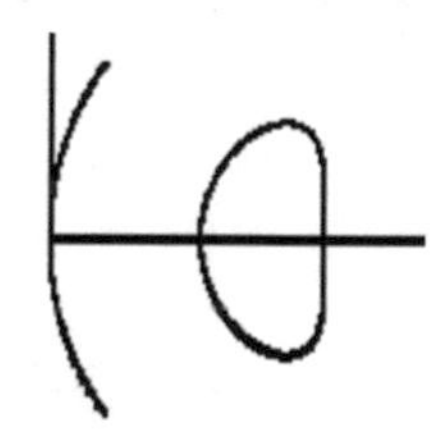

Fig. 13.96 The loci of M and N points for AB = 40; XD = 80

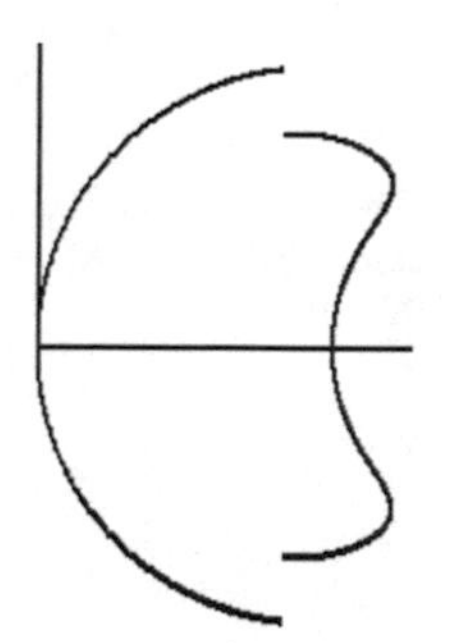

Fig. 13.97 The loci of M and N points for AB = 40; XD = 60

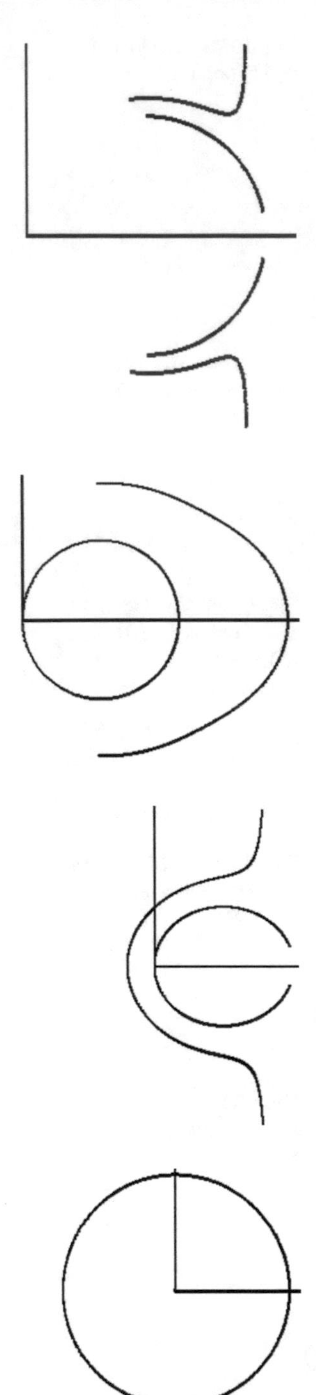

Fig. 13.98 The loci of M and N points for AB = 60; XD = 40

Fig. 13.99 The loci of M and N points for AB = 50; XD = 50

Fig. 13.100 The loci of M and N points for AB = 30; XD = 30

Fig. 13.101 The loci of M and N points for AB = 30; XD = 0

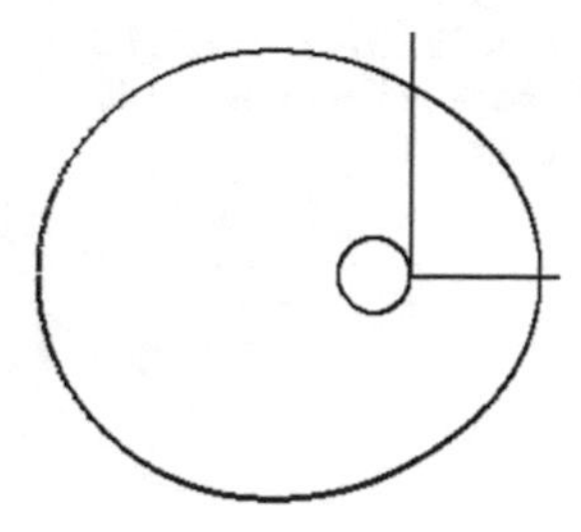

Fig. 13.102 The loci of M and N points for AB = 20; XD = −10

having the initial values and BC and CD sides were modified (Figs. 13.103, 13.104, 13.105 and 13.106).

There was obtained a wide palette of curves as loci, including lemniscated curves. Further different values from the ones above have been considered, for all lengths, including for particular cases. The results could be seen below (Figs. 13.107, 13.108, 13.109, 13.110, 13.111, 13.112, 13.113 and 13.114).

Mathematically interesting curves have been got. Other cases were also studied and the obtained curves were similar.

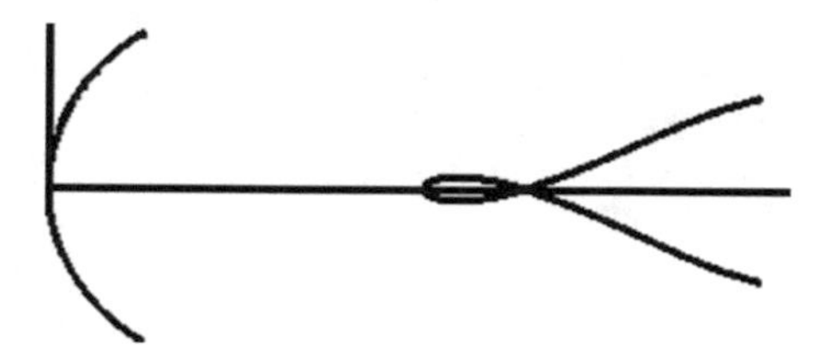

Fig. 13.103 The loci of M and N points for BC = 60; CD = 25

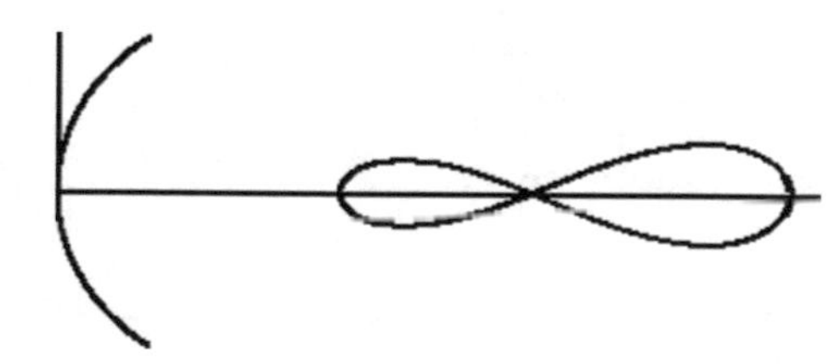

Fig. 13.104 The loci of M and N points for BC = 50; CD = 25

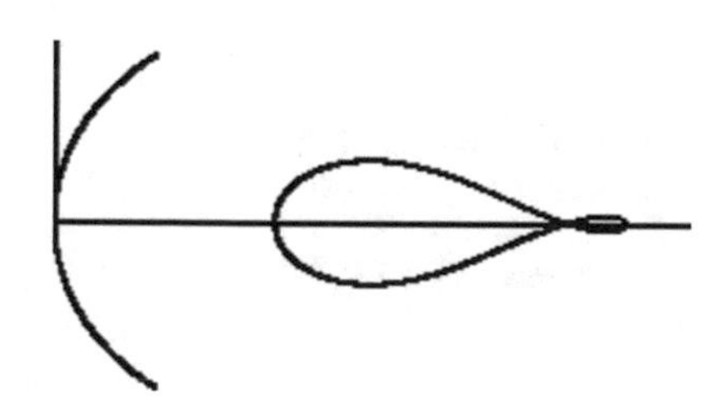

Fig. 13.105 The loci of M and N points for BC = 50; CD = 40

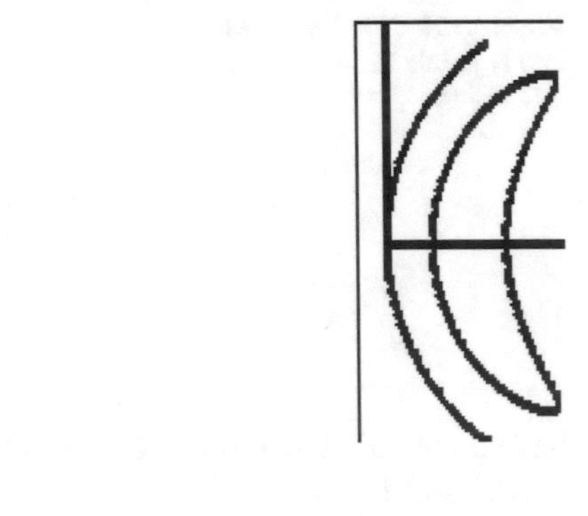

Fig. 13.106 The loci of M and N points for BC = 50; CD = 60

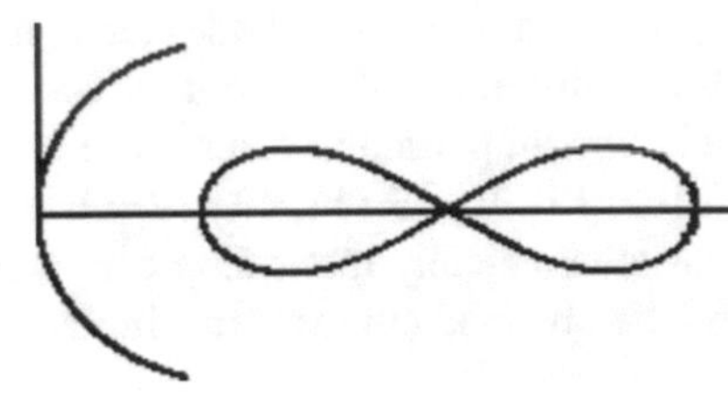

Fig. 13.107 The loci of M and N points for B = 30; BC = 50; CD = 30; XD = 50—parallelogram

Fig. 13.108 The loci of M and N points for AB = 50; BC = 50; CD = 30; XD = 30)—rhombus

Fig. 13.109 The loci of M and N points for AB = 30; BC = 50; CD = 30; XD = –50—parallelogram with XD < 0

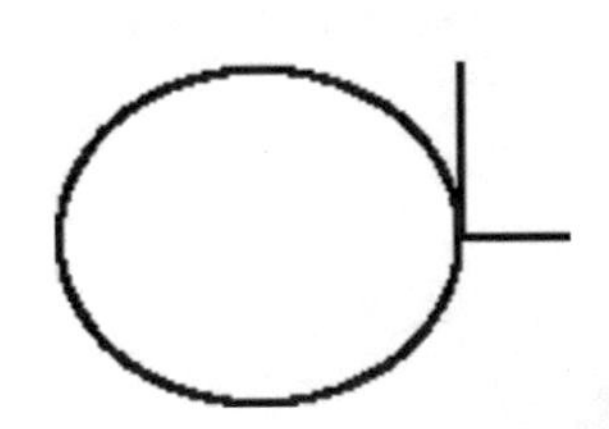

Fig. 13.110 The loci of M and N points for AB = 40; BC = 40; CD = 40; XD = –40—rhombus with XD < 0

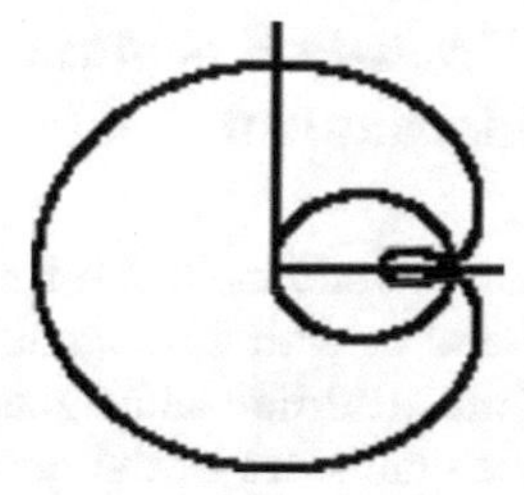

Fig. 13.111 The loci of M and N points for AB = 40; BC = 60; CD = 35; XD = 15—Pascal's snail

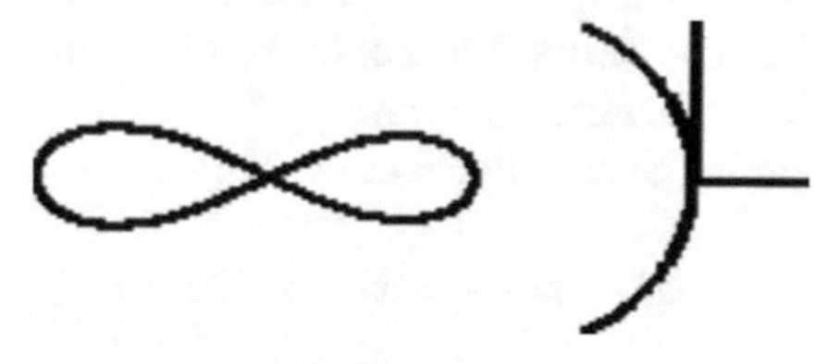

Fig. 13.112 The loci of M and N points for AB = 20; BC = 50; CD = 35; XD = −40

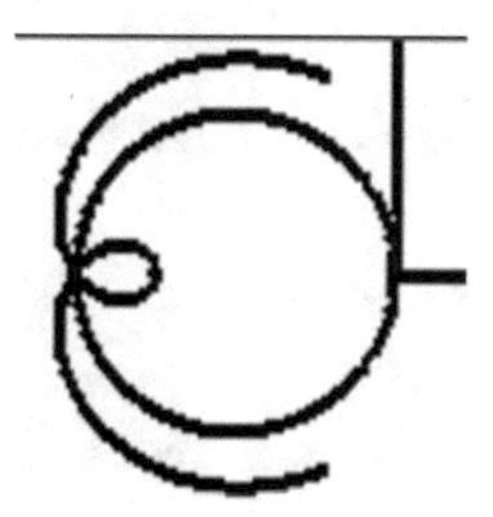

Fig. 13.113 The loci of M and N points For AB = 50; BC = 40; CD = 30; XD = −30

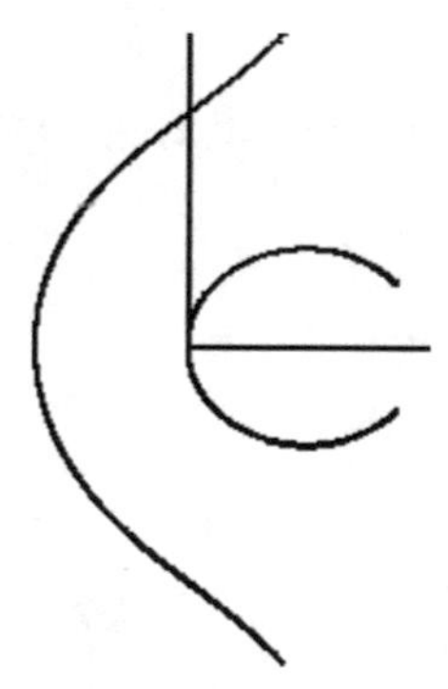

Fig. 13.114 The loci of M and N points for AB = 40; BC = 25; CD = 60; XD = 35

13.5 The Loci for the Cross-Points of the Heights Built in the Four-Bar Mechanism

In the previous chapters, we have found the loci for the cross-points of the heights with the sides of the quadrilateral or with its diagonals. Below, there were studied the loci for the crosspoints of the quadrilateral heights. First, all the perpendiculars were drawn from the corners to the sides of the quadrilateral. Many heights and cross-points were determined.

The calculation method was the same, so only the cross-points between two heights: DG and DF with the third one, AE (the E angle measures 90°) was studied below (Fig. 13.115). The kinematic schema of the mechanism was not given anymore as building it was similar to the previous case.

The following computing expressions are written:

$$DE/\sin\gamma = x_D/\sin(90) \tag{13.28}$$

$$x_E = x_D + DE\cos\beta \tag{13.29}$$

$$y_E = y_D + DE\sin\beta \tag{13.30}$$

$$AG/\sin(90-\varphi) = x_D/\sin(90) \tag{13.31}$$

$$x_G = AG\cos\varphi \tag{13.32}$$

$$y_G = AG\sin\varphi \tag{13.33}$$

$$x_K = AK\cos\gamma = x_D + DK\cos\lambda \tag{13.34}$$

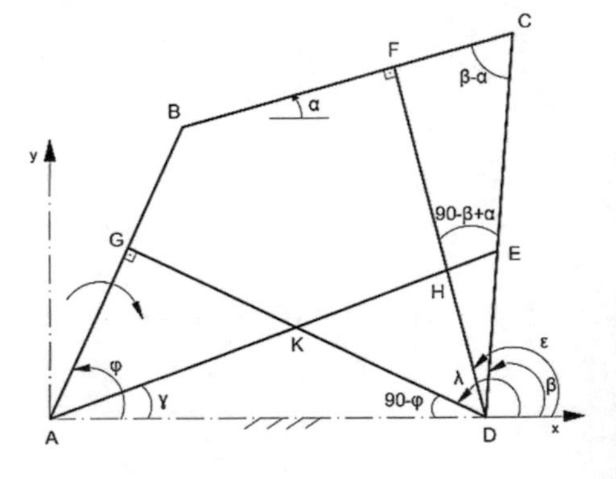

Fig. 13.115 The heights built in the four-bar mechanism

$$y_K = AK \sin\gamma = y_D + DK \sin\lambda \tag{13.35}$$

$$x_H = x_D + DH \cos\varepsilon = AH \cos\gamma \tag{13.36}$$

$$y_H = y_D + DH \sin\varepsilon = AH \sin\gamma \tag{13.37}$$

$$DF/\sin(\beta - \alpha) = CD/\sin(90) \tag{13.38}$$

$$\gamma = \beta - 90; \varepsilon = \alpha + 90; \lambda = \phi + 90. \tag{13.39}$$

In Fig. 13.116 the generated mechanism is shown in one position, and in Fig. 13.117, its successive positions were presented, with the initial data: AB = 20, BC = 30, CD = 40, XD = 45.

There are discontinuities in the operation of the mechanism.

In Fig. 13.118 the locus of the K point is shown, i.e. a four-branch curve.

The mechanism took the correct positions, i.e. the positions closest to the previous position, due to dynamic reasons. Below, the solutions that ensure the continuity of the motion were considered, but in this case the leaps to infinity also appeared, so the values were limited to 400 mm, otherwise the motion diagrams could not be read. In Fig. 13.119 it was given the trajectory of K point for this situation.

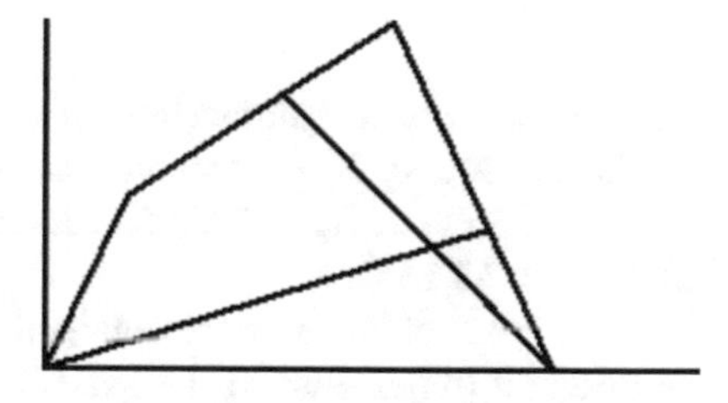

Fig. 13.116 The mechanism in one position

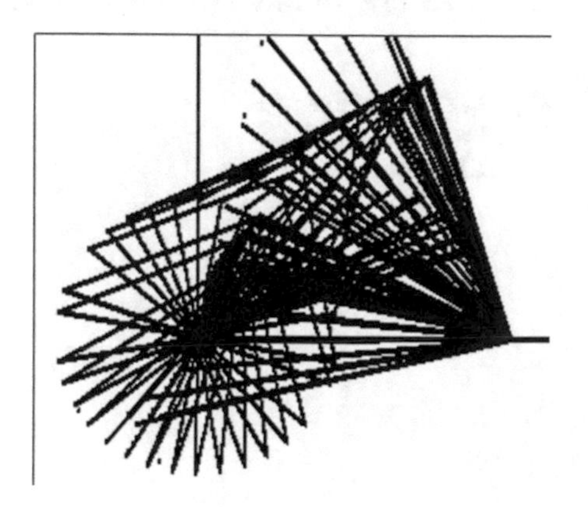

Fig. 13.117 Successive positions

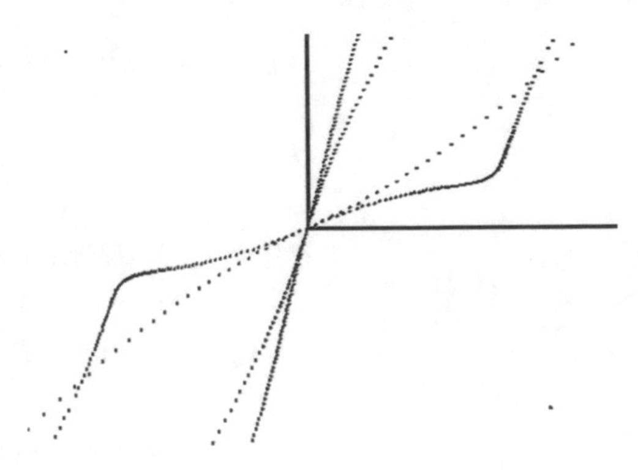

Fig. 13.118 The locus of K

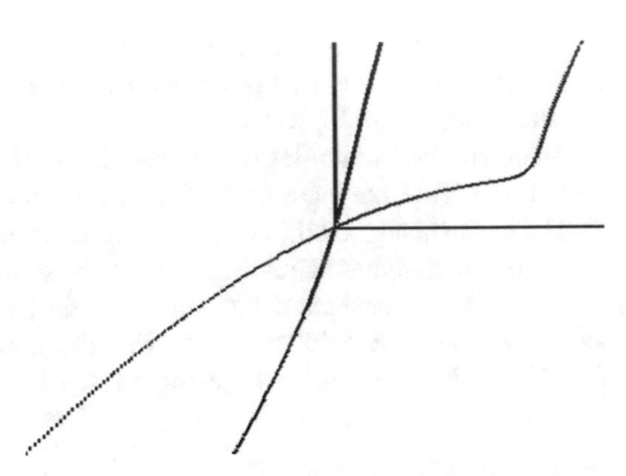

Fig. 13.119 The locus of K for the case of restricted dimensions

Next, some of the dimensions of the mechanism were changed, and the curves obtained as loci of K are presented in the following pictures (Figs. 13.120, 13.121, 13.122, 13.123, 13.124, 13.125, 13.126, 13.127, 13.128, 13.129, 13.130, 13.131, 13.132 and 13.133).

Many types of curves were obtained as loci. Next, the loci of H point were drawn, for different dimensions of the quadrilateral (Figs. 13.134, 13.135, 13.136, 13.137, 13.138, 13.139, 13.140, 13.141 and 13.142).

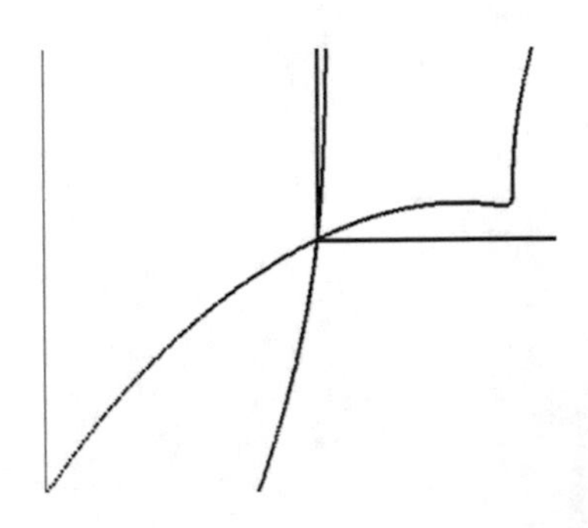

Fig. 13.120 The locus of K for AB = 25; BC = 30; CD = 40; XD = 45

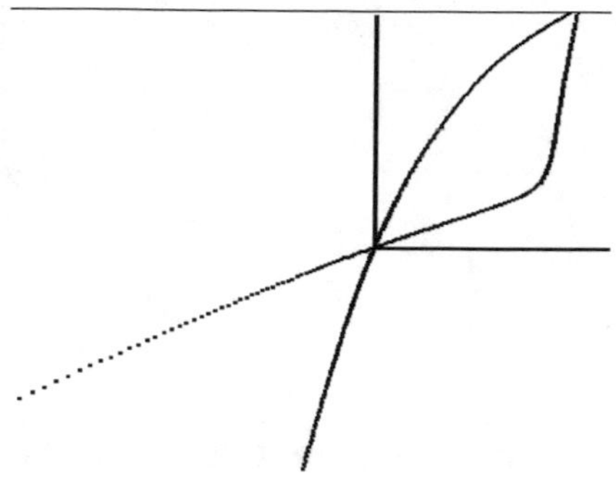

Fig. 13.121 The locus of K for AB = 20; BC = 50; CD = 70; XD = 45

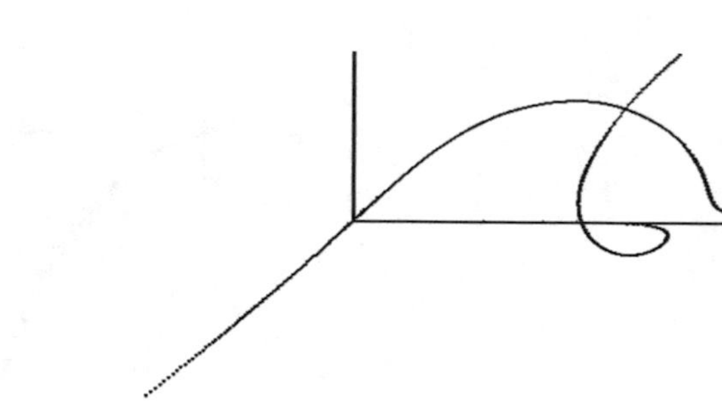

Fig. 13.122 The locus of K for AB = 20; BC = 60; CD = 60; XD = 45

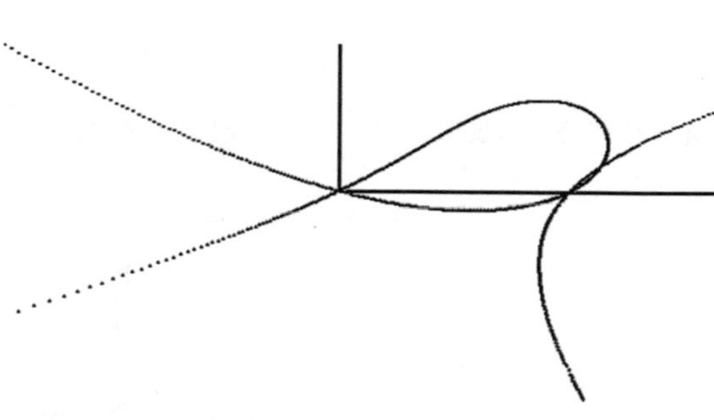

Fig. 13.123 The locus of K for AB = 20; BC = 60; CD = 40; XD = 45

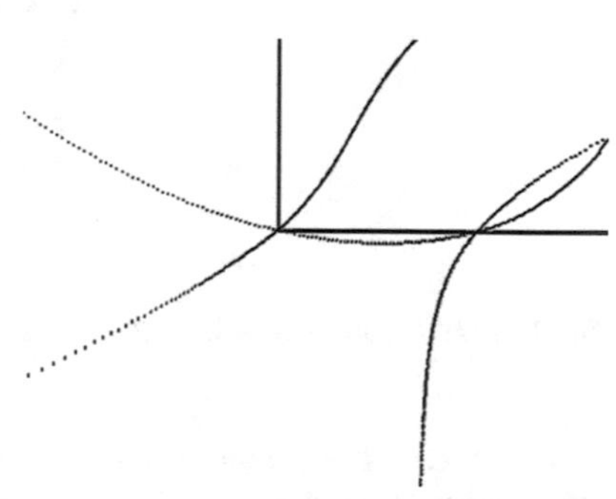

Fig. 13.124 The locus of K for AB = 20; BC = 50; CD = 25; XD = 45

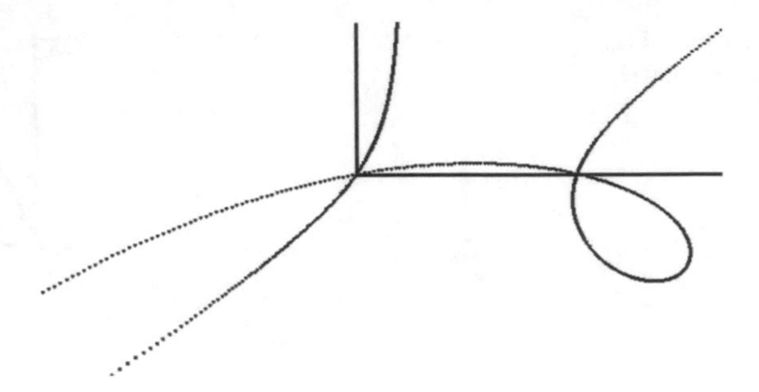

Fig. 13.125 The locus of K for AB = 20; BC = 40; CD = 25; XD = 45

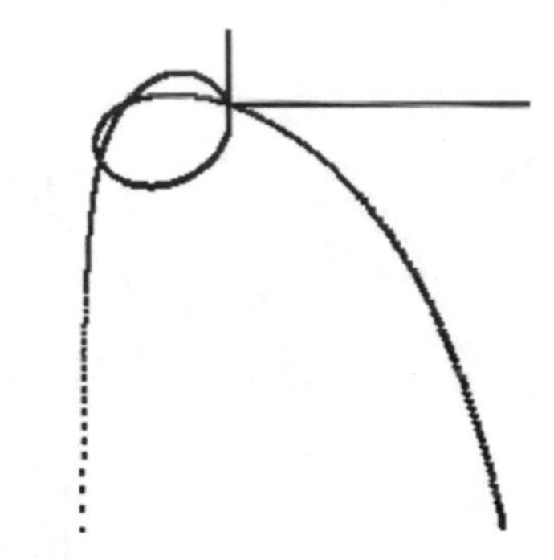

Fig. 13.126 The locus of K for AB = 20; BC = 30; CD = 40; XD = –10

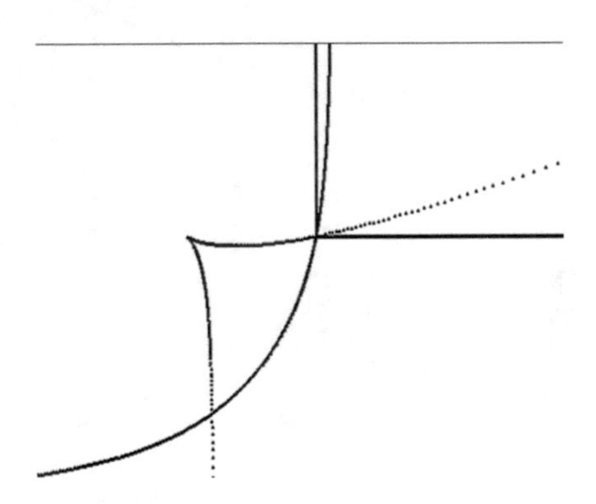

Fig. 13.127 The locus of K for AB = 20; BC = 30; CD = 40; XD = –30

In this case also, many shapes of loci were obtained, some curves being closed, others open. From the examples above, we could draw conclusions about the influence of certain lengths on the shapes of the resulting curves.

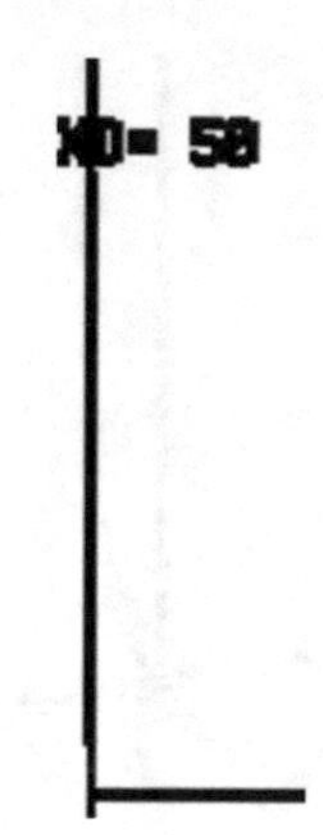

Fig. 13.128 The locus of K for AB = 50; BC = 50; CD = 50; XD = 50—rhombus

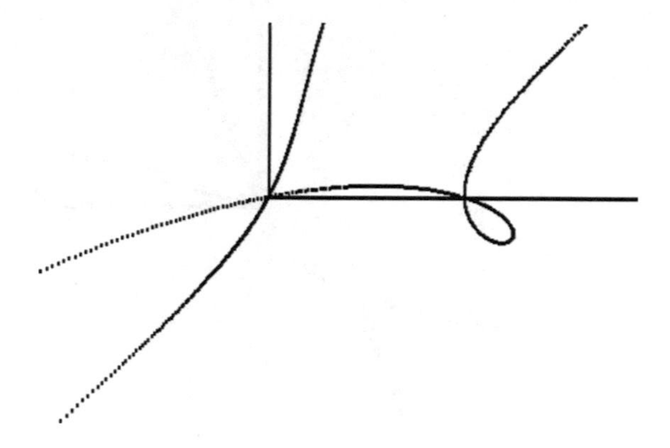

Fig. 13.129 The locus of K for AB = 35; BC = 60; CD = 50; XD = 70

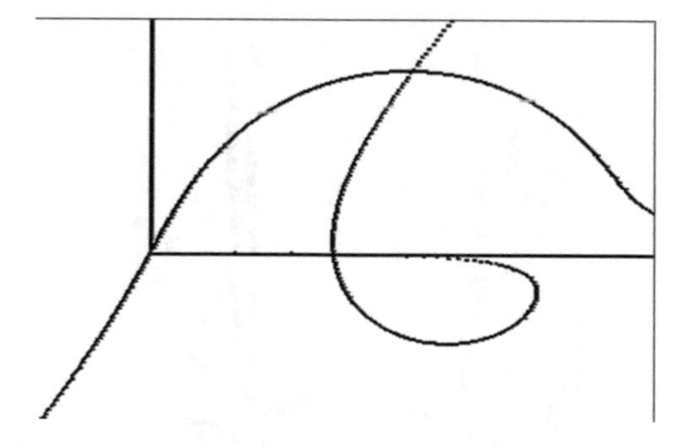

Fig. 13.130 The locus of K for AB = 35; BC = 60; CD = 60; XD = 55

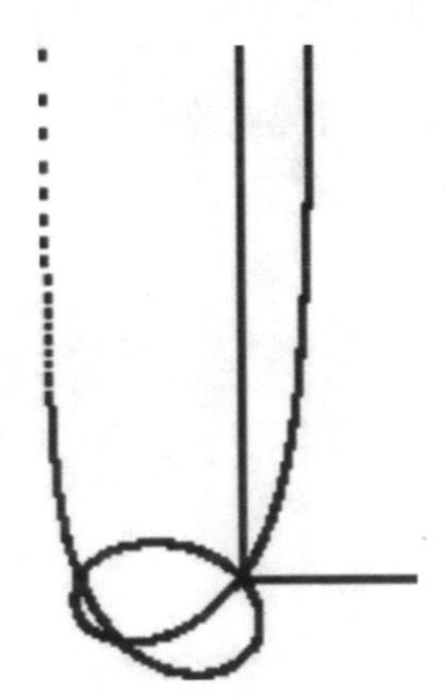

Fig. 13.131 The locus of K for AB = 30; BC = 60; CD = 50; XD = –20

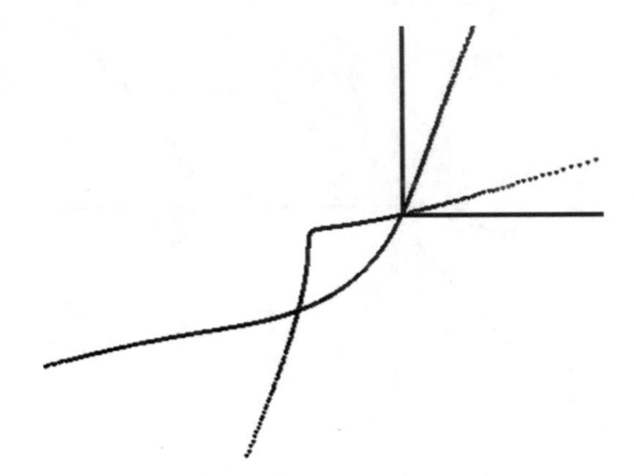

Fig. 13.132 The locus of K for AB = 20; BC = 45; CD = 60; XD = –40

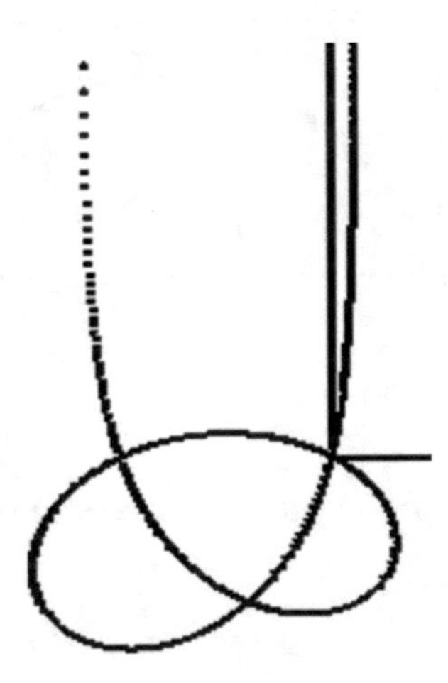

Fig. 13.133 The locus of K for AB = 50; BC = 60; CD = 45; XD = –35

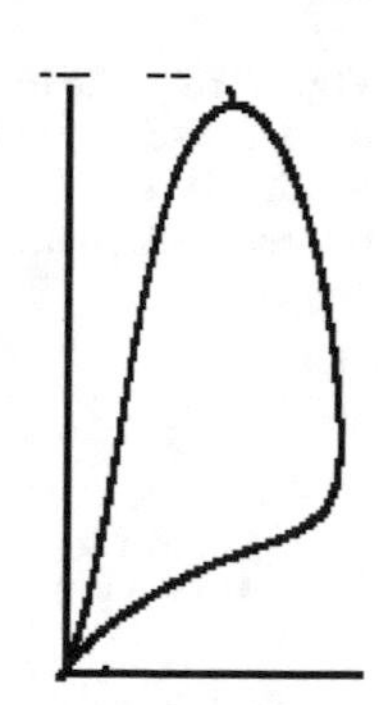

Fig. 13.134 The locus of H for AB = 20; BC = 35; CD = 50; XD = 60

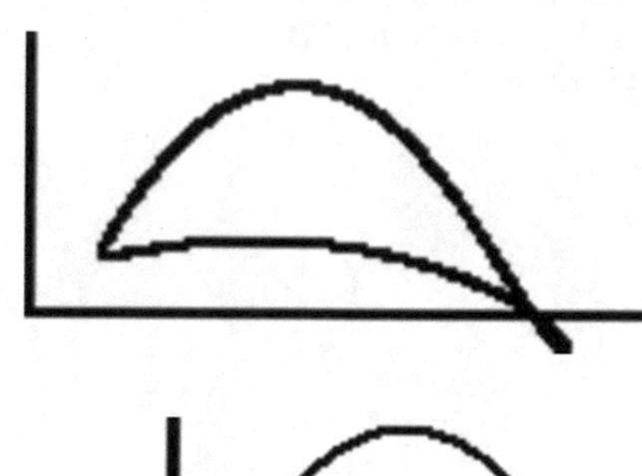

Fig. 13.135 The locus of H for AB = 20; BC = 50; CD = 45; XD = 50

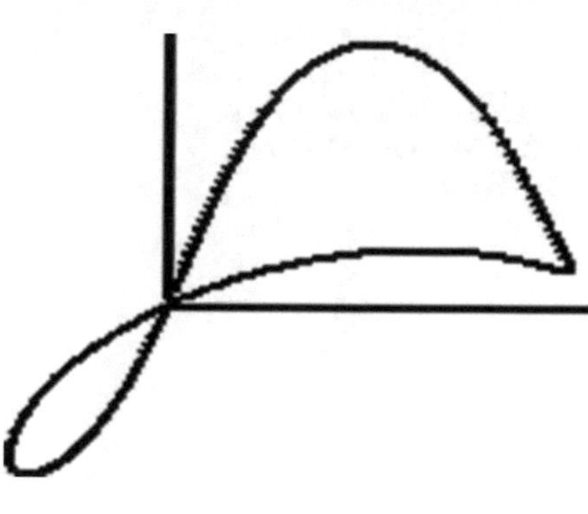

Fig. 13.136 The locus of H for AB = 30; BC = 50; CD = 60; XD = 60

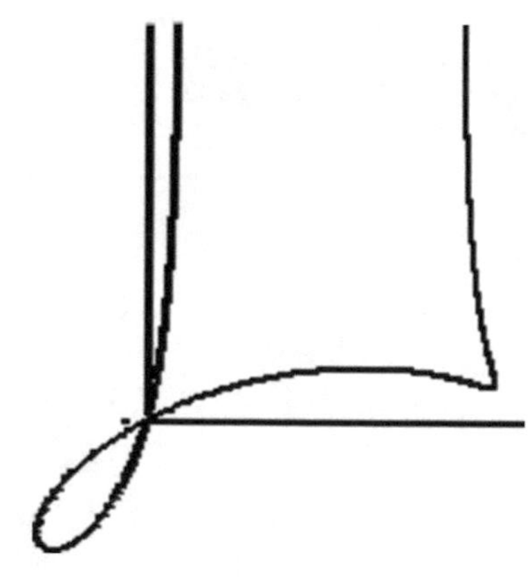

Fig. 13.137 The locus of H for AB = 35; BC = 45; CD = 55; XD = 65

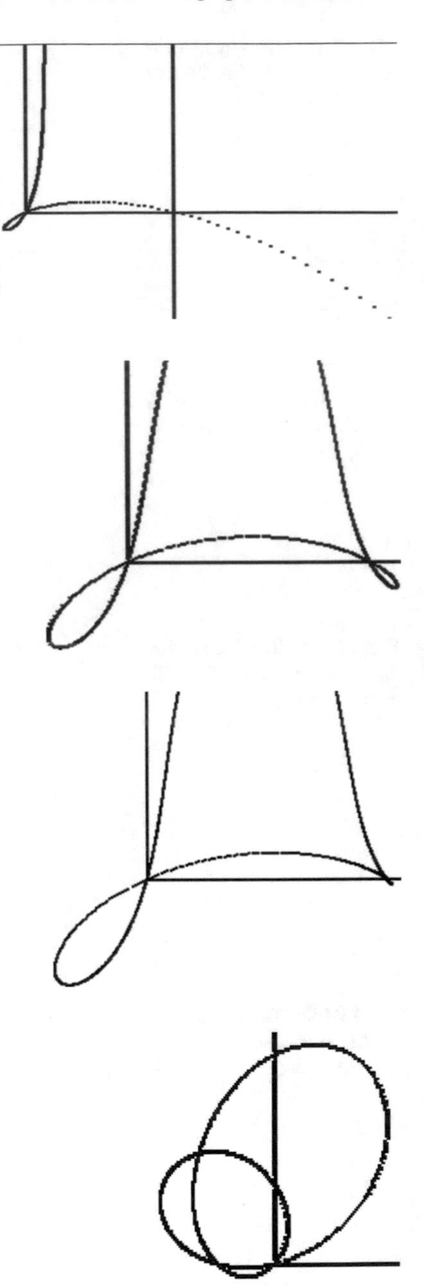

Fig. 13.138 The locus of H for AB = 45; BC = 60; CD = 45; XD = 60—parallelogram

Fig. 13.139 The locus of H for AB = 45; BC = 60; CD = 65; XD = 75

Fig. 13.140 The locus of H for AB = 50; BC = 65; CD = 75; XD = 85

Fig. 13.141 The locus of H for AB = 30; BC = 50; CD = 60; XD = −15

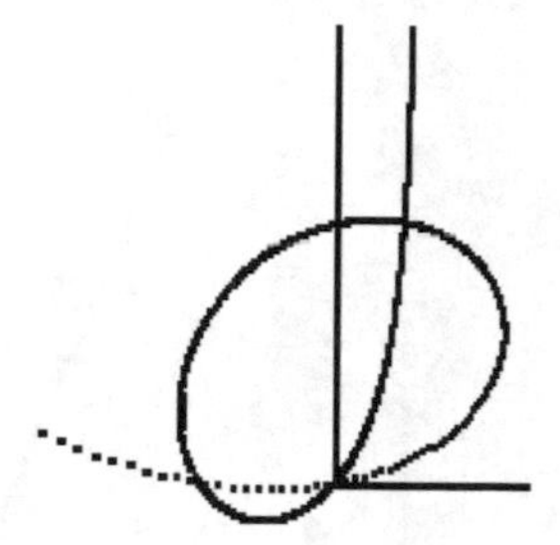

Fig. 13.142 The locus of H for AB = 30; BC = 55; CD = 45; XD = −20

13.6 The Loci for the Cross-Points of the Mid-Perpendiculars

For the ABCD quadrilateral in Fig. 13.143 the mid-perpendiculars of the four sides were drawn and we wanted to study the loci of their cross-points, two by two. Here, they were not all concurring like in the triangle. The N, F, E, L, points, representing the basis of the mid-perpendiculars, can be built on the generated mechanism by additional kinematic chains. For example, in order to find the position of E (the mid-point of the CD line), a circle arc with a center in D and a radius greater than or equal to CD/2 was drawn and then, another circle arc was draw, by the center in C and with the same radius, concurring in a point P, from which a perpendicular to the CD side was traced, which will pass through the E point. Similarly, the M and R points could

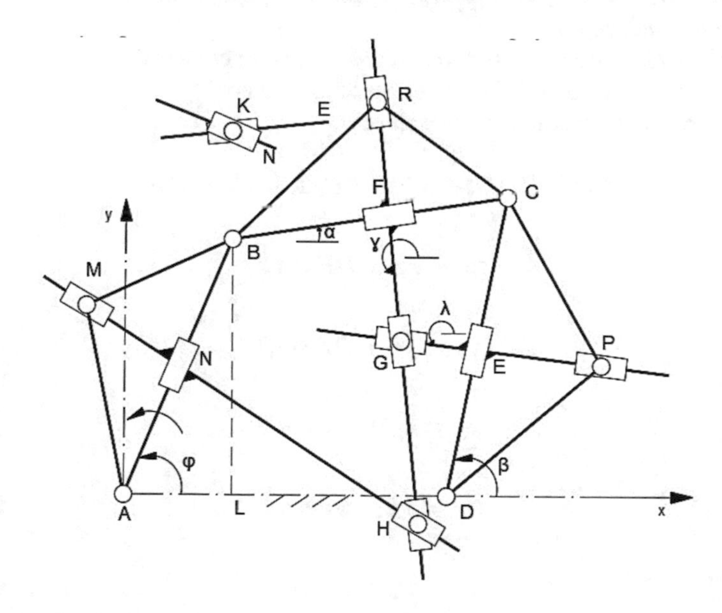

Fig. 13.143 The mechanism

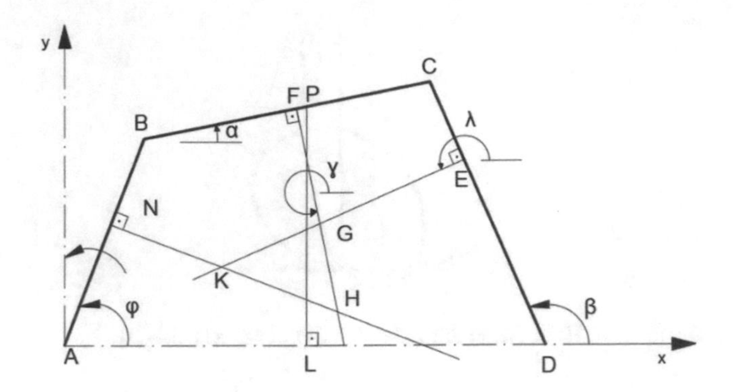

Fig. 13.144 The equivalent solution

be determined. The MN and RF mid-perpendiculars concur in the H point, MN and PE in K, and PE and RF in H. The mid-perpendicular built from the L point was not taken into account, since it is fixed, the AD side being the base. All its cross - points with other mid-perpendiculars will move on the mid-perpendicular from L, i.e. on a line.

Sliders were placed in the E, F, N, points, moving on the respective sides. These are connected with elements perpendicular to the sides of the mechanism. In the G, K, H points, there were introduced sliders that rotate to each other in order to materialize the points of the loci.

Below, in order to simplify the drawings of the mechanisms, they are not drawn anymore, but only the equivalent geometric figures they represent. Still, the calculations are kept fair.

The mechanism of Fig. 13.143 was equivalent to the geometric construction shown in Fig. 13.144, which would further be the study material.

The following expressions were written:

$$DE = CD/2; \quad BF = BC/2; \quad AN = AB/2 \tag{13.40}$$

$$x_E = x_D + DE\cos\beta \tag{13.41}$$

$$y_E = y_D + DE\sin\beta \tag{13.42}$$

$$x_F = x_B + BF\cos\alpha \tag{13.43}$$

$$y_F = y_B + BF\sin\alpha \tag{13.44}$$

$$x_G = x_E + EG\cos\lambda \tag{13.45}$$

$$y_G = y_E + EG \sin \lambda \tag{13.46}$$

$$\lambda = \beta + 90; \tag{13.47}$$

$$x_G = x_F + FG \cos \gamma \tag{13.48}$$

$$y_G = x_F + FG \sin \gamma \tag{13.49}$$

$$\gamma = \alpha + 270; \tag{13.50}$$

$$x_N = AN \cos \varphi \tag{13.51}$$

$$y_N = AN \sin \varphi \tag{13.52}$$

$$x_H = x_N + NH \cos(\varphi + 270) = x_F + FH \cos \gamma \tag{13.53}$$

$$y_H = y_N + NH \sin(\varphi + 270) = y_F + FH \sin \gamma \tag{13.54}$$

$$x_E + EG \cos \lambda = x_F + FG \cos \gamma \tag{13.55}$$

$$y_E + EG \sin \lambda = y_F + FG \sin \gamma \tag{13.56}$$

$$x_K = x_E + EK \cos \lambda = x_N + NK \cos(\varphi + 270) \tag{13.57}$$

$$y_K = y_E + EK \sin \lambda = y_N + NK \sin(\varphi + 270) \tag{13.58}$$

The following dimensions were taken into consideration: AB = 20, BC = 30, CD = 40, XD = 45. In Fig. 13.145 the mechanism was presented for one position, and in Fig. 13.146 the successive positions of the mechanism were drawn.

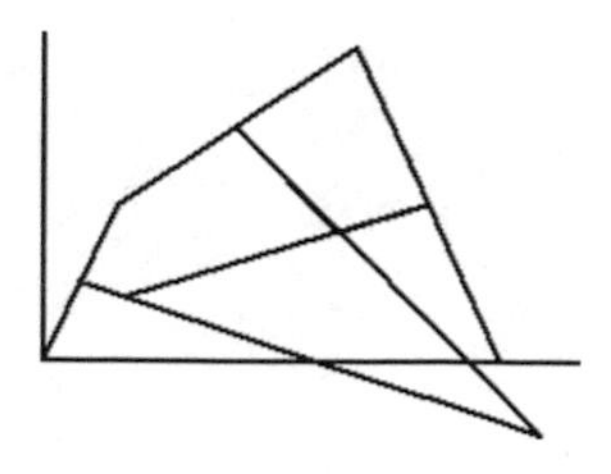

Fig. 13.145 The mechanism in one position

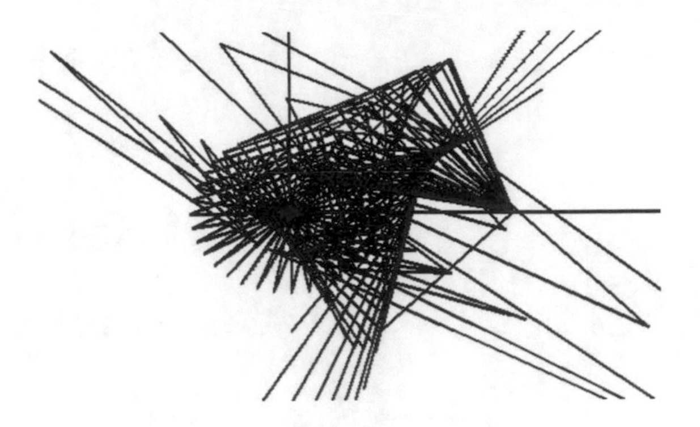

Fig. 13.146 The successive positions

It was noticeable that some lines intersect outside the drawing plan; in order to obtain a clear picture, the cross—points were limited to 400 mm, excluding the larger values.

The searched loci are given together in Fig. 13.147, the curves intersecting with each other. To make it clearer, these loci geometric places were separately represented below.

The locus of the G point is shown in Fig. 13.148 for the initial data of the mechanism.

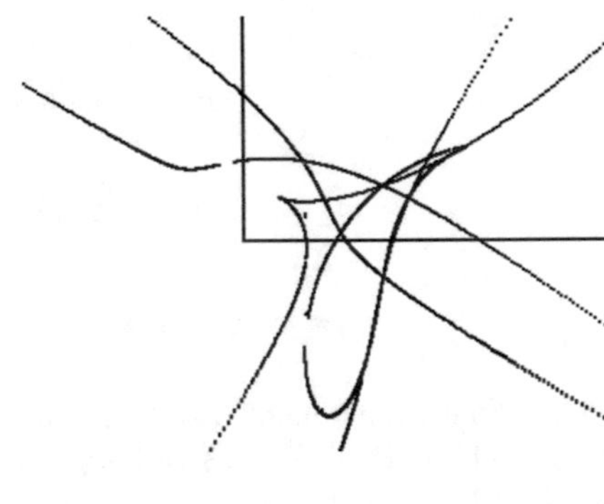

Fig. 13.147 All the loci

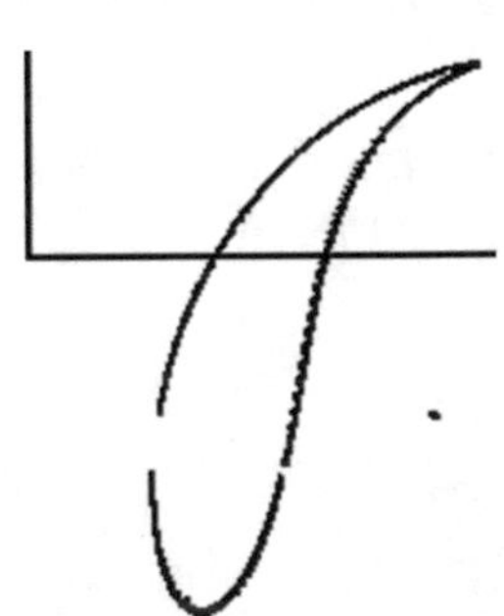

Fig. 13.148 The locus for G

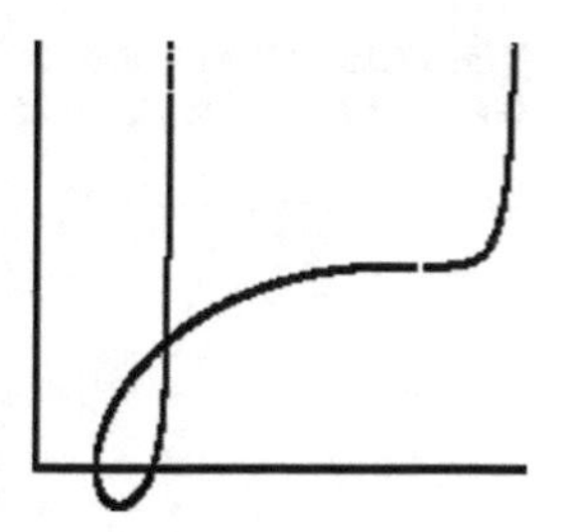

Fig. 13.149 The locus of G for AB = 20; XD = 30

Furthermore, the values for BC and CD links were kept constant, and the values for AB and XD were changed, obtaining the loci of G point shown below (Figs. 13.149, 13.150, 13.151, 13.152 and 13.153).

Next, different sets of values were considered for all sides (Figs. 13.154, 13.155 and 13.156).

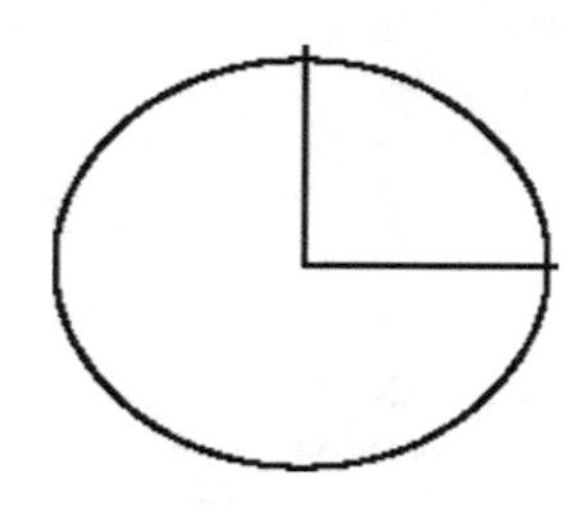

Fig. 13.150 The locus of G for AB = 20; XD = 0—circle

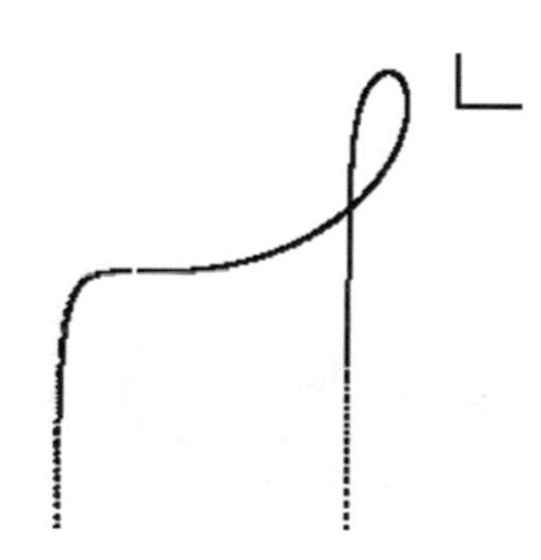

Fig. 13.151 The locus of G for AB = 20; XD = −30

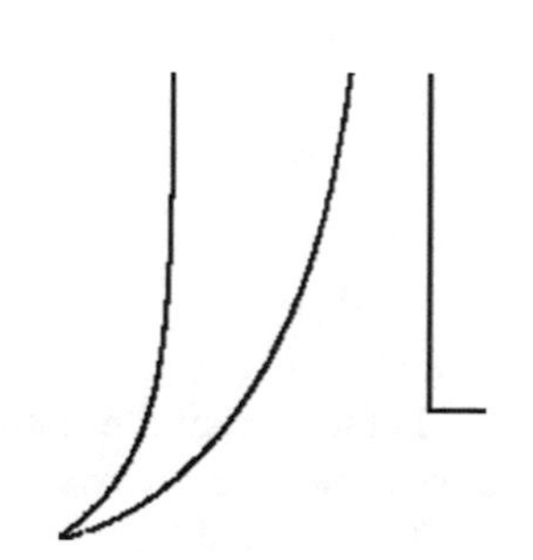

Fig. 13.152 The locus of G for AB = 20; XD = −50

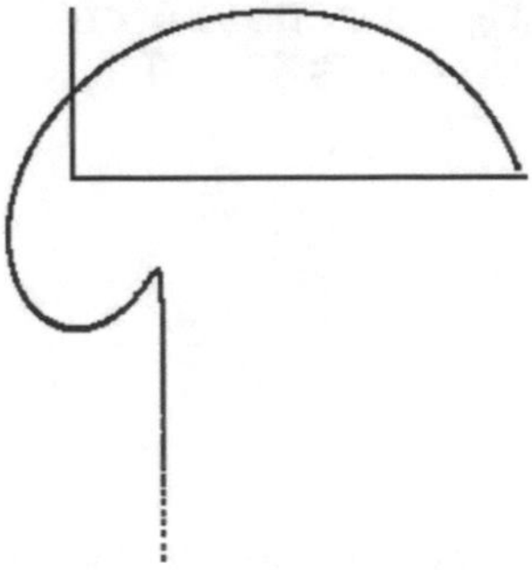

Fig. 13.153 The locus of G point for AB = 35; XD = 25

Fig. 13.154 The locus of G point for AB = 30; BC = 40; CD = 50; XD = −50

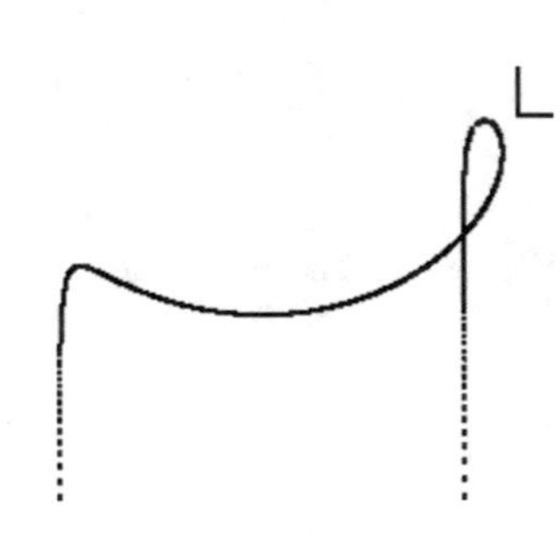

Fig. 13.155 The locus of G point for AB = 25; BC = 50; CD = 45; XD = –30

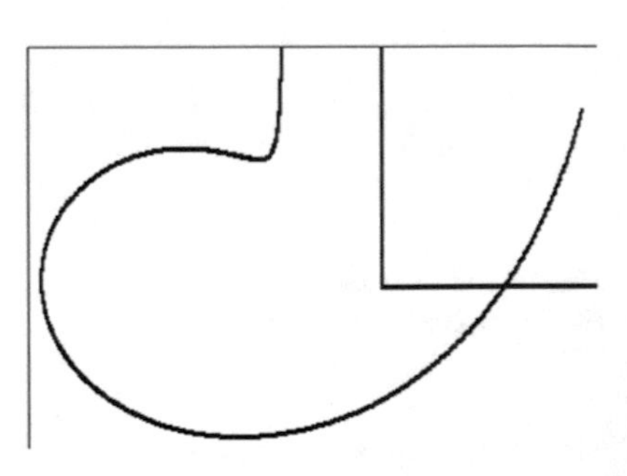

Fig. 13.156 The locus of G point for AB = 50; BC = 30; CD = 45; XD = –25

In the case of the rhombus there were no solutions of the algebric system, so no locus could be determined.

The locus of the K point for the initial data was shown in Figs. 13.157.

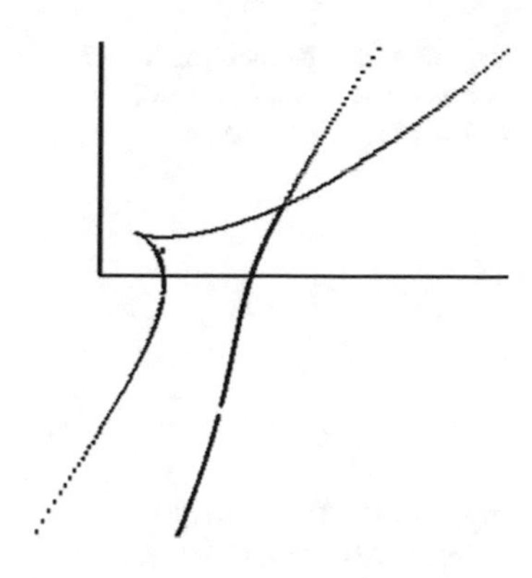

Fig. 13.157 The locus for K point

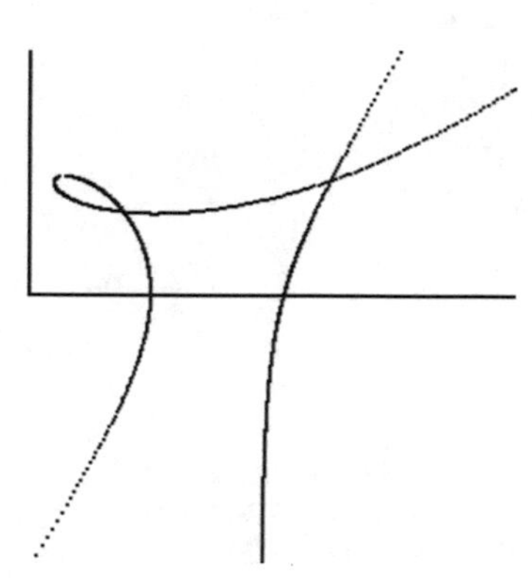

Fig. 13.158 The locus of K point for AB = 35; BC = 45; CD = 55; XD = 65

Next, different values were taken into consideration for the sides of the mechanism, thus obtaining the loci of K point in the following pictures (Figs. 13.158, 13.159 and 13.160).

For the case of the parallelogram, a parallel line with the Y-coordinate was determined, and in the case of the rhombus, the locus did not exist (Figs. 13.161, 13.162, 13.163, 13.164 and 13.165).

It was noticeable that there were obtained curves with two open branches.

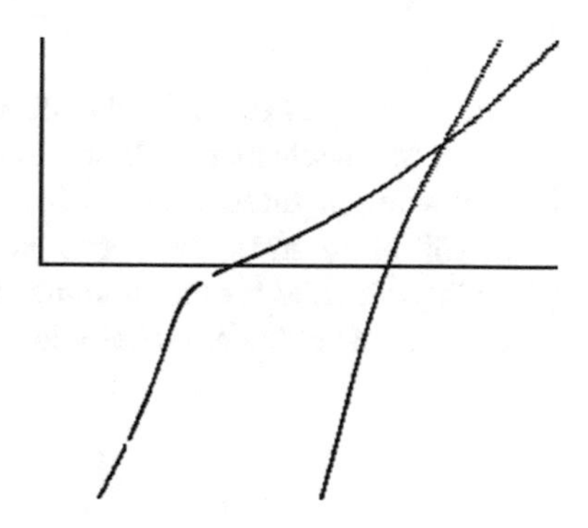

Fig. 13.159 The locus of K point for AB = 25; BC = 50; CD = 75; XD = 90

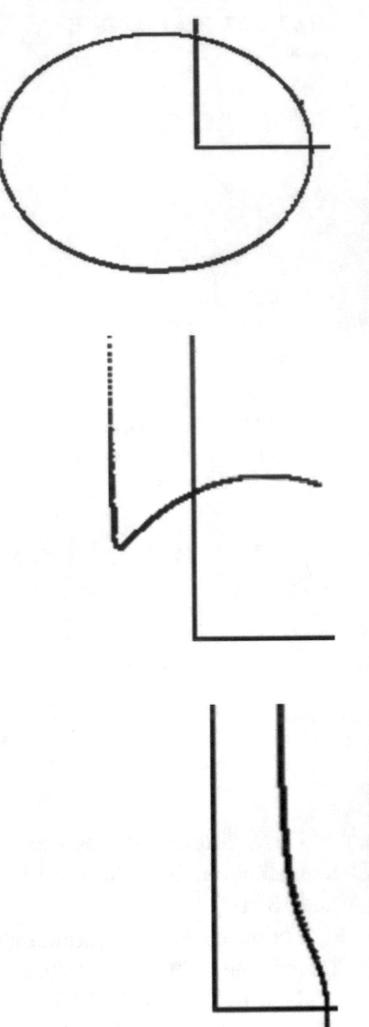

Fig. 13.160 The locus of K point or AB = 30; BC = 40; CD = 35; XD = −10

Fig. 13.161 The locus of K point for AB = 30; BC = 40; CD = 40; XD = −30

Fig. 13.162 The locus of K point for AB = 20; BC = 45; CD = 45; XD = −20

The locus of H point for the initial data was shown in Fig. 13.166.

Next, the mechanism data was modified in order to obtain the loci of H point in the following pictures (Figs. 13.167, 13.168 and 13.169).

In this case also, the locus is a line parallel to the Y-coordinate for the parallelogram, and for the rhombus, there is no locus.

The loci have two branches and more curvatures than the ones for the K point.

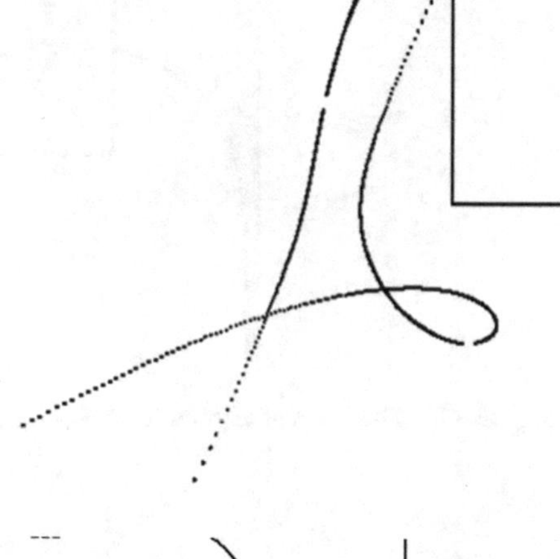

Fig. 13.163 The locus of K point for AB = 30; BC = 40; CD = 50; XD = −50

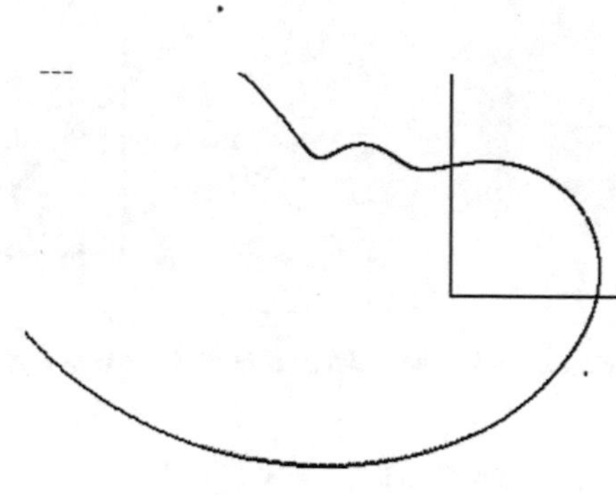

Fig. 13.164 The locus of K point for AB = 45; BC = 60; CD = 55; XD = −35

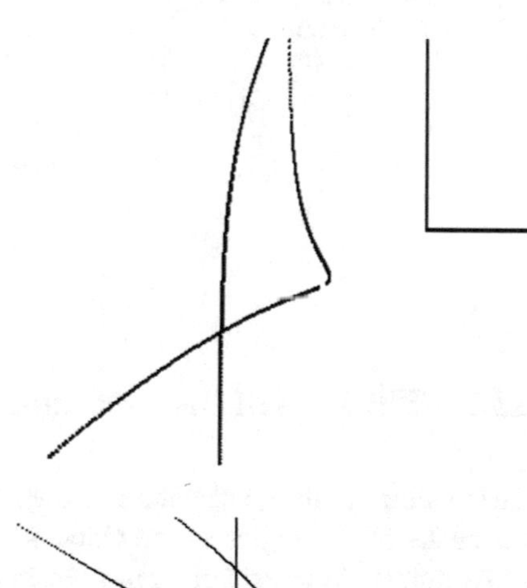

Fig. 13.165 The locus of K point for AB = 35; BC = 45; CD = 90; XD = −80

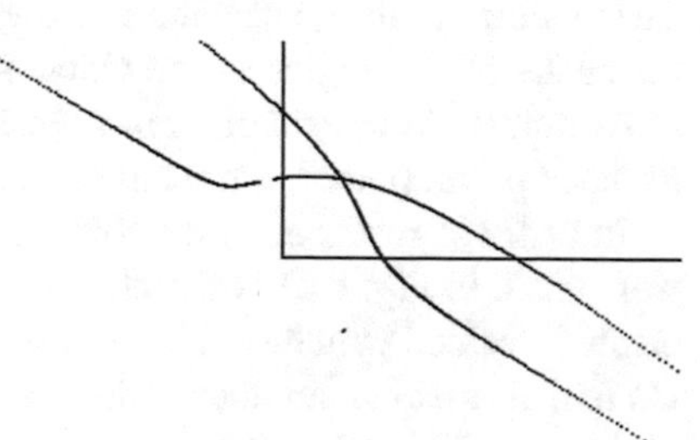

Fig. 13.166 The locus of H point

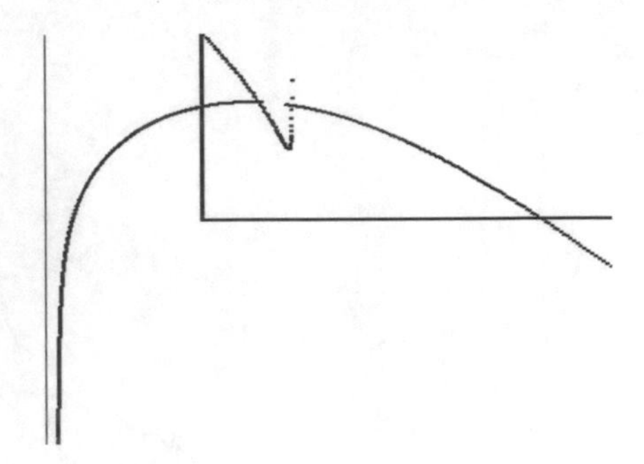

Fig. 13.167 The locus of H point for AB = 35; BC = 45; CD = 60; XD = 50

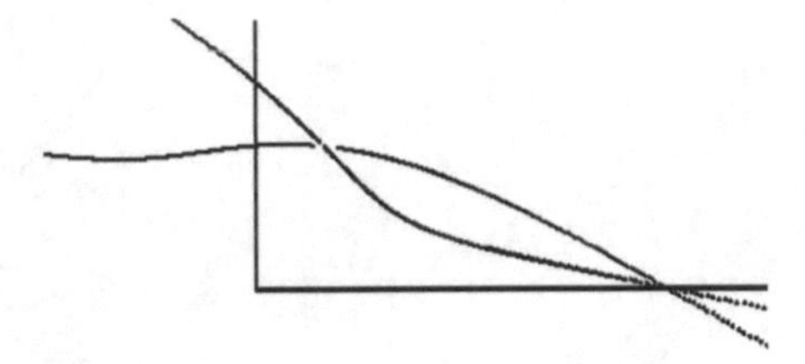

Fig. 13.168 The locus of H point for AB = 20; BC = 45; CD = 60; XD = 60

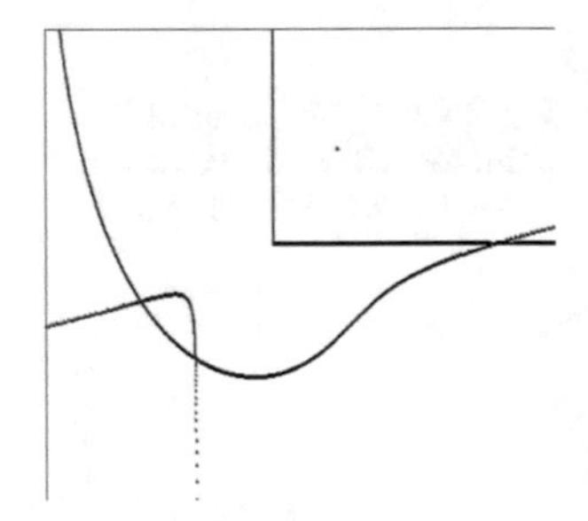

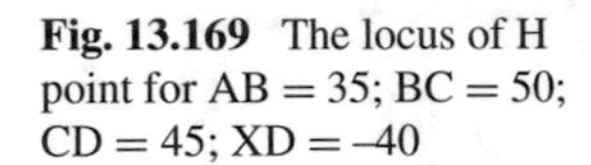

Fig. 13.169 The locus of H point for AB = 35; BC = 50; CD = 45; XD = –40

13.7 The Loci of the Medians Cross-Points

The medians of the quadrilateral shown in Fig. 13.170 were built by joining together the peaks of the angles of the sides with the middle of the opposite sides. Several cross-points of these medians have been obtained. Next, it was studied only the locus of the H point, placed at the intersection of the AE and DF medians.

In order to materialize the mid-points of the sides, additional kinematic chains were used. In Fig. 13.171 the chain for the CD side was shown, made up of the EM element welded with the slider glidig on the CD bar with a right angle in the E point. On the AE median, another slider was placed in the E point, connected to the slider gliding on CD bar by a revolute pair. With CM = MD, the E point should be at the middle of the CD, no matter what position the CD element will have.

The following expressions were written:

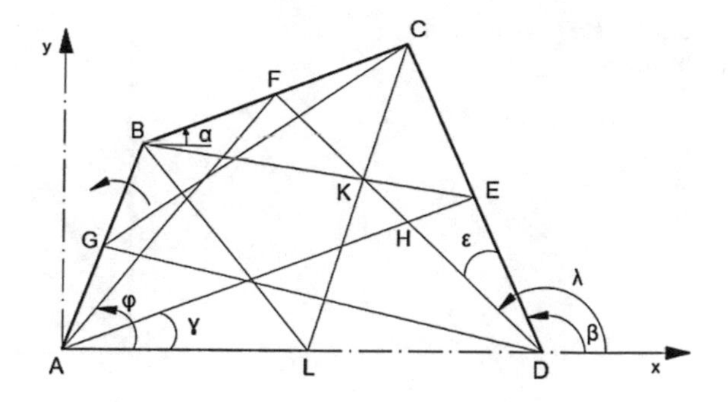

Fig. 13.170 The medians

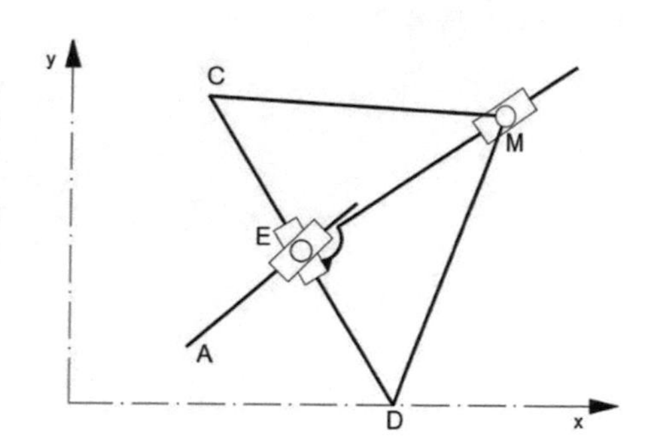

Fig. 13.171 The CMED cinematic chain

$$x_E = (CD \cos \beta)/2;\ y_E = (CD \sin \beta)/2 \tag{13.59}$$

$$AE^2 = x_E^2 + y_E^2 \tag{13.60}$$

$$\sin \gamma = y_E/AE; \quad \cos \gamma = x_E/AE \tag{13.61}$$

$$x_H = AH \cos \gamma = x_D + DH \cos \lambda; \quad \cos \lambda = (y_F - y_D)/DF \tag{13.62}$$

$$y_H = AH \sin \gamma = y_D + DH \sin \lambda \tag{13.63}$$

and we used the expressions below for the determination of λ:

$$x_F = x_B + (BC \cos \alpha)/2 \tag{13.64}$$

$$y_F = y_B + (BC \sin \alpha)/2 \tag{13.65}$$

$$DF^2 = (x_D - x_F)^2 + (y_D - y_F)^2 \tag{13.66}$$

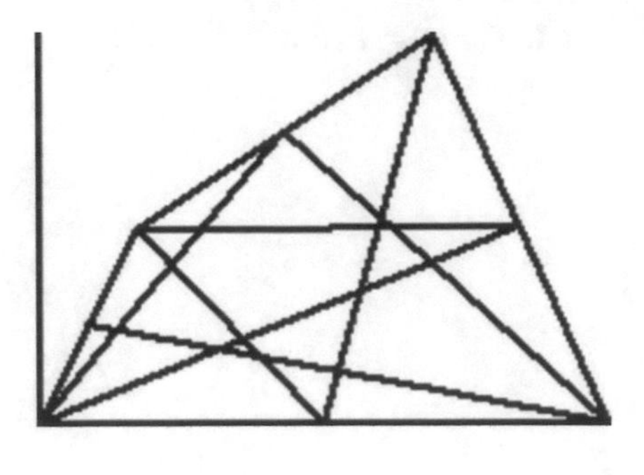

Fig. 13.172 The built-in medians

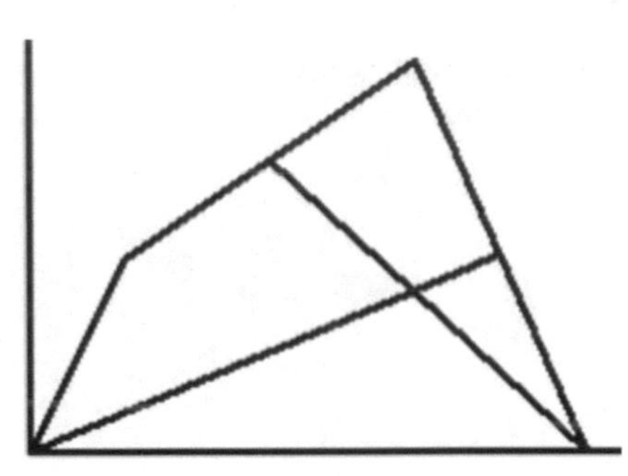

Fig. 13.173 Point H

$$\sin \lambda = (y_F - y_D)/DF \tag{13.67}$$

$$\cos \lambda = (x_F - x_D)/DF \tag{13.68}$$

Initial data: AB = 20, BC = 30, CD = 40, XD = 45. In Fig. 13.172 the mechanism was shown in one position with built-in medians.

Next, only the locus for the H cross point between the AE and DF medians (Fig. 13.173) was studied.

In Fig. 13.174 it is presented the successive positions of the mechanism, and in Fig. 13.175 the locus of H point for the initial data.

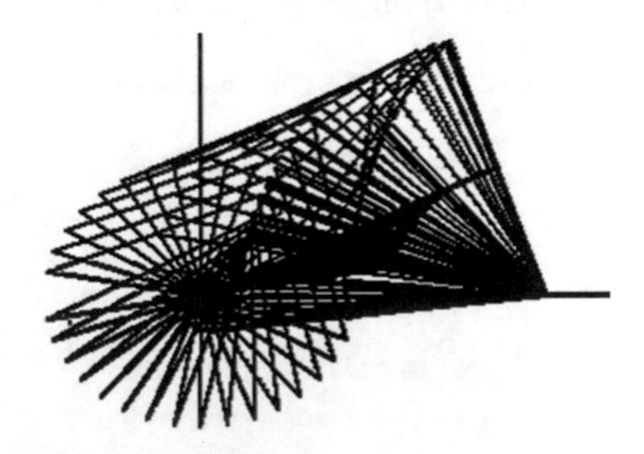

Fig. 13.174 The successive positions

Fig. 13.175 The locus of H point for AB = 20; BC = 30; CD = 40; XD = 45

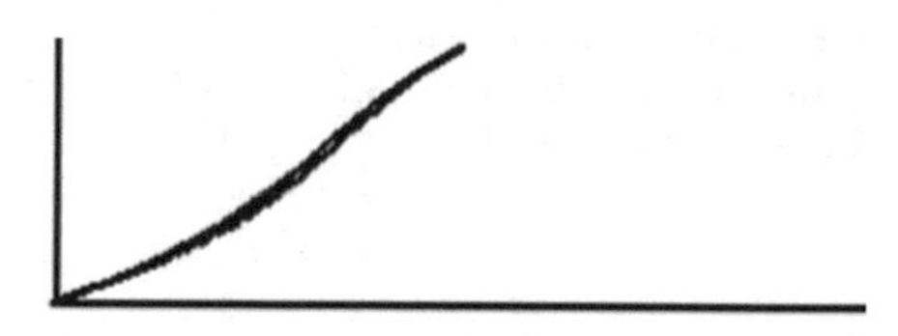

Fig. 13.176 The locus of H point for AB = 20; BC = 40; CD = 60; XD = 55

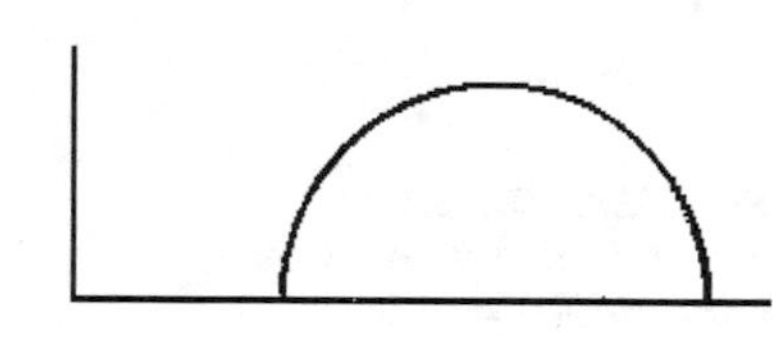

Fig. 13.177 The locus of H point for AB = 50; BC = 50; CD = 50; XD = 50)-rhombus

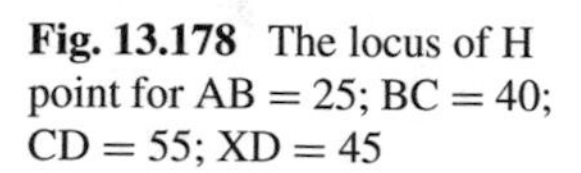

Fig. 13.178 The locus of H point for AB = 25; BC = 40; CD = 55; XD = 45

Furthermore, the lengths of the elements of the quadrilateral were modified, and different shapes of the locus of H point were obtained (Figs. 13.176, 13.177, 13.178, 13.179, 13.180, 13.181, 13.182, 13.183, 13.184, 13.185, 13.186 and 13.187).

A great variety of curves was obtained. By analizing it, the influence of the side lengths on the shape of the locus may be determined.

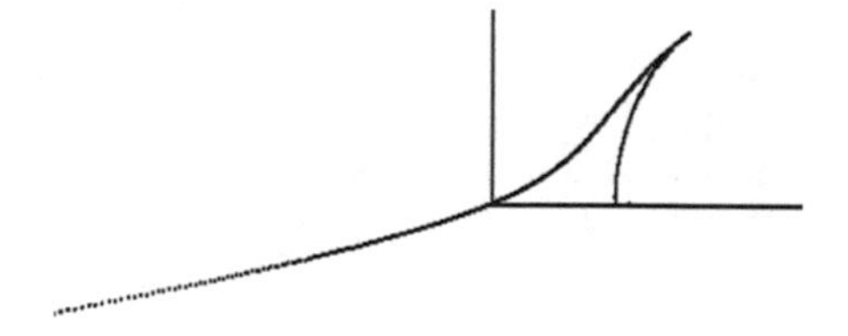

Fig. 13.179 The locus of H point for AB = 25; BC = 35; CD = 65; XD = 55

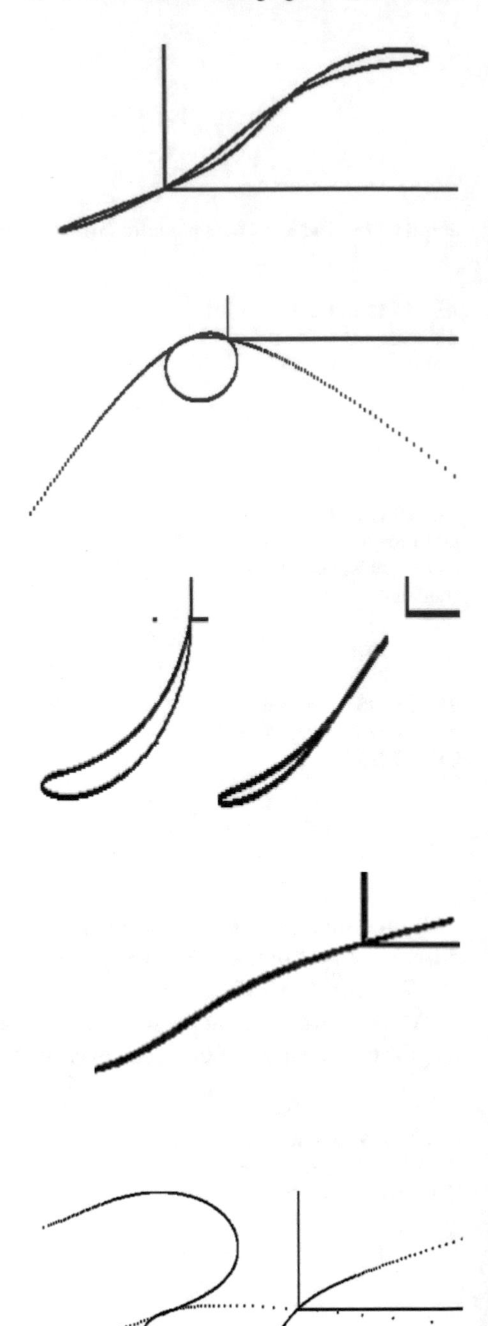

Fig. 13.180 The locus of H point for AB = 30; BC = 50; CD = 60; XD = 45

Fig. 13.181 The locus of H point for AB = 30; BC = 45; CD = 55; XD = −10

Fig. 13.182 The locus of H point for AB = 20; BC = 50; CD = 60; XD = −30

Fig. 13.183 The locus of H point for AB = 20; BC = 35; CD = 45; XD = −45

Fig. 13.184 The locus of H point for AB = 50; BC = 45; CD = 55; XD = −35

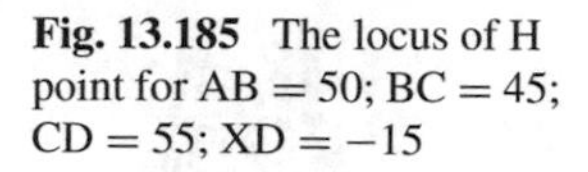
Fig. 13.185 The locus of H point for AB = 50; BC = 45; CD = 55; XD = −15

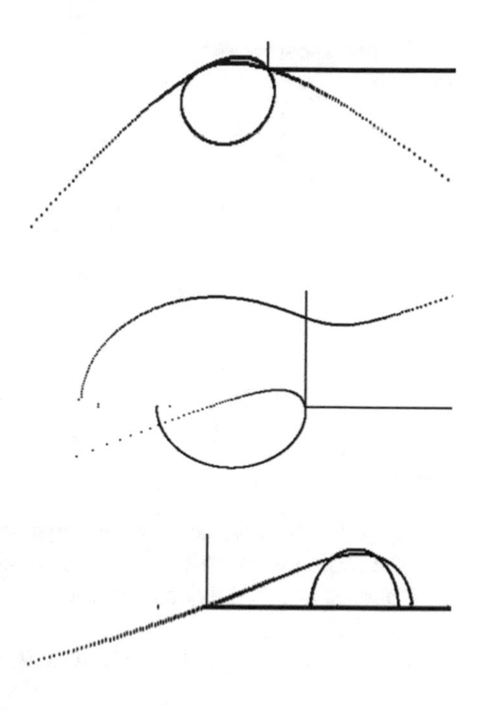

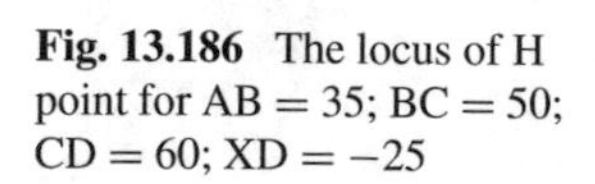
Fig. 13.186 The locus of H point for AB = 35; BC = 50; CD = 60; XD = −25

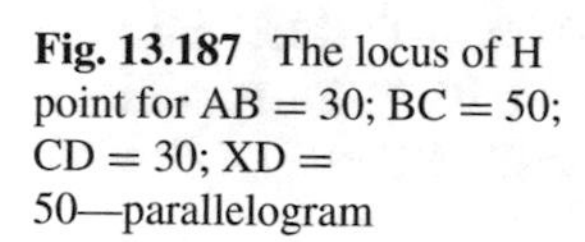
Fig. 13.187 The locus of H point for AB = 30; BC = 50; CD = 30; XD = 50—parallelogram

13.8 The Loci of the Bisecting Lines Cross-Points

The quadrilateral of Fig. 13.188 and the bisecting lines of the A, B, C, D angles were drawn. Within the mechanism, kinematic chains like the one drawn for D point were used in order to materialize the bisecting lines. By changing the position of the CD side, the position of the bisecting line of the D angle was also changed, so that the slider built in G moved along the CD side.

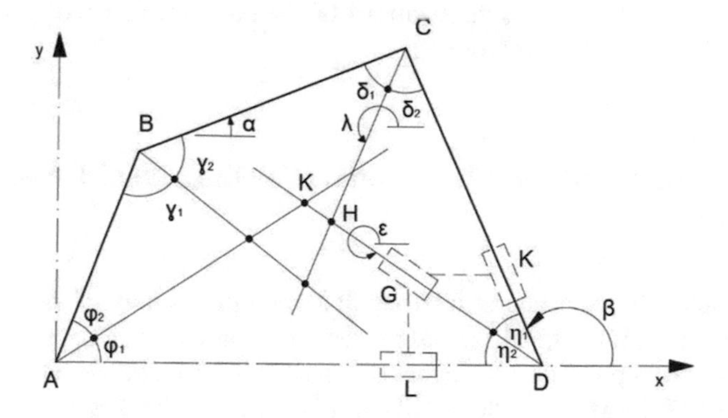

Fig. 13.188 The cross-points of the bisecting lines

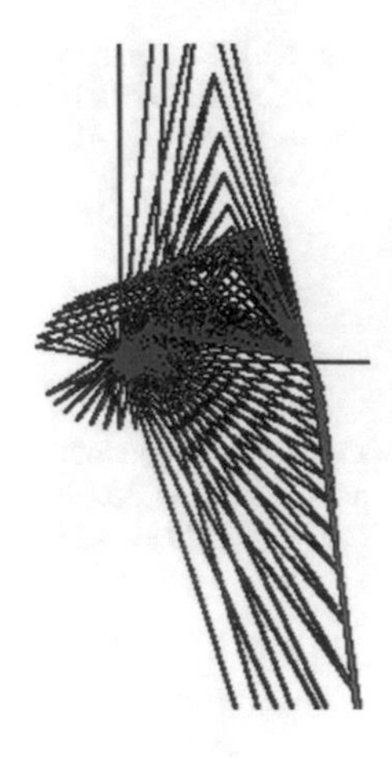

Fig. 13.189 The successive positions

The GK = GL elements were perpendicular to the CD and AD sides, ensuring that the DHK line is maintained as a bisecting line. It was similarly done for the A, B, C angles. The initial quadrilateral was maintained (AB = 20, BC = 30, CD = 40, XD = 45).

The following relationships were written:

$$x_K = AK\cos\varphi_1 = x_D + DK\cos(\varepsilon - \pi) \tag{13.69}$$

$$y_K = AK\sin\varphi_1 = y_D + DK\sin(\varepsilon - \pi) \tag{13.70}$$

$$\varphi_1 = \varphi/2 \tag{13.71}$$

It is noted that the bisecting lines concurr at several points. Below, only the K point is examined at the cross-point of the DK and AK bisecting line.

In Fig. 13.189 the successive positions of the mechanism were shown, and in Fig. 13.190 the locus was determined.

13.9 The Locus of the Quadrilateral's Diagonals' Cross Point

The diagonals of a quadrilateral intersect at a point that is fixed if the quadrilateral is fixed. However, if the quadrilateral moves, by a stationary element and a driving one, then the cross point of the diagonals describes a curve, that is a locus.

In Fig. 13.191, it is given the ABCD four-bar mechanism, for which we need to determine the locus of the M cross point of diagonals [2].

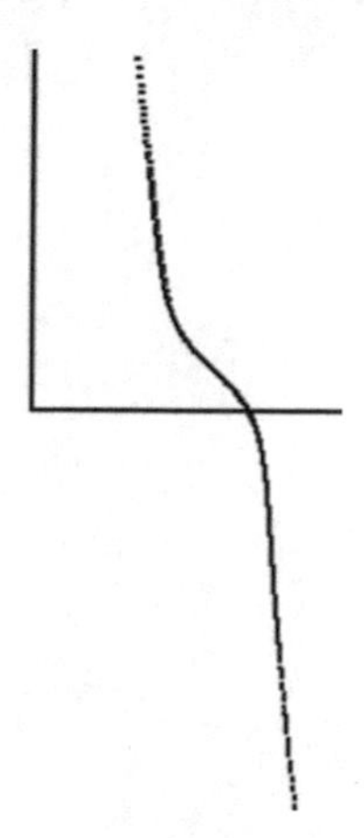

Fig. 13.190 The locus

The two diagonals, that have variable lengths, have been drawn, so that there were mounted sliders on the diagonals in C point and in D point. In M point, there were mounted two sliders rotating between them, ensuring that the point M slides over the two diagonals. It is cycled the φ angle and it is obtained the position of point C and the α, β, γ. angles

The following expressions are written:

$$x_B = a\cos\varphi \tag{13.72}$$

$$y_B = a\sin\varphi \tag{13.73}$$

$$(x_B - x_C)^2 + (y_B - y_C)^2 = b^2 \tag{13.74}$$

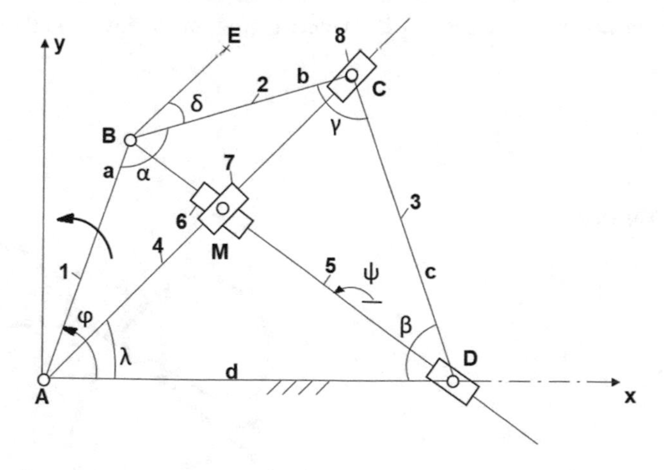

Fig. 13.191 Cross point of diagonals

$$(x_D - x_C)^2 + (y_D - y_C)^2 = c^2 \tag{13.75}$$

$$AC^2 = x_C^2 + y_C^2 \tag{13.76}$$

$$\sin\lambda = y_C/AC \tag{13.77}$$

$$\cos\lambda = x_C/AC \tag{13.78}$$

$$BD^2 = (x_B - x_D)^2 + (y_B - y_D)^2 \tag{13.79}$$

$$\sin\psi = (y_B - y_D)/BD \tag{13.80}$$

$$\cos\psi = (x_B - x_D)/BD \tag{13.81}$$

$$x_M = AM\cos\lambda = x_D + MD\cos\psi \tag{13.82}$$

$$y_M = AM\sin\lambda = y_D + MD\sin\psi \tag{13.83}$$

that determine AM, MD, XM, YM.

For the values: a = 28; e = 40; b = 60; c = 45, d = 70 and $\gamma = 180$, the angle φ was cycled, i.e. the AB crank was rotated, resulting the trajectory of the E point shown in Fig. 13.192 and the locus we needed to ascertain, which is *a lemniscate type curve*, with uneven branches. It is also drawn one position of the mechanism. The trajectory of E point is plotted in order to compare the two curves.

For other values (a = 22; e = 40; b = 41; c = 45, d = 50 şi $\gamma = 100$), there were obtained the cuves shown in Fig. 13.193 where the lemniscate is more flattened than before.

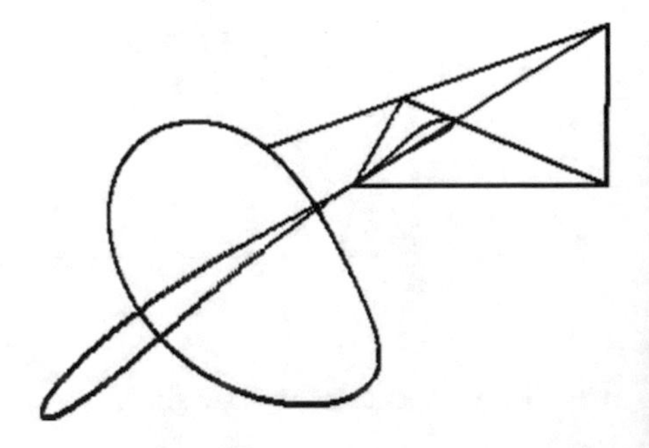

Fig. 13.192 The locus for γ = 180

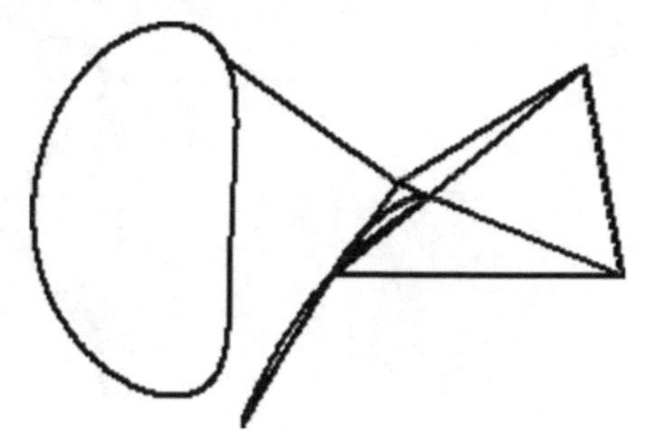

Fig. 13.193 The locus for $\gamma = 100$

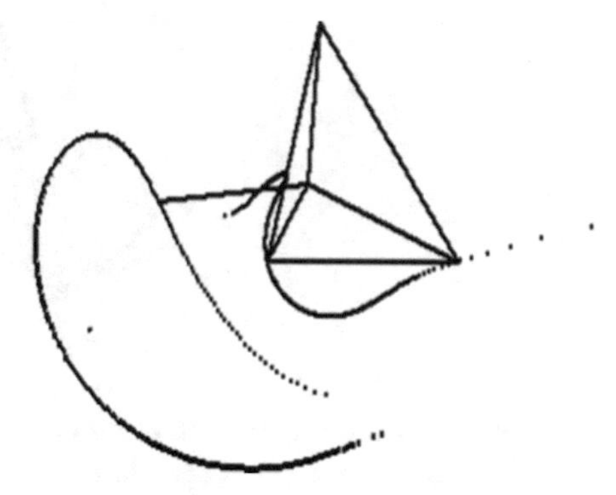

Fig. 13.194 The locus for c = 60

Further, values a = 22; e = 40; b = 41; d = 50 and $\gamma = 100$ remained constant and c was modified as in the pictures below: in Fig. 13.194, the locus is a circular arc, traversed in both directions, and in Fig. 13.195 it is a complex curve.

Furthermore, c = 37 was maintained and b modified (Fig. 13.196). One can ascertain that the locus degenerates into two branches because the geometrical necessary

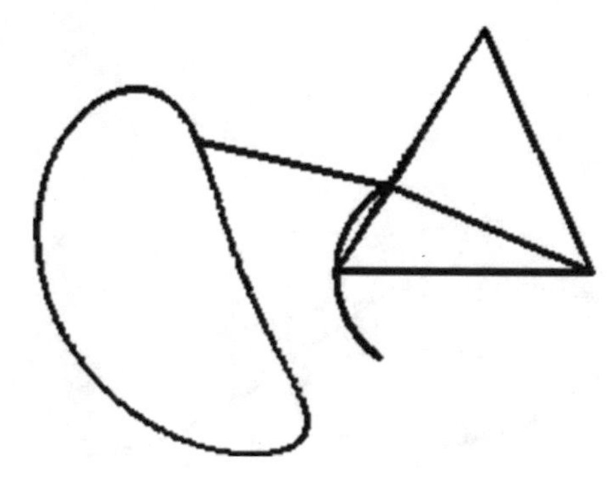

Fig. 13.195 The locus for c = 70

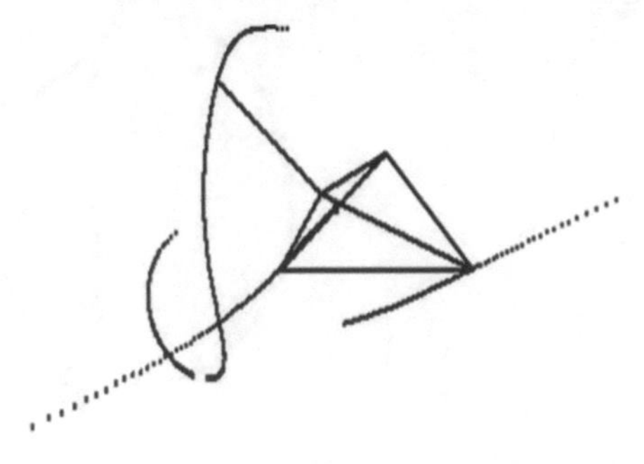

Fig. 13.196 The locus for b = 20

Fig. 13.197 The locus for d = –35

conditions for the AB to perform complete rotation are no longer met (Grashof). It can be seen that neither the trajectory of E point is a closed curve.

It was also studied the possibility of d being negative (Fig. 13.197), i.e. the point D is on the left side of the y-coordinate, having as result curves distinct from the ones above. They are presented in the following figures (Figs. 13.198 and 13.199).

When the quadrangle is a square (the sides are equal to 22), during the motion it becomes a lozenge and the locus is a circular arc (Fig. 13.200), as in the case of the four-bar mechanism (Fig. 13.201).

The trapeze case was also studied (Fig. 13.202). The locus is a two-branch open curve, as the Grashof conditions are not met.

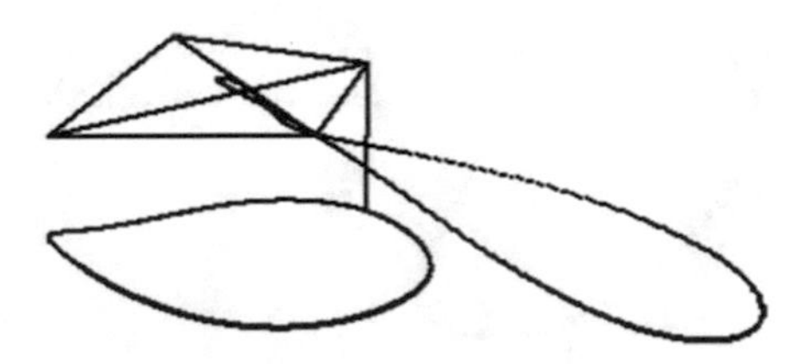

Fig. 13.198 The locus for d = –55

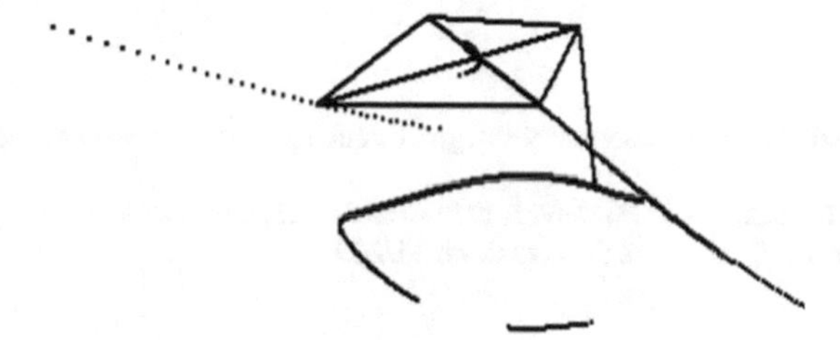

Fig. 13.199 The locus for d = –60

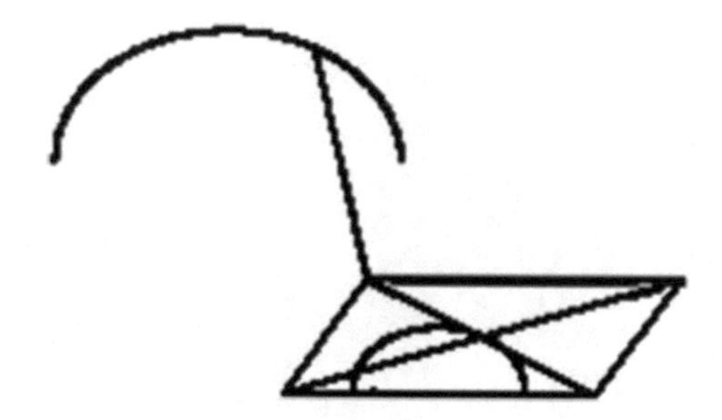

Fig. 13.200 The locus for AB = c = d

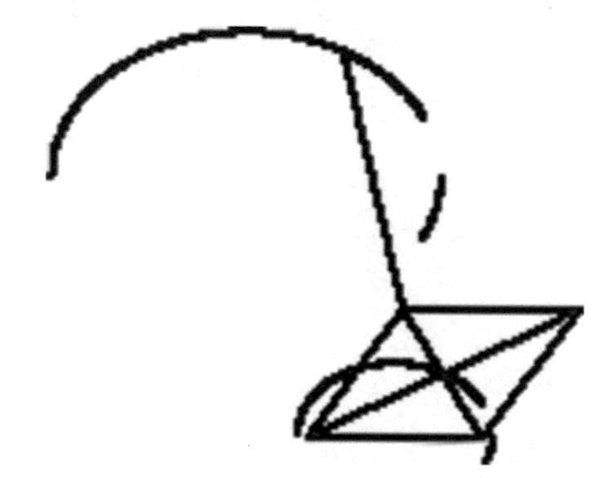

Fig. 13.201 The locus for a = c = 22; b = d = 40

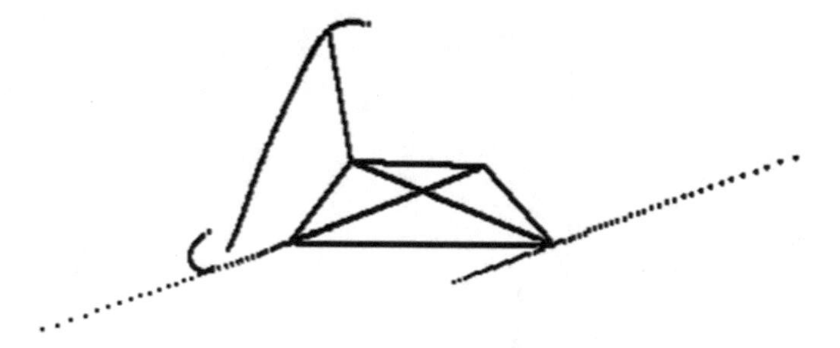

Fig. 13.202 The locus for a = 28; e = 40; b = 30; c = 28, d = 60 and $\gamma = 100$

References

1. Popescu I (2016) Locuri geometrice şi imagini estetice generate cu mecanisme. Editura Sitech, Craiova
2. Sass L, Popescu I, Luca L (2003) Determinarea traiectoriei punctului de intersecţie a diagonalelor unui patrulater. În: Construcţia de maşini, nr. 11/2003

Chapter 14
The Locus for the Cross-Point of the Diagonals in a Pentagon

Abstract Here we study the trajectories drawn by the points of intersection of the diagonals in a pentagon. When one side is fixed, and the remaining are mobile, there are two conducting elements. Taking two intersecting diagonals, the result is 5 trajectory-generating points. Mechanisms are built for all these cases and the trajectories found are drawn. Many resulting curves are given for each value of q. Successive positions of the created mechanisms are also given. The resulting curves are generally open, with several branches [1]. The length of the sides of the mechanisms did not correlate with conditions like Grashof, so that in general the mechanisms do not work throughout the cycle, but only in certain subintervals of the cycle [2]. The discontinuities that appear are indicated below in the diagrams where the jumps are seen.

14.1 The Loci of the K Point

For the pentagon shown in Fig. 14.1, all possible diagonals were drawn. Note that the following cross-points were obtained: G, H, F, K, L.

One could imagine a mechanism that could draw all the loci specific to these cross-points, but the Fig. 14.1 would be very complicated, which is why the locus for each point of intersection was determined.

For the K point, the mechanism of Fig. 14.2 was determined by installing in K two sliders that rotate between each other, and other sliders in D and E points allowing the variation in the lengths of the CE and BD diagonals. *The mechanism had two driving elements, AB and ED,* and the revolute pair is connected to the frame in E point.

The following relationships were written based on Fig. 14.2:

$$x_B = AB\cos\ \varphi \tag{14.1}$$

I. Popescu et al., *Problems of Locus Solved by Mechanisms Theory*,
Springer Tracts in Mechanical Engineering,
https://doi.org/10.1007/978-3-030-63079-9_14

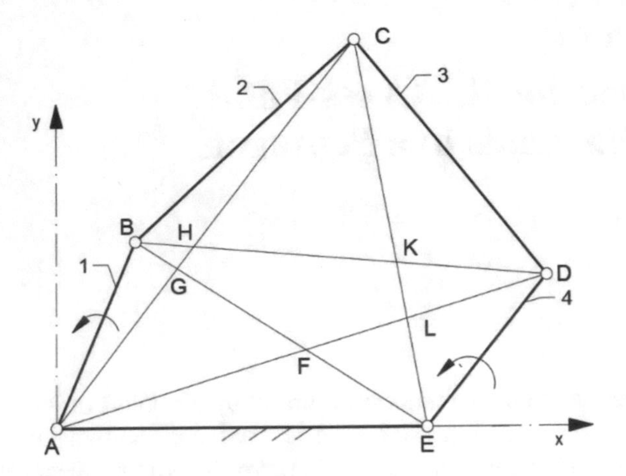

Fig. 14.1 The diagonals

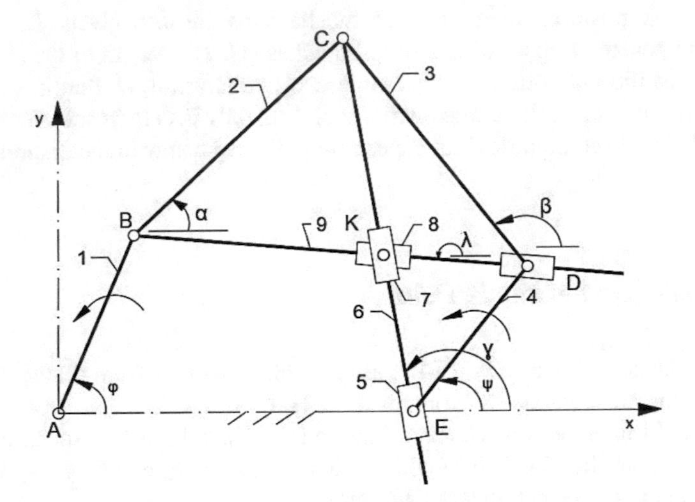

Fig. 14.2 The mechanism for the K point

$$y_B = AB \sin \varphi \tag{14.2}$$

$$x_D = x_E + ED \cos \psi \tag{14.3}$$

$$y_D = ED \sin \psi \tag{14.4}$$

$$x_C = x_B + BC \cos \alpha = x_D + DC \cos \beta \tag{14.5}$$

$$y_C = y_B + BC \sin\alpha = y_D + DC \sin\beta \tag{14.6}$$

$$BD^2 = (x_B - x_D)^2 + (y_B - y_D)^2 \tag{14.7}$$

$$CE^2 = (x_C - x_E)^2 + (y_C - y_E)^2 \tag{14.8}$$

$$x_K = x_E + EK \cos\gamma = x_D + DK \cos\lambda \tag{14.9}$$

$$y_K = y_E + EK \sin\gamma = y_D + DK \sin\lambda \tag{14.10}$$

$$\cos\gamma = (x_C - x_E)/CE \tag{14.11}$$

$$\sin\gamma = (y_C - y_E)/CE \tag{14.12}$$

$$\cos\lambda = (x_B - x_D)/BD \tag{14.13}$$

$$\sin\lambda = (y_B - y_D)/BD \tag{14.14}$$

The calculated mechanism was shown in one position in Fig. 14.3, for the following initial data (measured on the drawing used for verification): AB = 58; BC = 75; DC = 85; ED = 55; XE = 105; $\varphi = 60°$; $\psi = 40°$. These data were also kept for the following mechanisms based on Fig. 14.1. The motion of the driving element ED depended on the motion of the driving element AB, i.e. $\psi = q\varphi$. Different values were chosen for q.

For q = 1 the successive positions of the mechanism were determined in Fig. 14.4 and the locus of K point in Fig. 14.5.

The non-linear variation of the K point coordinates and the linear variation of ψ were presented in Fig. 14.6.

Next, the successive positions and loci of the K point for different values of q were presented (Figs. 14.7, 14.8, 14.9 and 14.10).

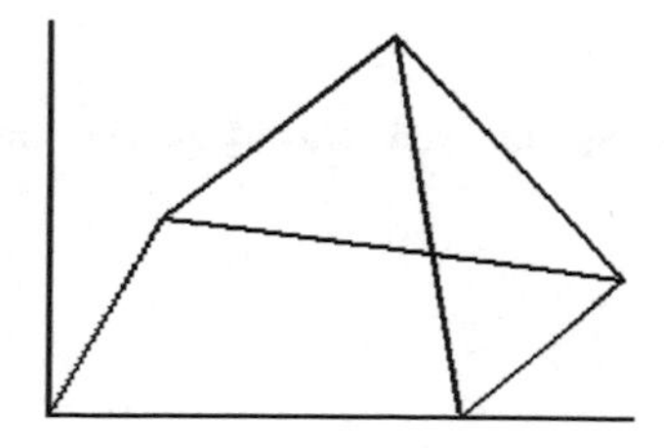

Fig. 14.3 The mechanism in one position

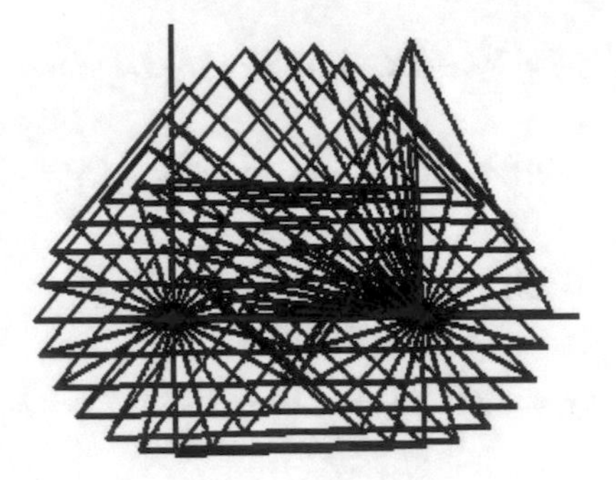

Fig. 14.4 Successive positions

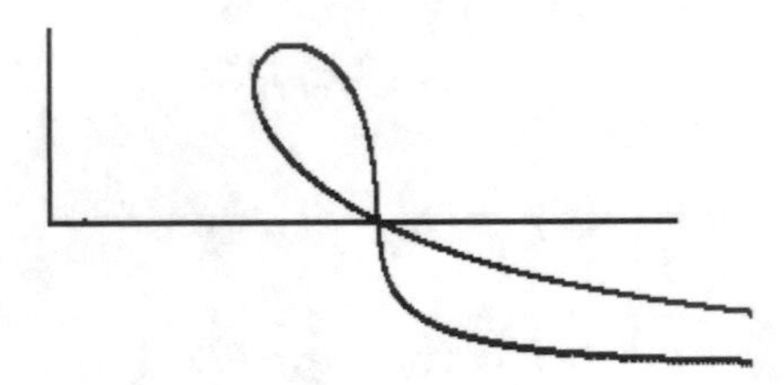

Fig. 14.5 The locus of K point for q = 1

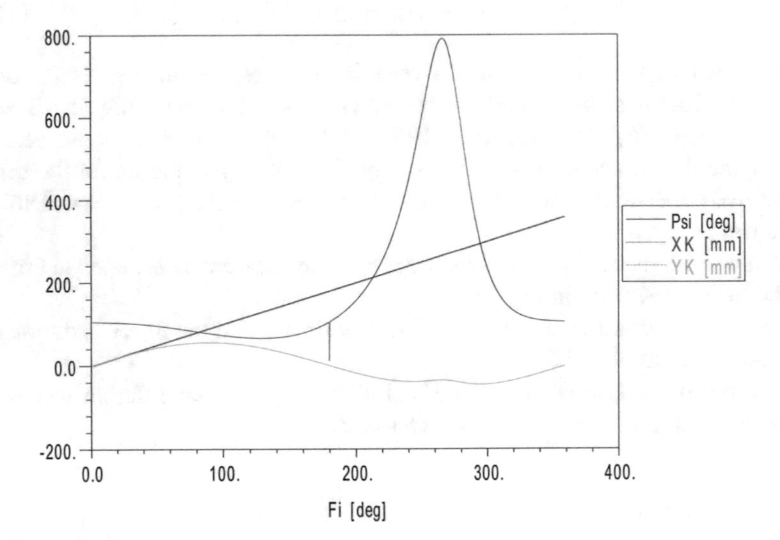

Fig. 14.6 Variation of K point coordinates for q = 1

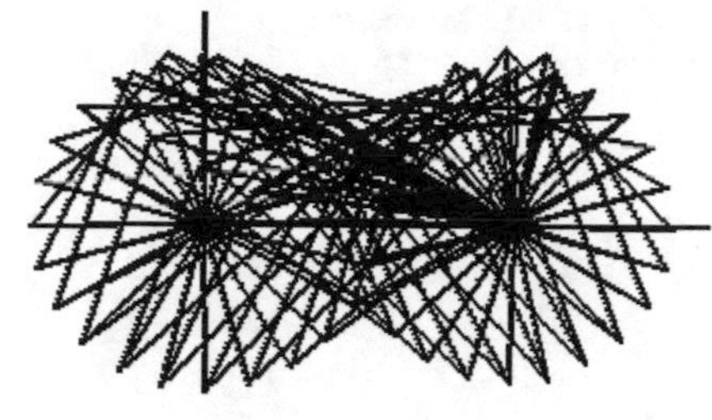

Fig. 14.7 The successive positions for $q = -1$

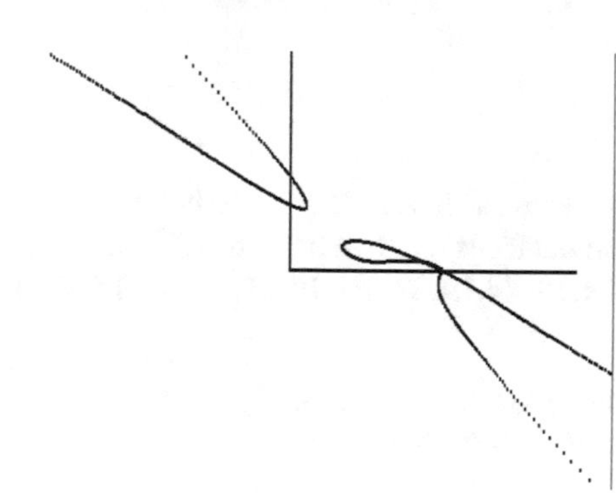

Fig. 14.8 The locus of K point for $q = -1$

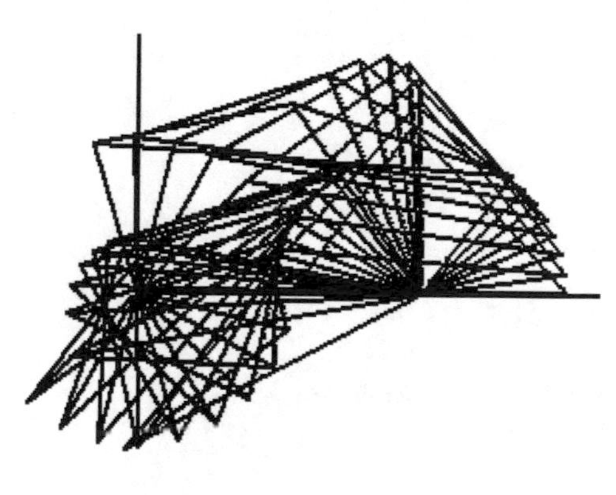

Fig. 14.9 The successive positions for $q = 0.5$

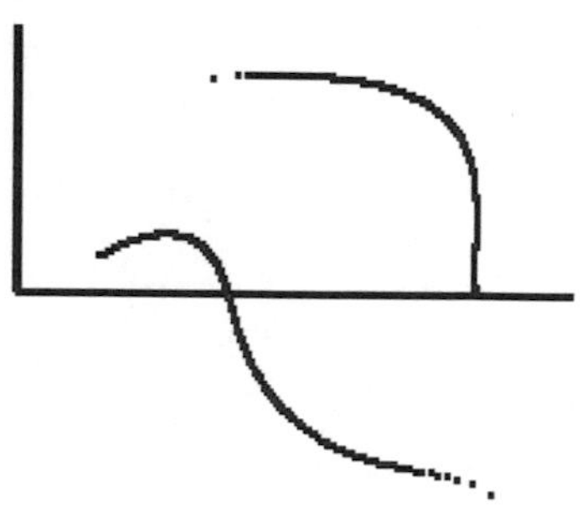

Fig. 14.10 The locus of K point for $q = 0.5$

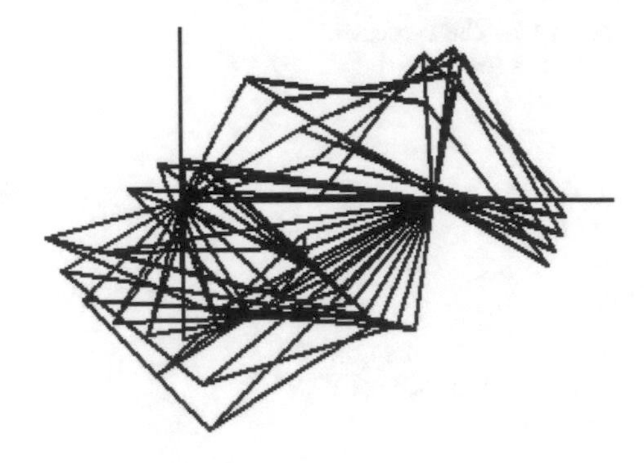

Fig. 14.11 The successive positions for q = −0.5

Here, as in the other cases below, the ED link did not perform complete rotations, so the locus had two branches (Figs. 14.11, 14.12, 14.13, 14.14, 14.15, 14.16, 14.17, 14.18, 14.19, 14.20, 14.21, 14.22, 14.23, 14.24, 14.25, 14.26, 14.27 and 14.28).

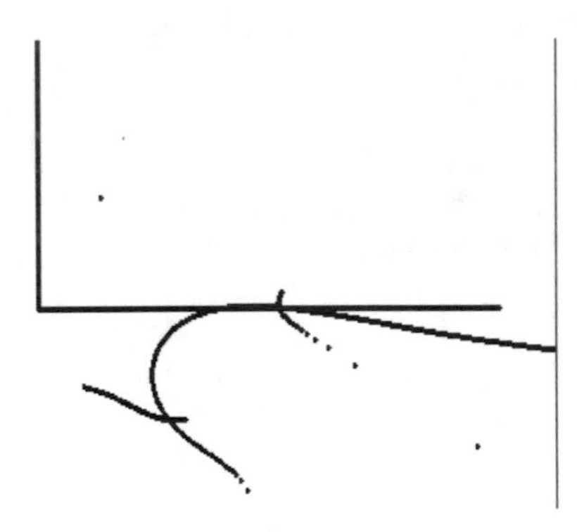

Fig. 14.12 The locus of K point for q = −0.5

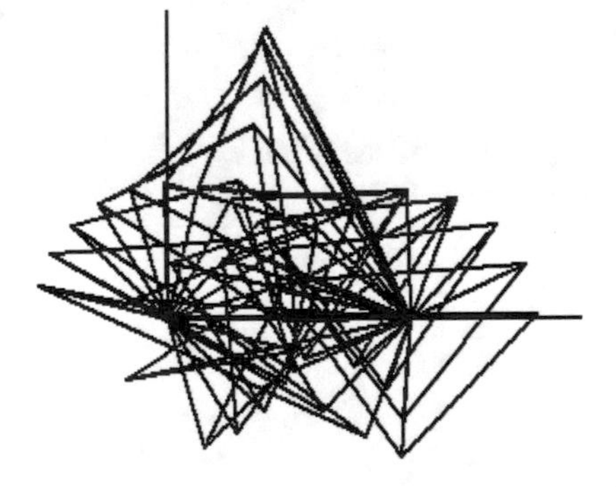

Fig. 14.13 The successive positions for q = 1.5

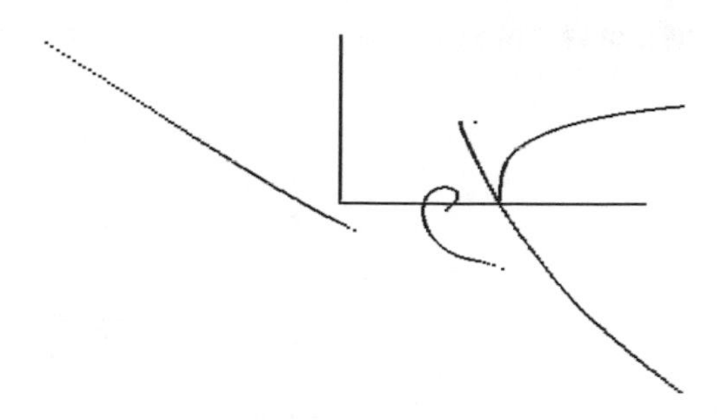

Fig. 14.14 The locus of K point for q = 1.5

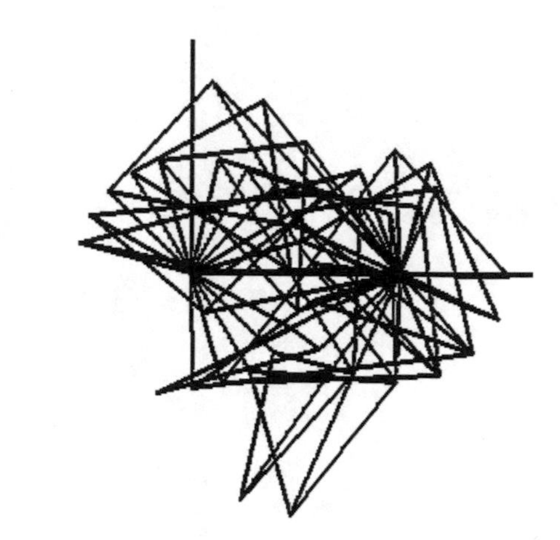

Fig. 14.15 The successive positions for q = −1.5

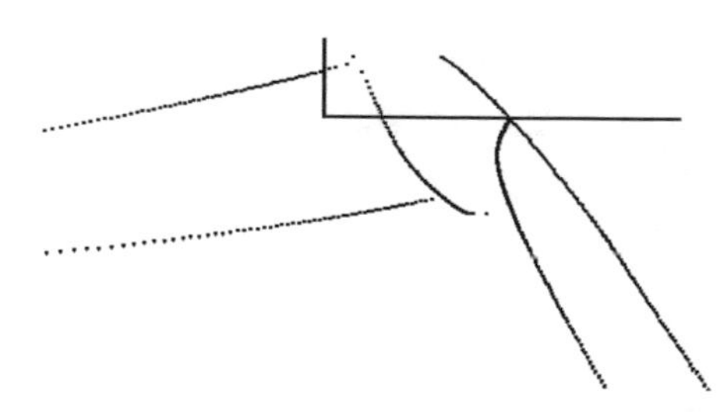

Fig. 14.16 The locus of K point for q = −1.5

14.2 The Loci of the G Point

The mechanism that determines the locus of the G point was presented in Fig. 14.29.

The following relationships were written:

$$AC^2 = x_C^2 + y_C^2 \tag{14.15}$$

$$EB^2 = (x_E - x_B)^2 + (y_E - y_B)^2 \tag{14.16}$$

Fig. 14.17 The successive positions for q = 2

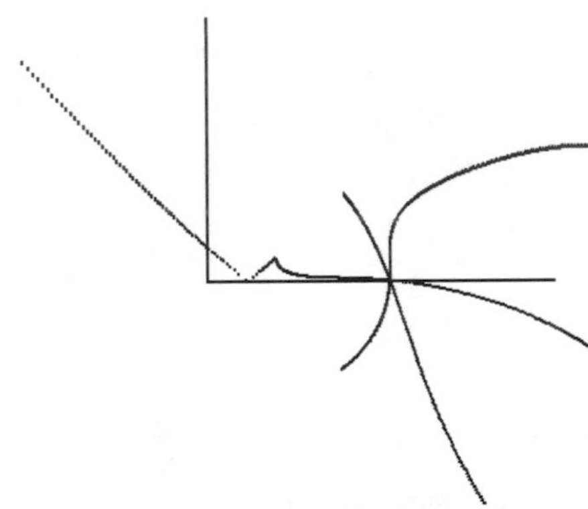

Fig. 14.18 The locus of K point for q = 2

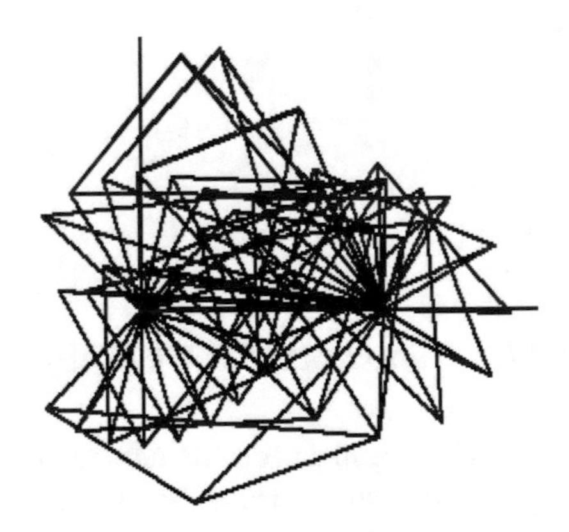

Fig. 14.19 The successive positions for q = −2

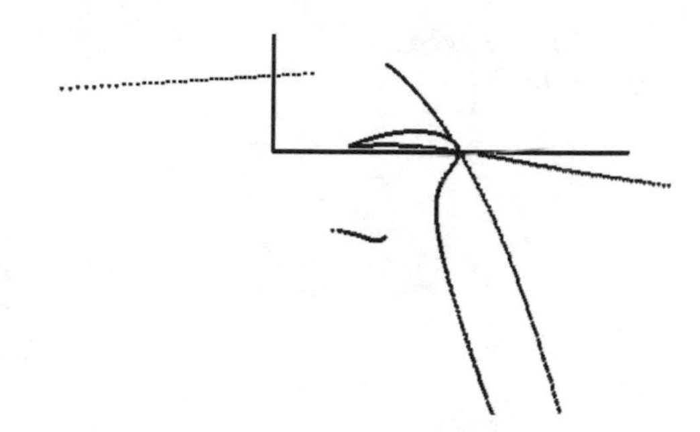

Fig. 14.20 The locus of K point for q = −2

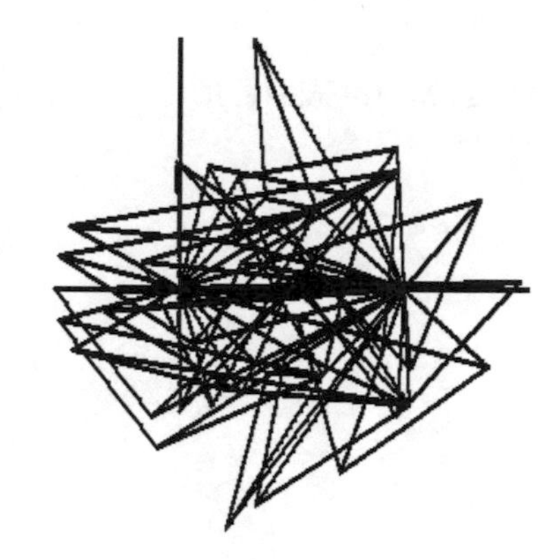

Fig. 14.21 The successive positions for q = 3

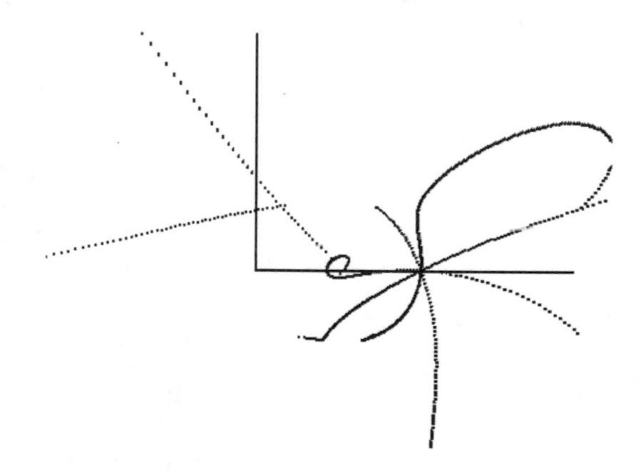

Fig. 14.22 The locus of K point for q = 3

$$x_G = AG\cos\lambda = x_E + EG\cos\gamma \tag{14.17}$$

$$y_G = AG\sin\lambda = y_E + EG\sin\gamma \tag{14.18}$$

$$\cos\lambda = x_C/AC \tag{14.19}$$

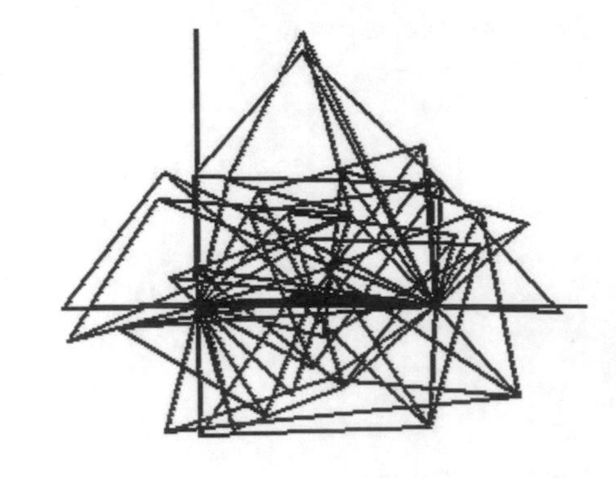

Fig. 14.23 The successive positions q = −3

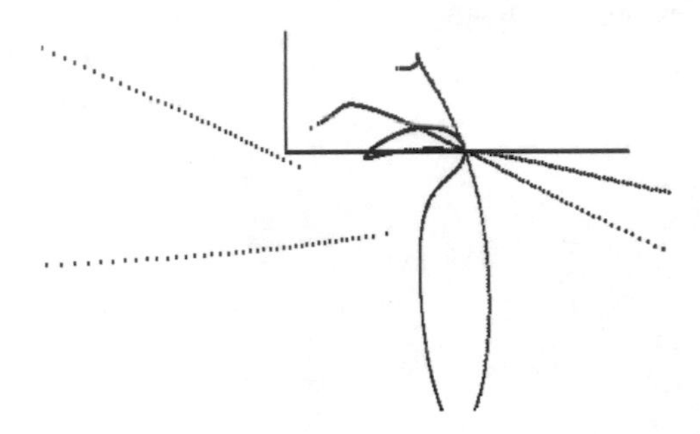

Fig. 14.24 The locus of K point for q = −3

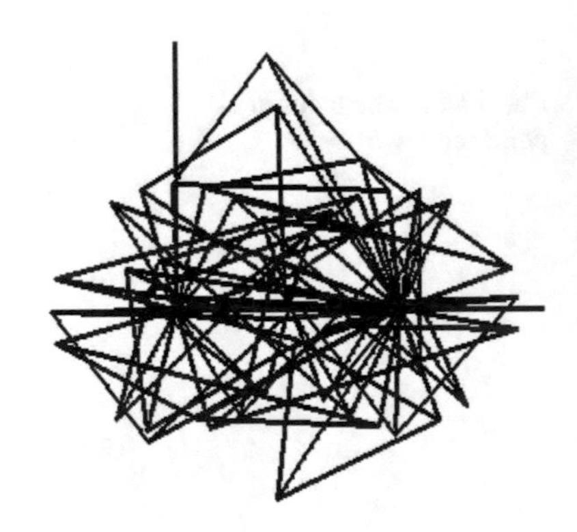

Fig. 14.25 The successive positions q = 5

$$\sin \lambda = y_C / AC \tag{14.20}$$

$$tg\gamma = (y_B - y_E)/(x_B - x_E) \tag{14.21}$$

In Fig. 14.30 the mechanism was shown for one position, having as initial data those of the mechanism in the previous chapter. The successive positions of the mechanism, for q = 1, could be seen in Fig. 14.31.

The locus for q = 1 was shown in Fig. 14.32.

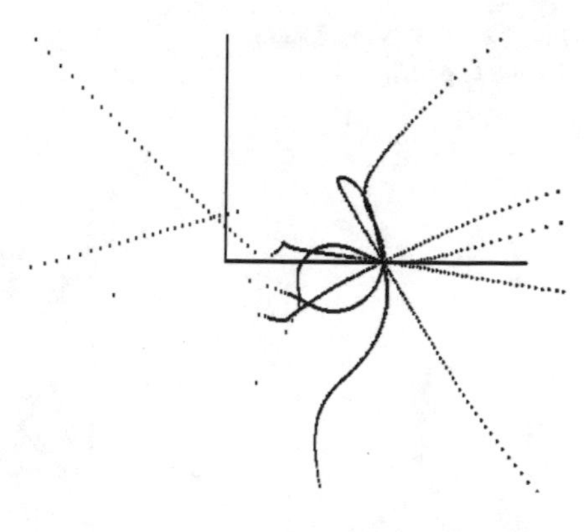

Fig. 14.26 The locus of K point for q = 5

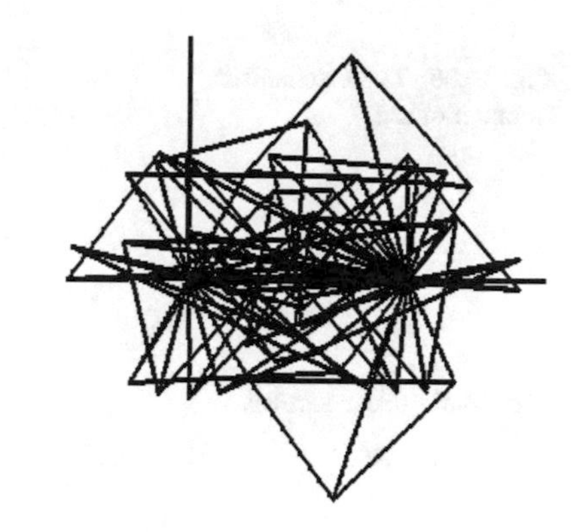

Fig. 14.27 The successive positions q = −5

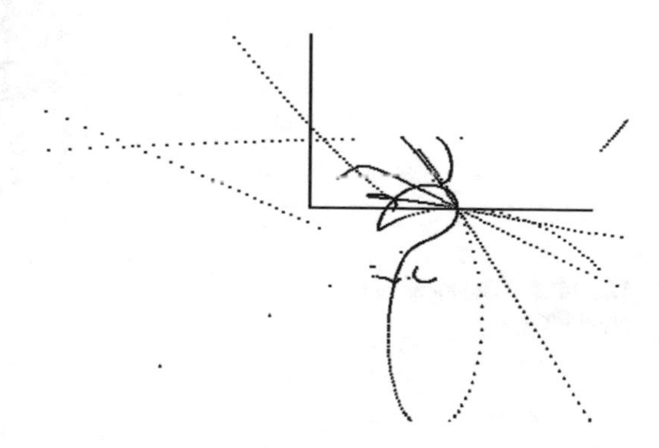

Fig. 14.28 The locus of K point for q = −5

The curve is not continuous. There were two discontinuities in the mechanism operation, as shown in Fig. 14.33.

There were presented only the loci of G point for the values of q shown below the pictures (Figs. 14.34, 14.35, 14.36, 14.37, 14.38, 14.39, 14.40, 14.41, 14.42, 14.43 and 14.44).

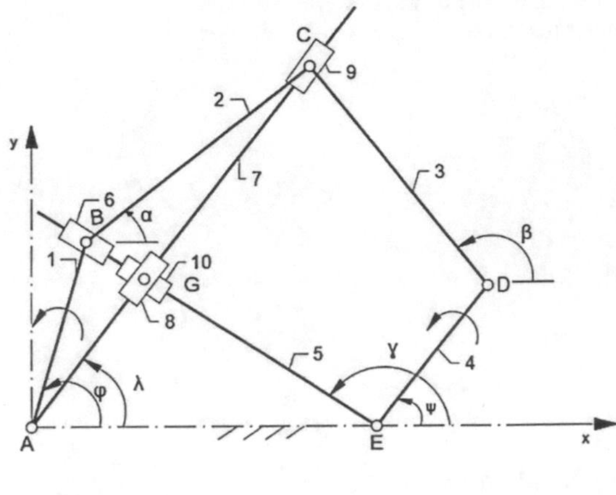

Fig. 14.29 The mechanism for the G point

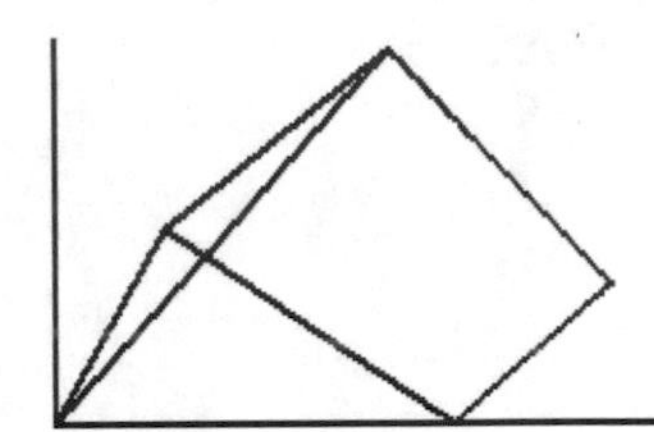

Fig. 14.30 The mechanism in one position

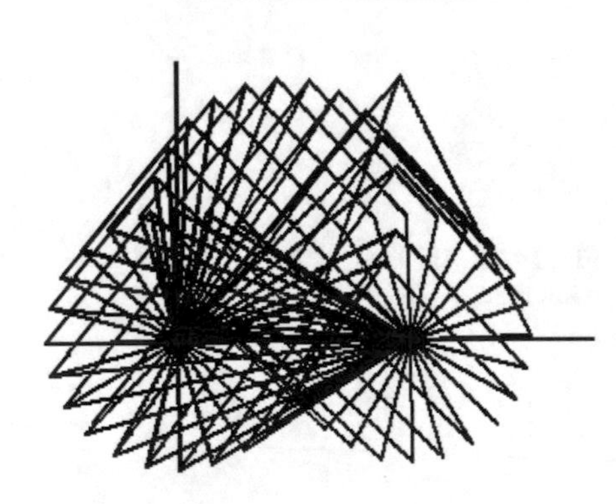

Fig. 14.31 The successive positions

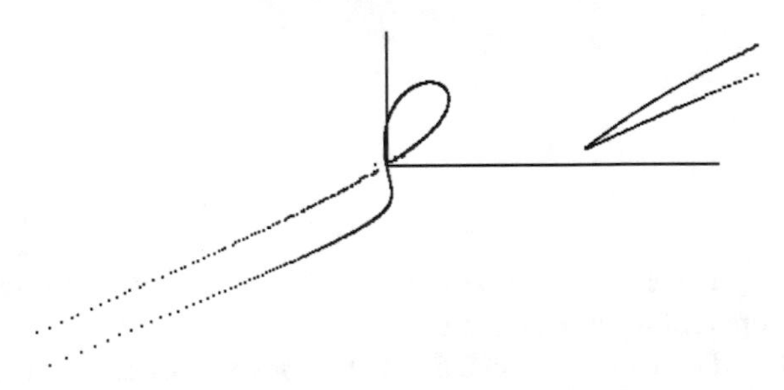

Fig. 14.32 The locus of G point for q = 1

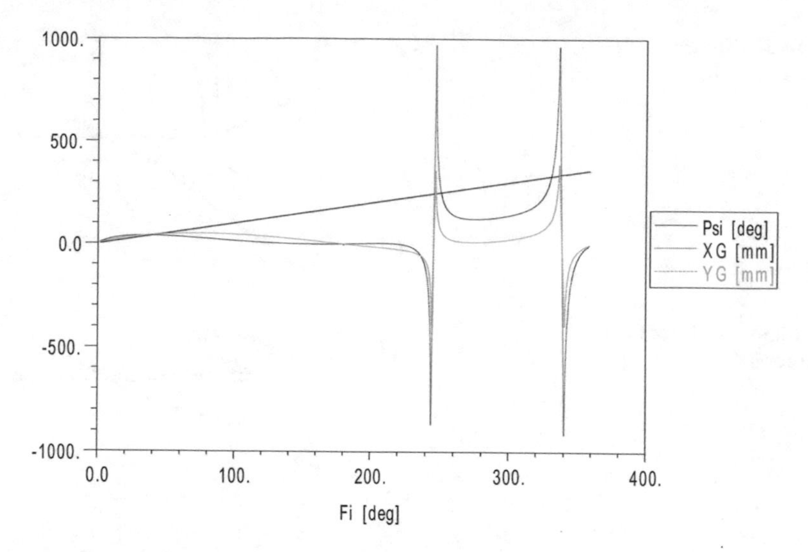

Fig. 14.33 Discontinuities of XG şi YG for q = 1

Fig. 14.34 The locus of G point for q = −1

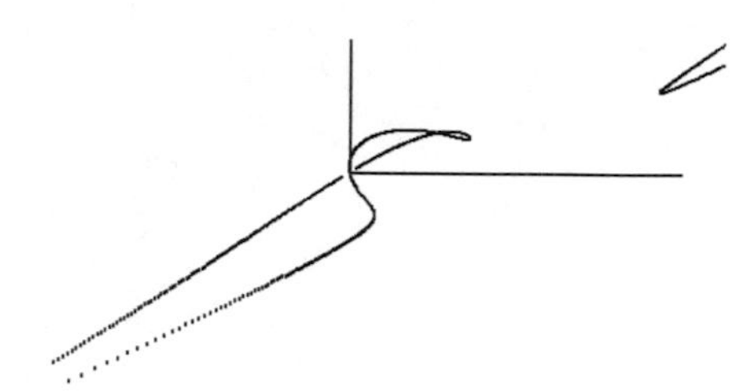

Fig. 14.35 The locus of G point for q = 0.5

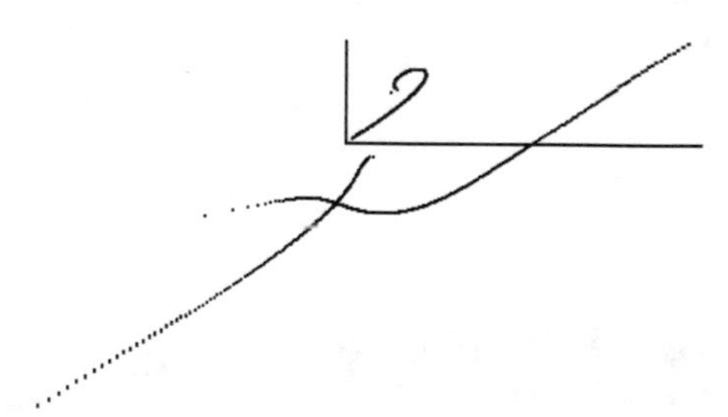

Fig. 14.36 The locus of G point for q = −0.5

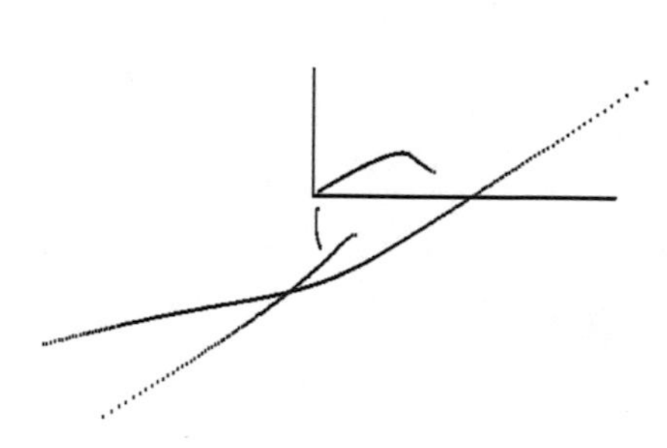

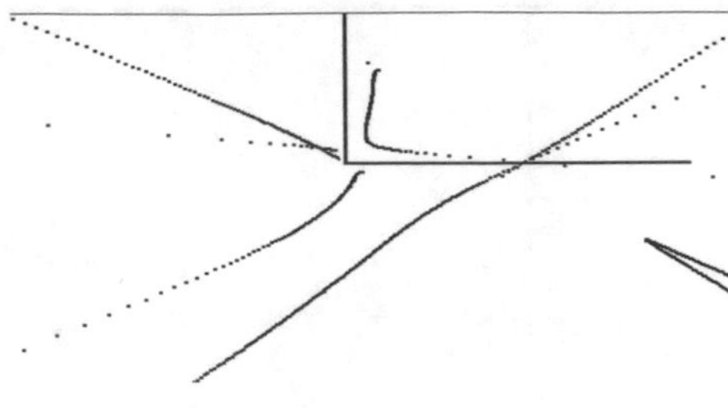

Fig. 14.37 The locus of G point for $q = 1.5$

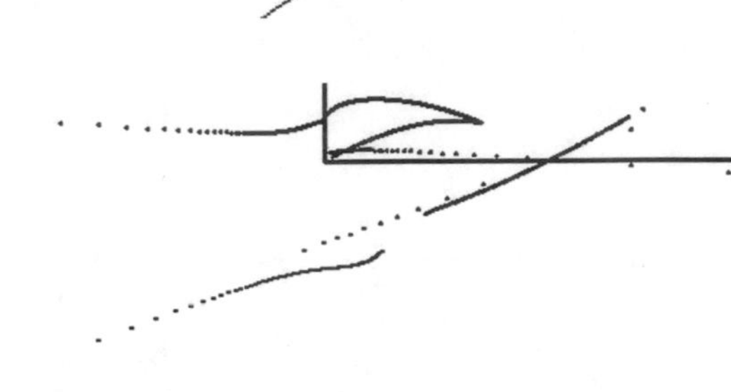

Fig. 14.38 The locus of G point for $q = -1.5$

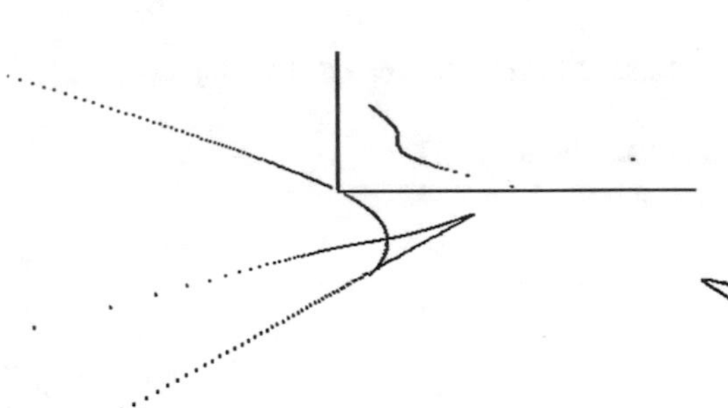

Fig. 14.39 The locus of G point for $q = 2$

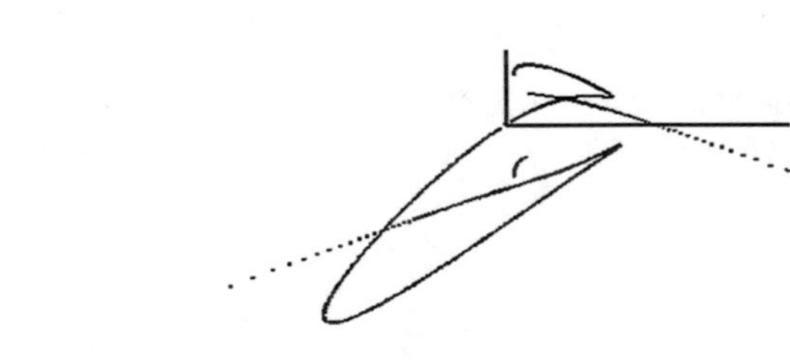

Fig. 14.40 The locus of G point for $q = -2$

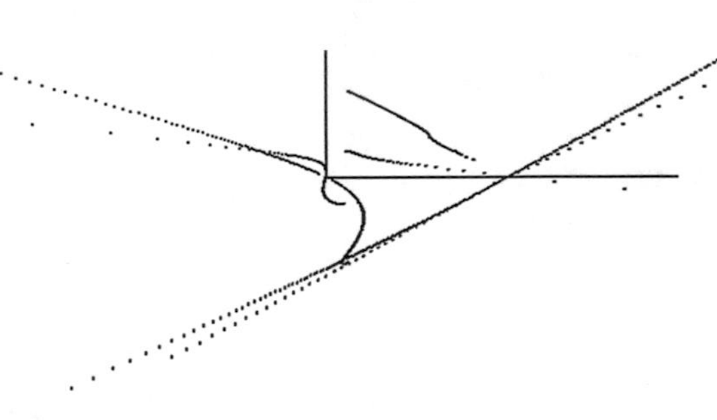

Fig. 14.41 The locus of G point for $q = 3$

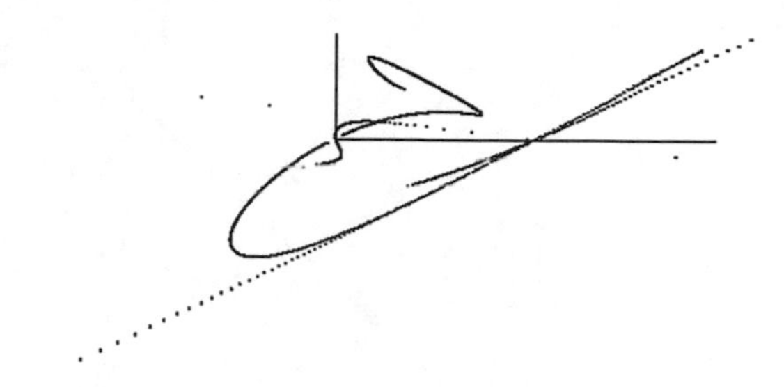

Fig. 14.42 The locus of G point for q = −3

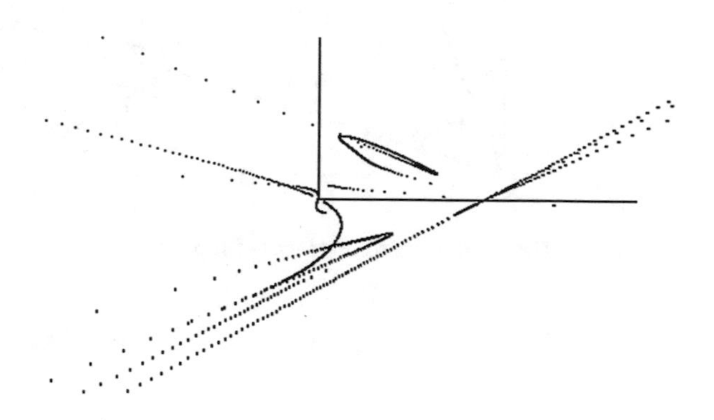

Fig. 14.43 The locus of G point for q = 5

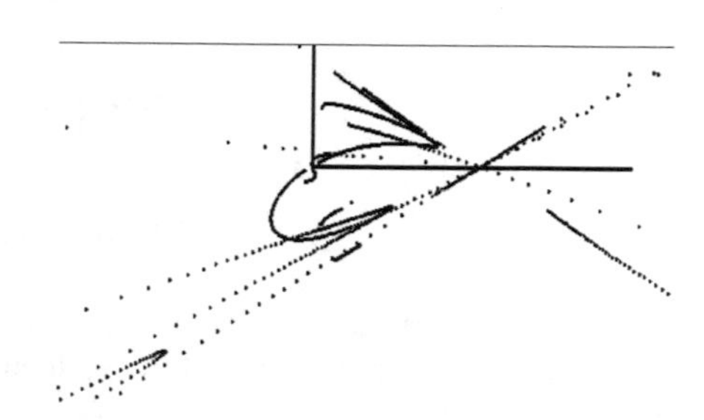

Fig. 14.44 The locus of G point for q = −5

14.3 The Loci of the L Point

The mechanism for determining the loci of the L point was given in Fig. 14.45.

The following expressions were written:

$$AD^2 = x_D^2 + y_D^2 \tag{14.22}$$

$$EC^2 = (x_E - x_C)^2 + (y_E - y_C)^2 \tag{14.23}$$

$$x_L = x_E + EL\cos\gamma = AL\cos\lambda \tag{14.24}$$

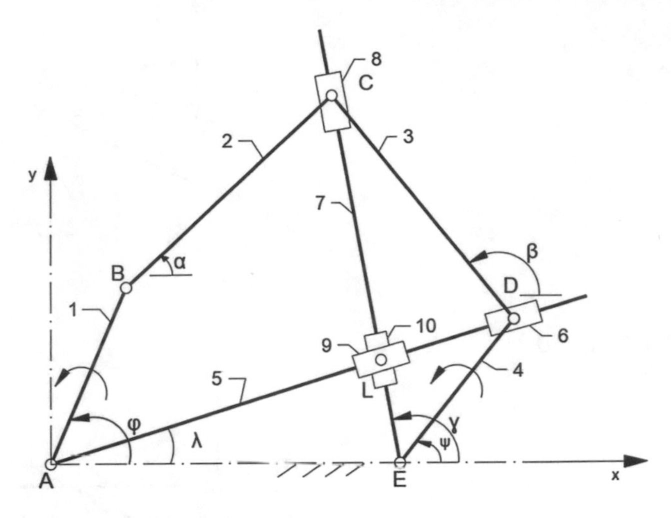

Fig. 14.45 The mechanism for the locus of L point

$$y_L = y_E + EL \sin\gamma = AL \sin\lambda \tag{14.25}$$

$$\cos\lambda = x_D/AD \tag{14.26}$$

$$\sin\lambda = y_D/AD \tag{14.27}$$

$$tg\gamma = (y_C - y_E)/(x_C - x_E) \tag{14.28}$$

The above data was used for the mechanism in one position presented in Fig. 14.46.

The successive positions of the mechanism for q = 1 were shown in Fig. 14.47. It can be observed in Fig. 14.48 that X_L and Y_L have two zones with discontinuities.

Below, there are given the loci of L point for different values of q (Figs. 14.49, 14.50, 14.51, 14.52, 14.53, 14.54, 14.55, 14.56, 14.57, 14.58 and 14.59).

Fig. 14.46 The mechanism in one position

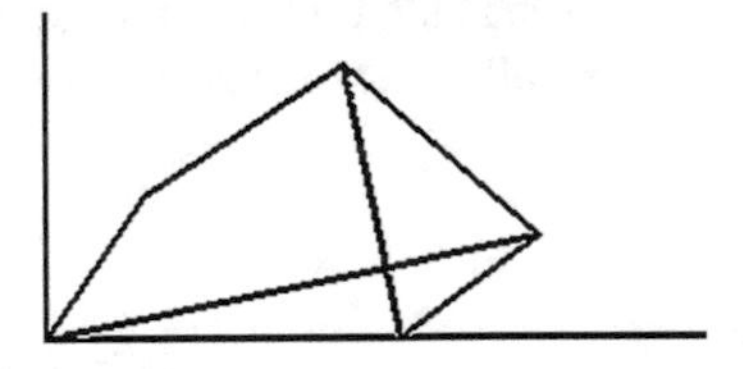

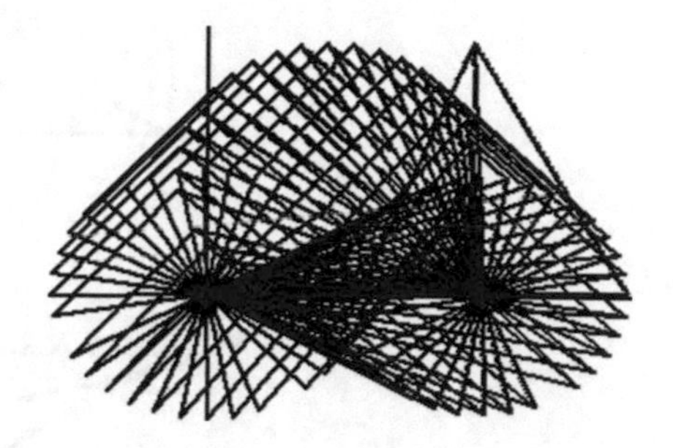

Fig. 14.47 The successive positions

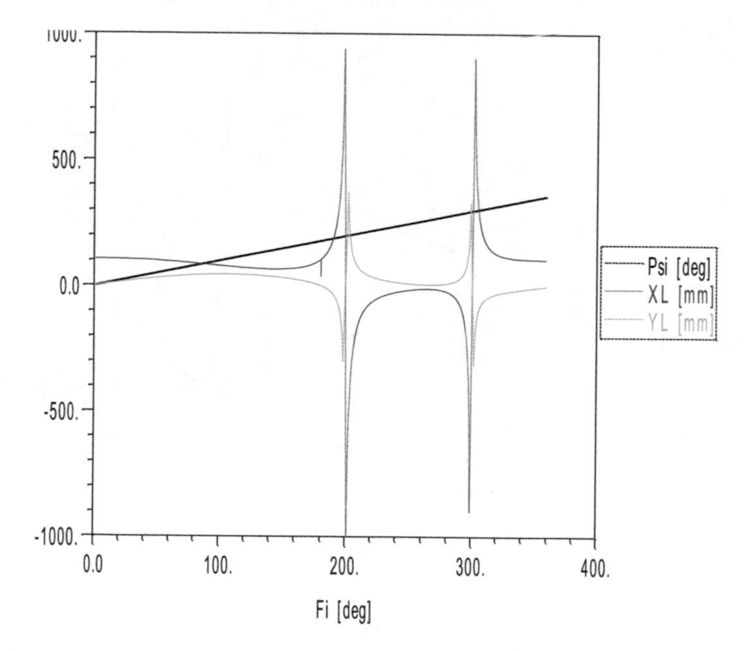

Fig. 14.48 Discontinuities for q = 1

Fig. 14.49 The locus of L point for q = −1

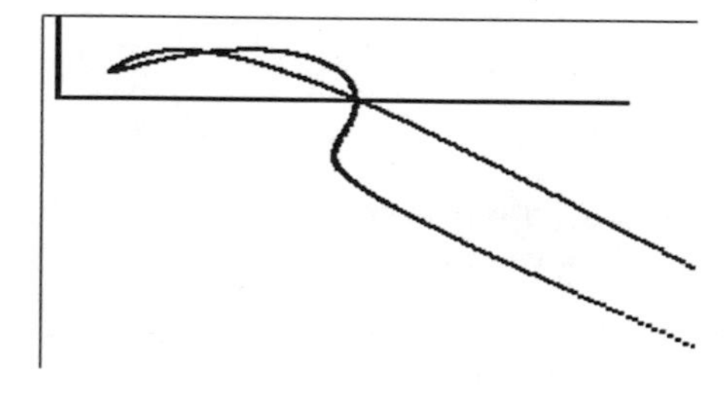

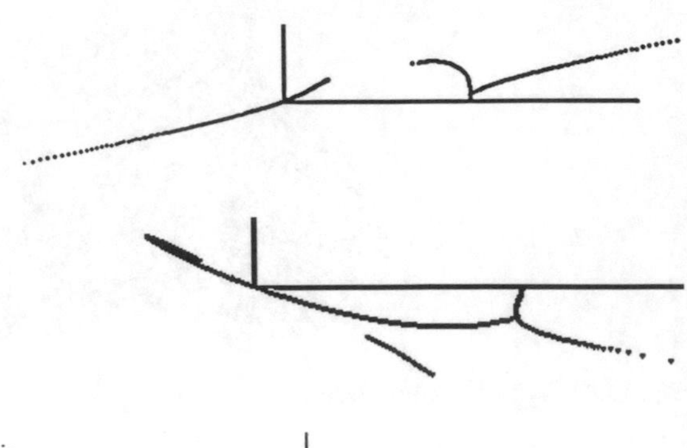

Fig. 14.50 The locus of L point for q = 0.5

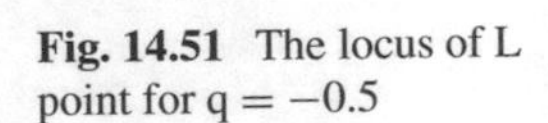

Fig. 14.51 The locus of L point for q = −0.5

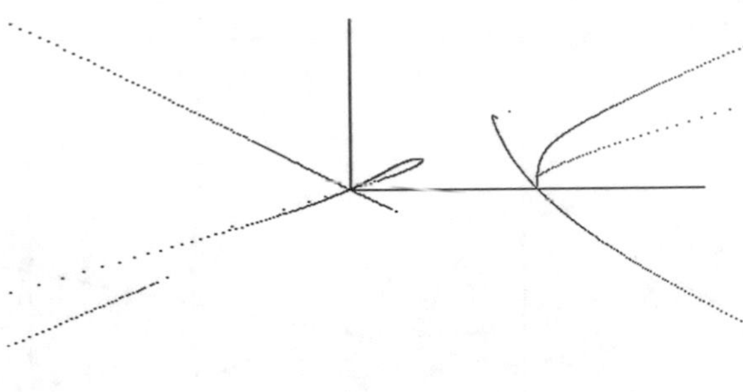

Fig. 14.52 The locus of L point for q = 1.5

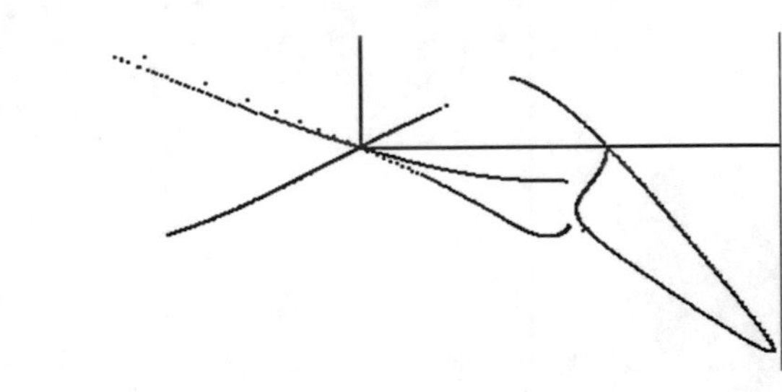

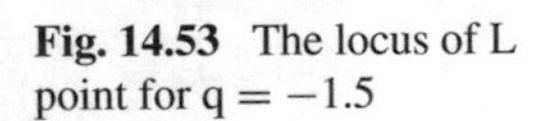

Fig. 14.53 The locus of L point for q = −1.5

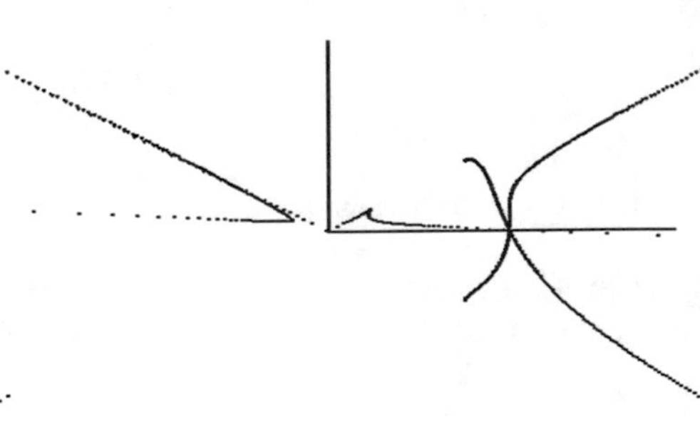

Fig. 14.54 The locus of L point for q = 2

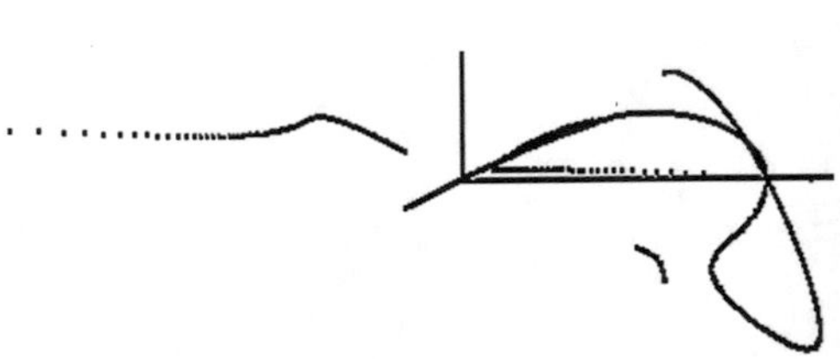

Fig. 14.55 The locus of L point for q = −2

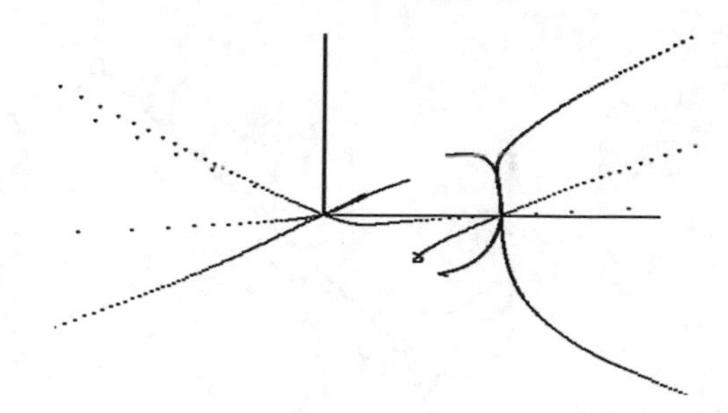

Fig. 14.56 The locus of L point for q = 3

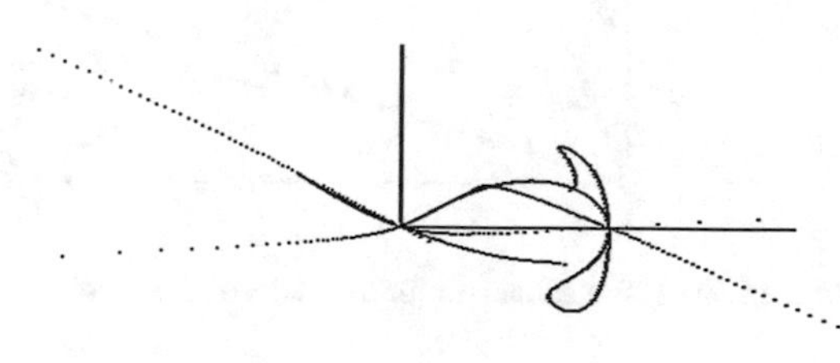

Fig. 14.57 The locus of L point for q = −3

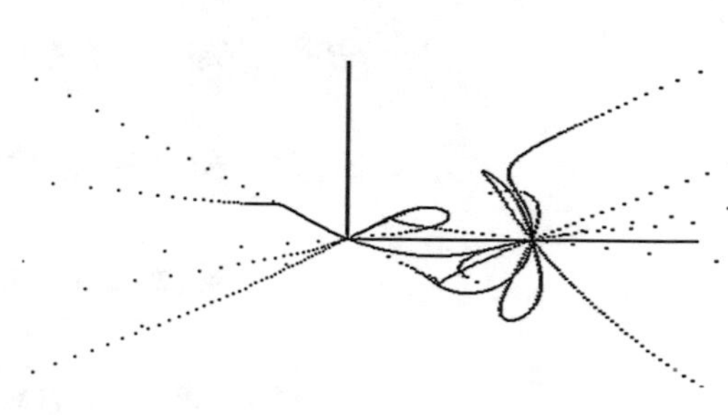

Fig. 14.58 The locus of L point for q = 5

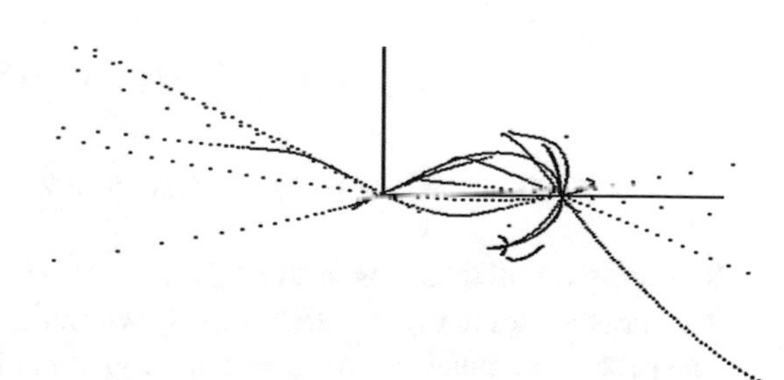

Fig. 14.59 The locus of L point for q = −5

14.4 The Loci of the F Point

The mechanism describing the locus of the F point was shown in Fig. 14.60.
The following expressions were written:

$$AD^2 = x_D^2 + y_D^2 \quad (14.29)$$

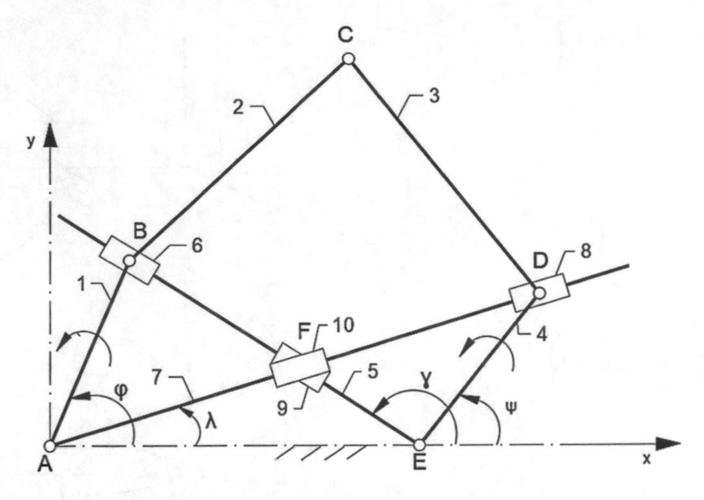

Fig. 14.60 The mechanism for the locus of the F point

$$BE^2 = (x_B - x_E)^2 + (y_B - y_E)^2 \tag{14.30}$$

$$\cos\lambda = x_D/AD \tag{14.31}$$

$$\sin\lambda = y_D/AD \tag{14.32}$$

$$tg\gamma = (y_B - y_E)/(x_B - x_E) \tag{14.33}$$

$$x_F = x_E + EF\cos\gamma = AF\cos\lambda \tag{14.34}$$

$$y_F = y_E + EF\sin\gamma = AF\sin\lambda \tag{14.35}$$

The same initial data was used as above.
The mechanism in one position was shown in Fig. 14.61.
The successive positions for q = 1 were given in Fig. 14.62.
For q = 1, the locus of F point is a circle (Fig. 14.63).

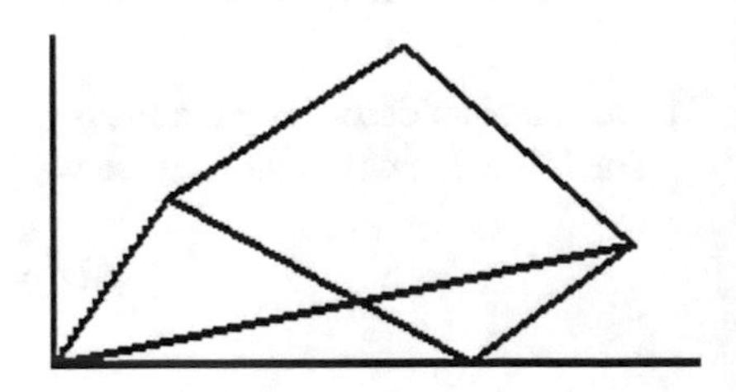

Fig. 14.61 The mechanism in one position

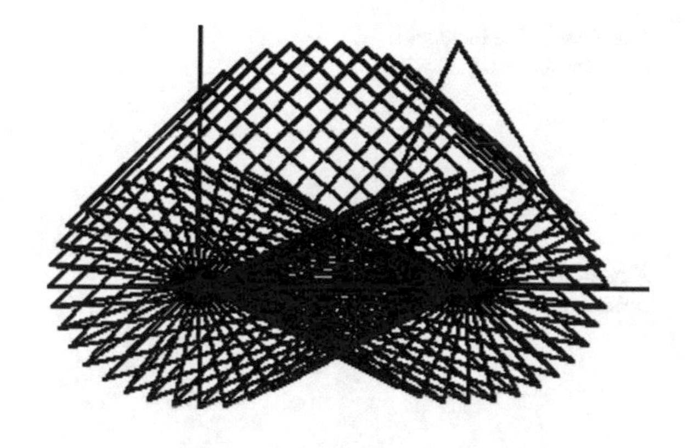

Fig. 14.62 The successive positions for q = 1

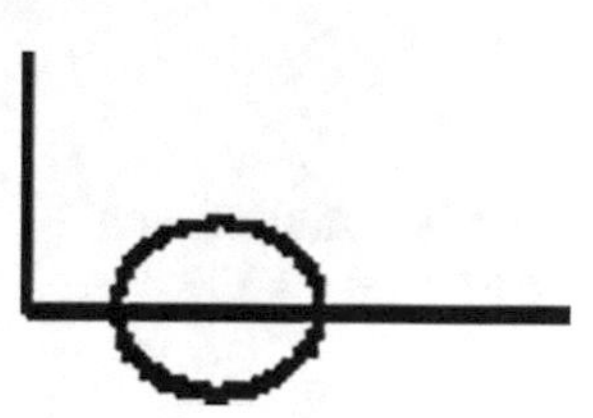

Fig. 14.63 The locus of F for q = 1

Below, there are presented other loci of F point for different values of q (Figs. 14.64, 14.65, 14.66, 14.67, 14.68, 14.69, 14.70, 14.71, 14.72, 14.73 and 14.74).

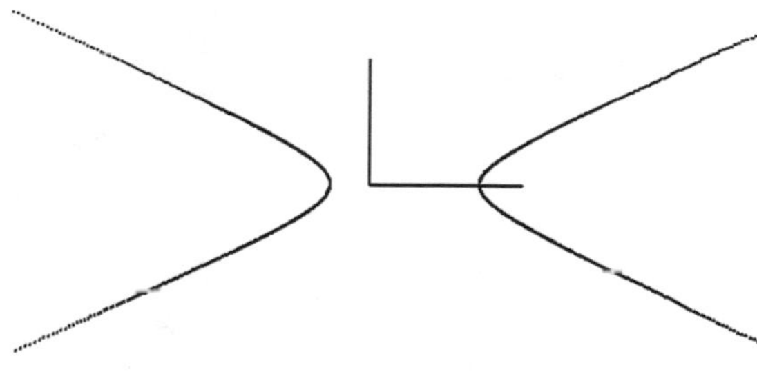

Fig. 14.64 The locus of F point for q = −1

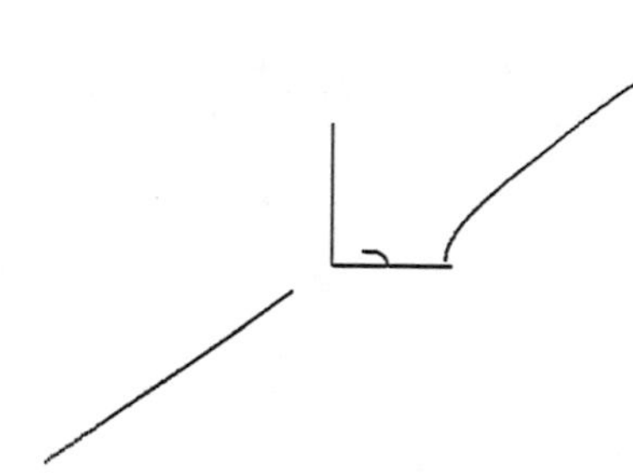

Fig. 14.65 The locus of F point for q = 0.5

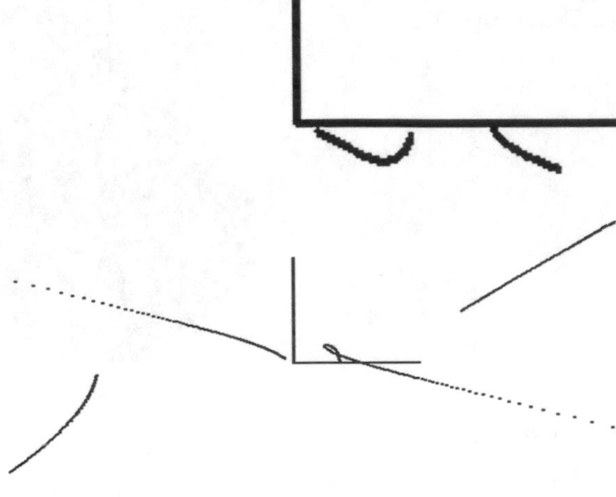

Fig. 14.66 The locus of F point for q = −0.5

Fig. 14.67 The locus of F point for q = 1.5

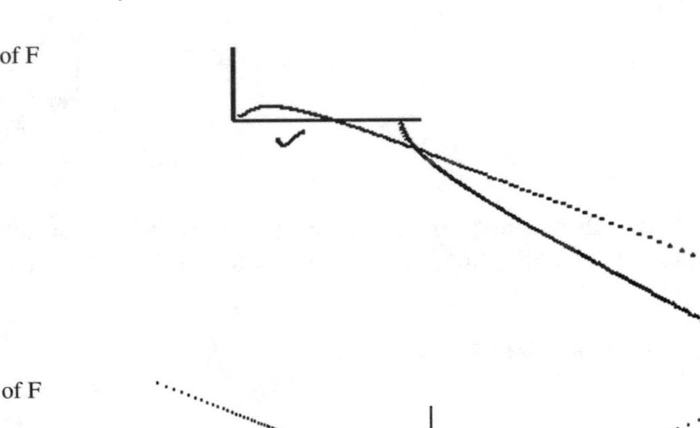

Fig. 14.68 The locus of F point for q = −1.5

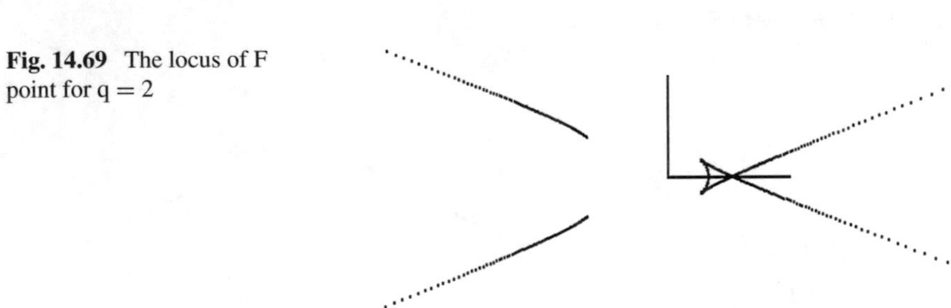

Fig. 14.69 The locus of F point for q = 2

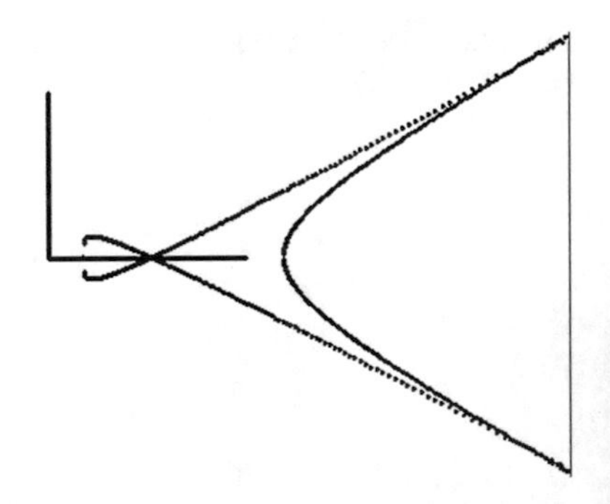

Fig. 14.70 The locus of F point for q = −2

Fig. 14.71 The locus of F point for q = 3

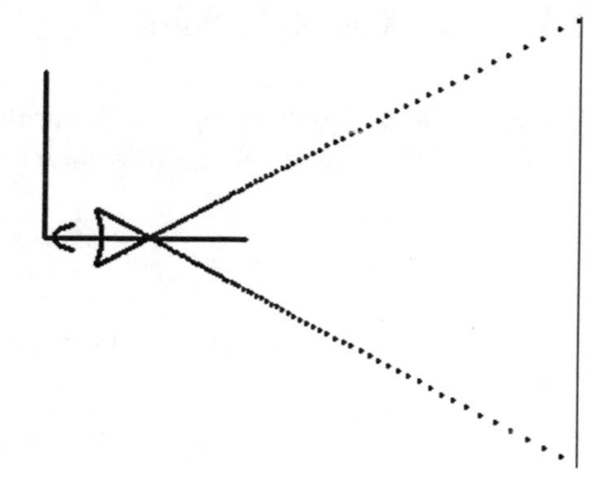

Fig. 14.72 The locus of F point for q = −3

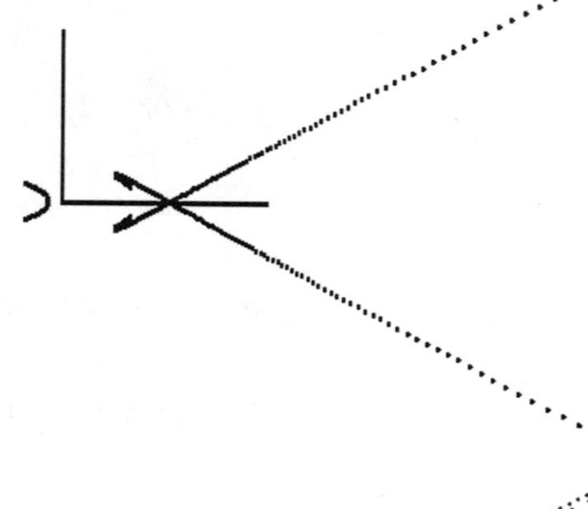

Fig. 14.73 The locus of F point for q = 5

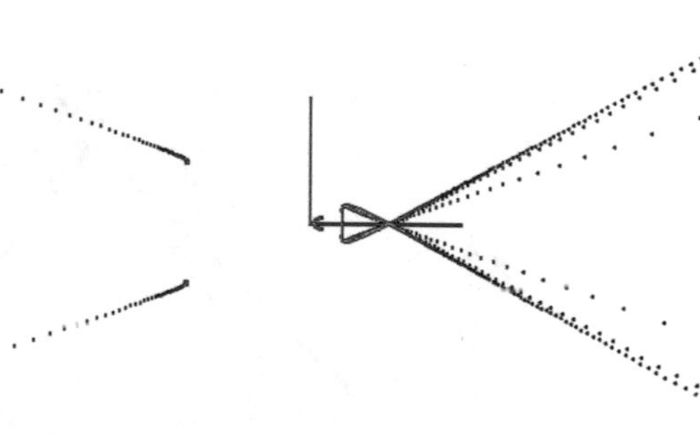

Fig. 14.74 The locus of F point for q = −5

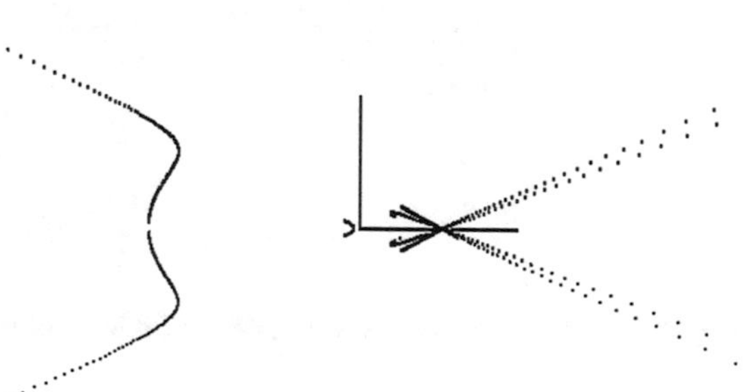

14.5 The Loci of H Point

The mechanism shown in Fig. 14.75 was used for drawing the loci of the H point.

The following expressions were written:

$$AC^2 = x_C^2 + y_C^2 \tag{14.36}$$

$$BD^2 = (x_B - x_D)^2 + (y_B - y_D)^2 \tag{14.37}$$

$$\cos\lambda = x_C/AC \tag{14.38}$$

$$\sin\lambda = y_C/AC \tag{14.39}$$

$$sin\gamma = (y_H - y_D)/HD \tag{14.40}$$

$$cos\gamma = (x_H - x_D)/HD \tag{14.41}$$

$$x_H = x_D + DH\cos\gamma = AH\cos\lambda \tag{14.42}$$

$$y_H = y_D + DH\sin\gamma = AH\sin\lambda \tag{14.43}$$

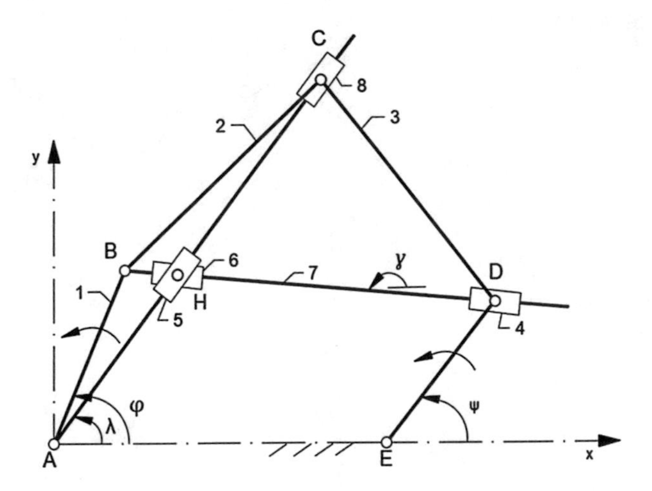

Fig. 14.75 The mechanism used for drawing the locus of H point

In Fig. 14.76 it is shown the mechanism in one position, in Fig. 14.77 it is shown the successive positions for $q = 1$ and, in Fig. 14.78 it is shown the successive positions for $q = -1$.

In Fig. 14.79 a jump of YH at $\varphi = 270$ was shown. Thc locus for $q = 1$ was shown in Fig. 14.80.

In Figs. 14.81, 14.82, 14.83, 14.84, 14.85, 14.86, 14.87, 14.88 and 14.89 there are presented other loci of H point for different values of q.

The loci for the cross-points of other pentagon lines like heights, bisecting-lines, mediana, mid-perpendiculars can be found similarly.

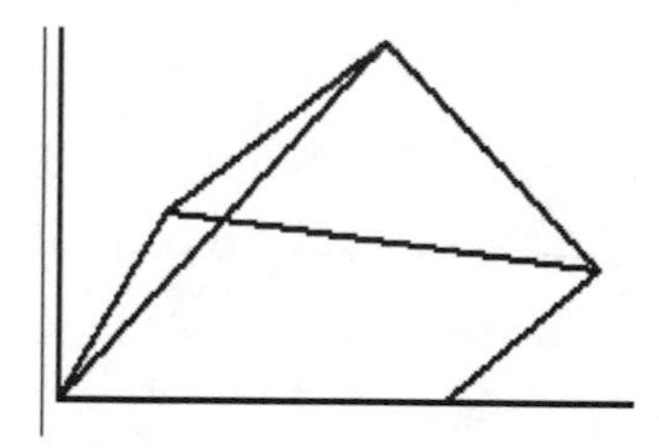

Fig. 14.76 The mechanism in one position

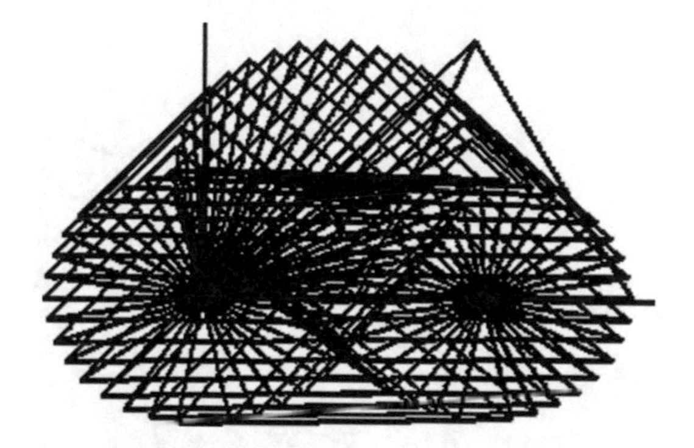

Fig. 14.77 The successive positions for $q = 1$

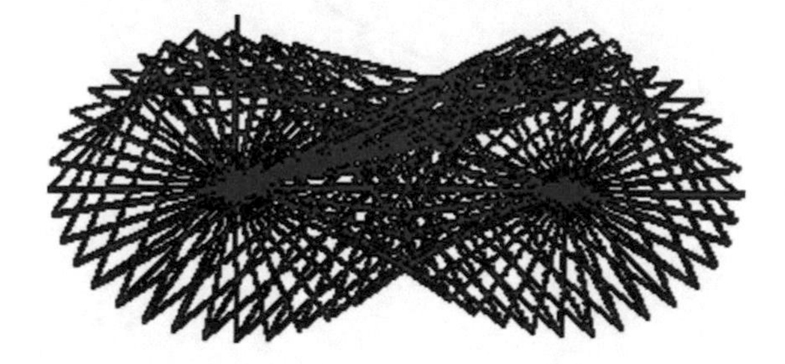

Fig. 14.78 The successive positions for $q = -1$

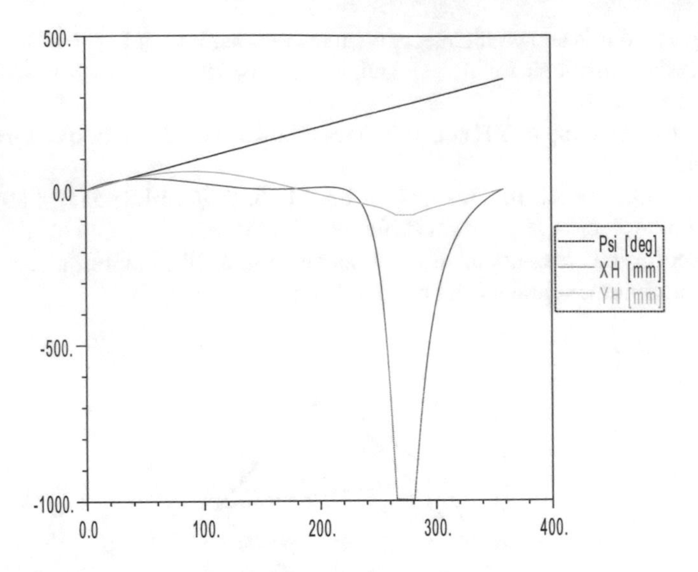

Fig. 14.79 The coordinates of H point for q = 1

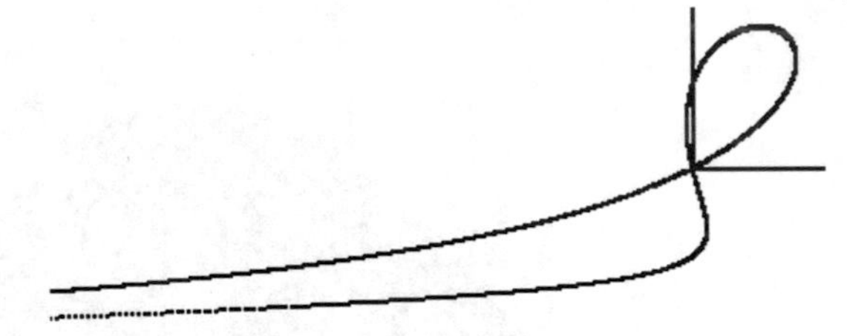

Fig. 14.80 The locus of H point for q = 1

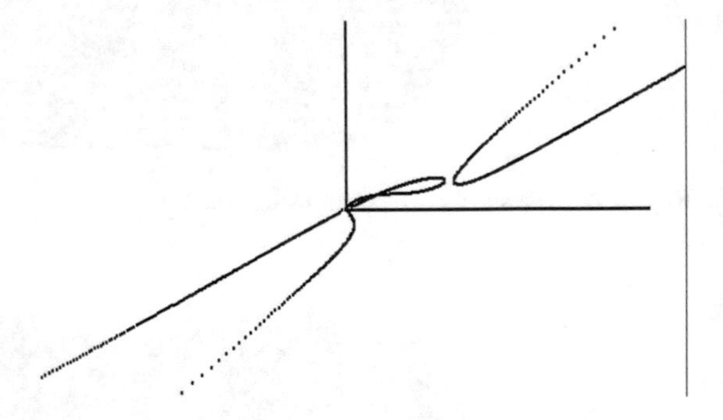

Fig. 14.81 The locus of H point for q = −1

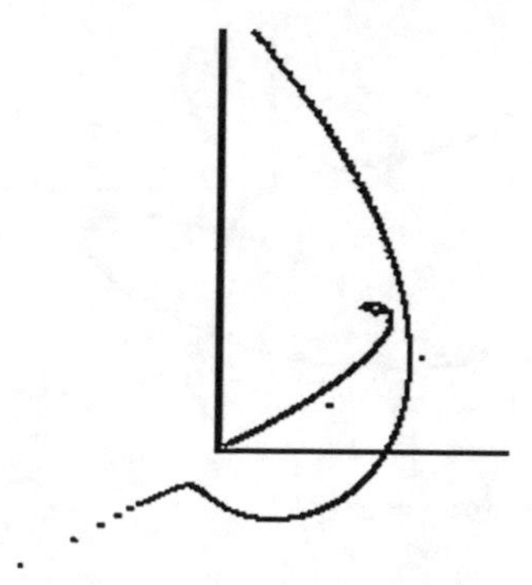

Fig. 14.82 The locus of H point for q = 0.5

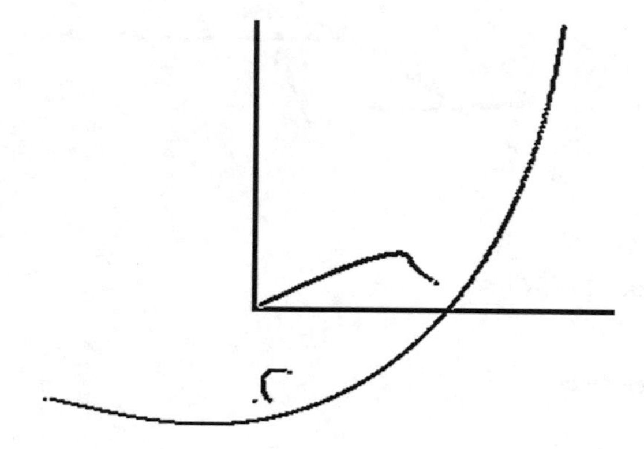

Fig. 14.83 The locus of H point for q = −0.5

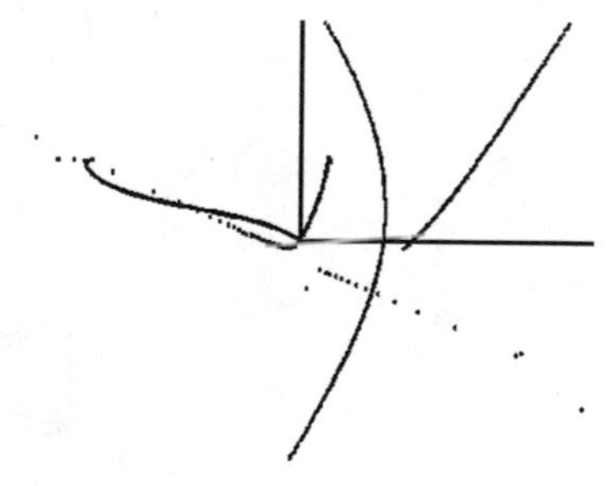

Fig. 14.84 The locus of H point for q = 1.5

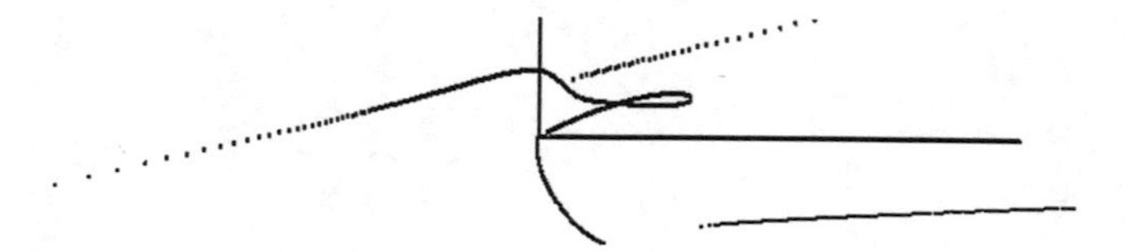

Fig. 14.85 The locus of H point for q = −1.5

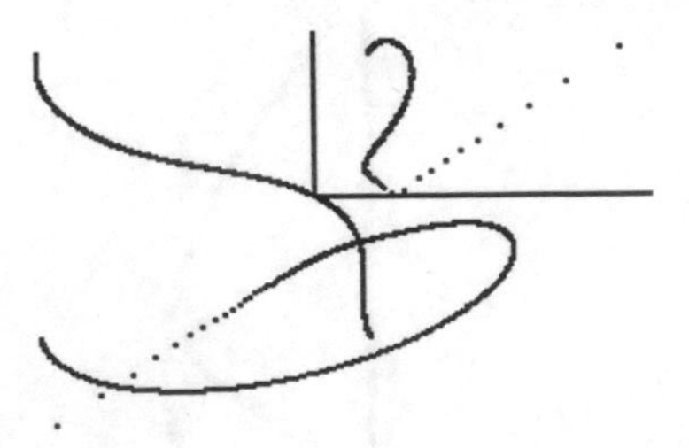

Fig. 14.86 The locus of H point for q = 2

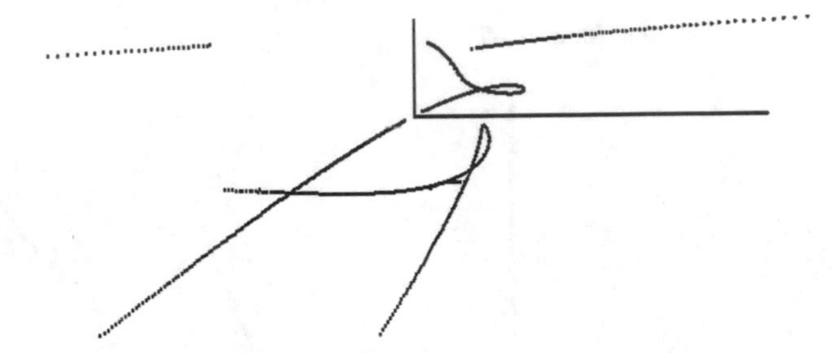

Fig. 14.87 The locus of H point for q = −2

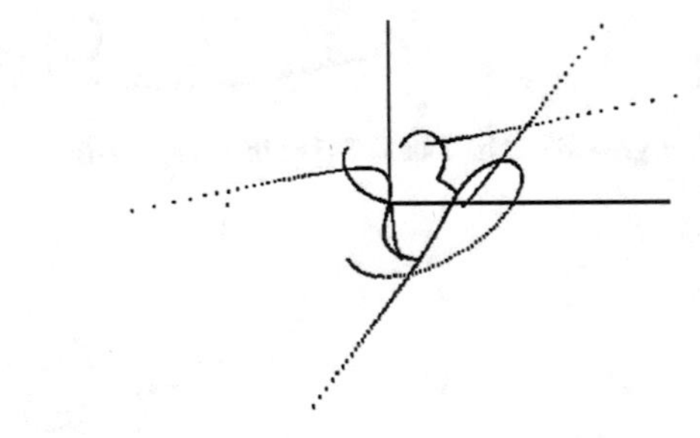

Fig. 14.88 The locus of H point for q = 3

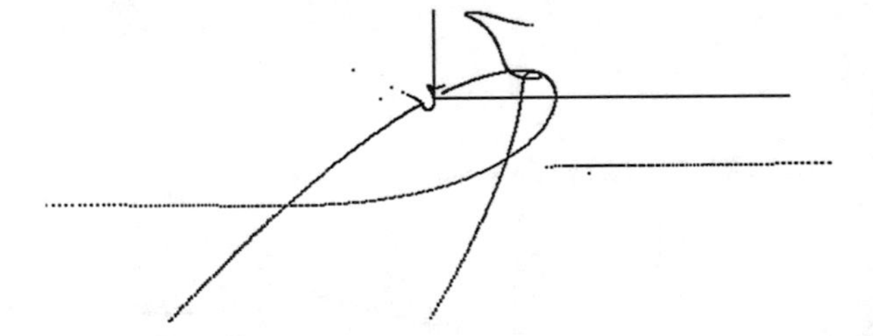

Fig. 14.89 The locus of H point for q = −3

References

1. Popescu I (2016) Locuri geometrice şi imagini estetice generate cu mecanisme. Editura Sitech, Craiova
2. Popescu I (2017) Curbe de bielă în diferite plane şi locuri geometrice generate de mecanisme. Editura Sitech, Craiova

Chapter 15
Correlation Between Track Generation and Synthesis of Mechanisms

Abstract A correlation is made between the generation of trajectories by mechanisms and the synthesis of those mechanisms when the trajectory is given. It is exemplified with the articulated quadrilateral mechanism, finding a nonlinear algebraic system with 9 equations and 9 constant unknowns, which are the parameters that define the mechanism. It is shown that this system cannot be solved with numerical analysis methods to provide exact solutions, so that approximate methods are used, given that the measurement of coordinates on the drawing introduces errors, so that the measured points are not sure on that curve. The methods used to roughly solve the system are indicated, but the solutions found cannot be applied because some negative or zero values result. This leads to the solution of the optimal synthesis, i.e. to optimization problems with nonlinear constraints, indicating approximate methods of solving. An example is also given of finding the mechanism that draws a straight-line segment imposed by 6 points, finding several mechanisms with different precisions. The Newton–Raphson algorithm was used, applicable to a nonlinear algebraic system with only 6 equations and 6 unknowns.

The analysis of the mechanisms deals with the study of some given mechanisms, i.e. their dimensions are known and the movement of the driving element is given, so that the positions, speeds and linear and angular accelerations can be calculated. With the relations from the positions, by the method of the contours connecting rod curves (trajectories) can be generated. For example the case of Fig. 1.8 Chap. 1, where the trajectory of a point on the connecting rod of an articulated four bar mechanism is shown. In the synthesis this is reversed, i.e. the connecting rod curve in Fig. 1.8 Chap. 1 the dimensions of the tracing mechanism are required.

In the case of the articulated fourbar we start from Fig. 15.1 in which point E draws a trajectory whose equation must be determined.

Based on Fig. 15.1 we have the equations [1]:

$$x_0 + a \cos \varphi + b \cos \alpha = x \tag{15.1}$$

I. Popescu et al., *Problems of Locus Solved by Mechanisms Theory*, Springer Tracts in Mechanical Engineering,
https://doi.org/10.1007/978-3-030-63079-9_15

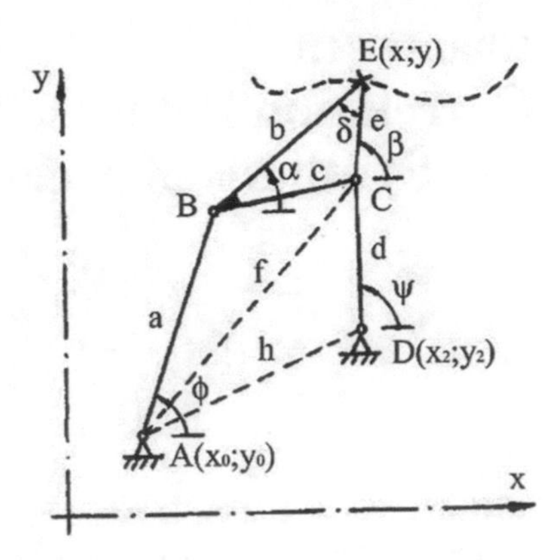

Fig. 15.1 Articulated four bar mechanism

$$y_0 + a \sin \varphi + b \sin \alpha = y \tag{15.2}$$

$$x_2 + d \cos \psi + e \cos \delta = x \tag{15.3}$$

$$y_2 + d \sin \psi + e \sin \delta = y \tag{15.4}$$

$$\beta = \alpha + \delta \tag{15.5}$$

The system has 9 unknown data in Eqs. 15.1 to 15.5, therefore for solving, 9 points are chosen on the connecting rod curve, whose coordinates x (i) and y (i), i = 1… 9, are the main data for the problem.

Removing φ and ψ, we obtain:

$$(x - x_0 - b \cos \alpha)^2 + (y - y_0 - b \sin \alpha)^2 = a^2 \tag{15.6}$$

$$[x - x_2 - e \cos(\alpha + \delta)]^2 = [y - y_2 - e \sin(\alpha + \delta)]^2 = d^2 \tag{15.7}$$

The system of Eqs. 15.6 and 15.7, in which α is variable with position, can be written using the following equations:

$$A_1 \cos \alpha + B_1 \sin \alpha = C_1 \tag{15.8}$$

$$A_2 \cos \alpha + B_2 \sin \alpha = C_2 \tag{15.9}$$

where A_1, B_1, C_1, A_2, B_2, C_2, are expressions depending on the parameters that define the mechanism.

From system formed by Eqs. 15.8 and 15.9 we obtain:

$$E_i = (C_1B_2 - C_2B_1)^2 + (A_1C_2 - A_2C_1)^2 - (A_1B_2 - A_2B_1)^2 = 0 \quad (15.10)$$

with $i = 1\ldots9$, where x = x(i), y = y(i), i being the number of thc chosen point on the connecting rod curve.

This is a nonlinear algebric system with 9 equations and 9 constant unknowns. The connecting rod curve equation is:

$$\begin{aligned} &Q_6x^6 + Q_3y^6 + Q_{22}xy^5 + Q_1x^4y^2 + Q_8x^2y^4 + +Q_{14}x^3y^3 \\ &\quad + Q_{25}x^5 + Q_{21}y^5 + Q_{11}x^3y^2 + Q_{12}x^2y^3 + Q_{13}x^4y + Q_{17}xy^4 \\ &\quad + Q_7x^4 + Q_4y^4 + Q_2x^2y^2 + Q_{15}x^3y + Q_{18}xy^3 \\ &\quad + Q_{26}x^3 + Q_{23}y^3 + Q_{16}x^2y + Q_{19}xy^2 + Q_9x^2 \\ &\quad + Q_5y^2 + Q_{20}xy + Q_{27}x + Q_{24}y + Q_{10} = 0 \end{aligned} \quad (15.11)$$

where Qi are nonlinear expressions depending on the parameters of the mechanism.

For example, for the mechanism with:

$$\begin{aligned} &a = 30;\ b = 50;\ d = 40;\ e = 40 \\ &x_A = 10;\ y_A = 10;\ x_D = 100;\ y_D = 100;\ \delta = 60 \end{aligned} \quad (15.12)$$

the equation of the connecting rod curve is:

$$\begin{aligned} &(2.272069459180975' \ast^\wedge 6 - 403917.81839644327'\mathrm{x} \\ &\quad + 1.0291177647383088' \ast^\wedge 6\mathrm{x}^\wedge 2 - 320794.25163305853'\mathrm{x}^\wedge 3 \\ &\quad + 43316.058002125144'\mathrm{x}^\wedge 4 - 2957.431961543907'\mathrm{x}^\wedge 5 \\ &\quad + 83.99987743596876'\mathrm{x}^\wedge 6 - 3.5109850529900985' \ast^\wedge 6\mathrm{y} \\ &\quad + 668102.9952181454'\mathrm{xy} - 304175.05080927734'\mathrm{x}^\wedge 2\mathrm{y} \\ &\quad + 44460.26475634218'\mathrm{x}^\wedge 3\mathrm{y} - 2039.0748415112719'\mathrm{x}^\wedge 4 + 0.'\mathrm{x}^\wedge 5\mathrm{y} \\ &\quad + 1.2677451705825222' \ast^\wedge 6\mathrm{y}^\wedge 2 - 358975.97026955767'\mathrm{xy}^\wedge 2 \\ &\quad + 68747.01091271496'\mathrm{x}^\wedge 2\mathrm{y}^\wedge 2 - 5914.863923087814'\mathrm{x}^\wedge 3\mathrm{y}^\wedge 2 \\ &\quad + 251.99963230790624'\mathrm{x}^\wedge 4\mathrm{y}^\wedge 2 - 201799.7687668981'\mathrm{y}^\wedge 3 \\ &\quad + 44460.26475634218'\mathrm{xy}^\wedge 3 - 4078.149683022544'\mathrm{x}^\wedge 2\ \mathrm{y}^\wedge 3 \\ &\quad + 0.'\mathrm{x}^\wedge 3\mathrm{y}^\wedge 3 + 25430.95291058982'\mathrm{y}^\wedge 4 - 2957.4319615439076'\mathrm{xy}^\wedge 4 \\ &\quad + 251.99963230790624'\mathrm{x}^\wedge 2\mathrm{y}^\wedge 4 - 2039.074841511272'\mathrm{y}^\wedge 5 \\ &\quad + 0.'\mathrm{xy}^\wedge 5 + 83.99987743596876'\mathrm{y}^\wedge 6 = 0. \end{aligned} \quad (15.13)$$

The data were divided by 10 (except for the angle), thus we will not have to deal with equations with large numbers (the equation is of degree 6). If Eq. 15.11

is expressed depending to the parameters of the mechanism, not to the Qi coefficients which are nonlinear expressions of these parameters, then, by solving with the Mathematica programs, 22 pages in the form below resulted:

$$
\begin{aligned}
164b^2d^4x^28b^2d^2e^2x^2 + 4b^2e^4x^28b^2d^2x^4 + 8b^2e^2x^4 \\
+ 4b^2x^68b^2d^4xxa + 16b^2d^2e^2xxa8b^2e^4xxa \\
+ 16b^2d^2x^3xa16b^2e^2x^3xa8b^2x^5xa \\
+ 4b\text{^}2d\text{^}4xa\text{^}28b\text{^}2d\text{^}2
\end{aligned}
\tag{15.14}
$$

To find the dimensions of the mechanism we reach a nonlinear algebric system impossible to solve.

The solution consists in solving the nonlinear algebric system Eq. 15.10 with 9 equations and 9 constant unknowns.

For solving, 9 points will be chosen on the given curve and their coordinates will be measured, as initial data. The problem is solved only if the coordinates of the measured points check the given connecting rod curve, so that Eq. 15.13 is reached. But what will they be measured with? Even if it is measured with the finest apparatus, errors still result in decimals, only Eq. 15.13 has numbers with many decimals. If some points do not verify the equation of the given connecting rod curve, then they are not on the curve, the solutions being incorrect. It is therefore clear that the problem does not have an exact solution. For this reason, we will work with approximate methods to solve the system Eq. 15.10. In [1] many of our researches are given for solving the system Eq. 15.10. By direct methods it was possible to solve for 5 points imposed on the connecting rod curve, obtaining satisfactory results [2]. Next we tried for 9 points on the connecting rod curve with the Newton–Raphson algorithm, then with the 2nd order Newton–Raphson algorithm, with Hessian matrix, with the Gauss–Seidel algorithm, with the Broyden method, with cubic spline functions of two variables, with iterative methods.

Other drawbacks have arisen, for example, point D in Fig. 15.1 tends to approach point A, line "d" of "a", and line "e" of "b", so that the kinematic chain ABE that runs through the curve results, but is no longer a mechanism. In other cases, some negative sides are reached, mathematically correct, but technically unacceptable.

Thus, it has become necessary to impose restrictions in the algorithms, regarding acceptable values for the parameters of the mechanism, the observance of the Grashof conditions so that the element AB can make complete rotations and other restrictions. The synthesis of the trajectories was thus solved as a nonlinear optimization problem, with linear and nonlinear constraints.

In [3–8] the results obtained using several methods are given: linearization and application of the Simplex algorithm, relaxation method, 2nd order Newton–Raphson method, Monte-Carlo method, genetic algorithms and others.

Acceptable results were obtained. The best algorithm turned out to be Monte-Carlo, with random numbers, although the calculation time is too long.

Returning to Fig. 1.8 of Chap. 1, if we give only the curve we want to be drawn by a mechanism, but not knowing which mechanism, we must try several mechanisms, finally choosing the simplest one.

All the methods mentioned above, tried by us, have appeared in other researchers over time.

Thus, [9] shows the classical methods of analytical and graphical synthesis for generating trajectories. In [10, 11] methods from mathematical statistics are used to solve some synthesis problems. To solve nonlinear algebric systems the gradient method in [3, 4, 7, 12, 13] is used. The optimal synthesis of the mechanisms generating trajectories is detailed in [1, 12, 14–22]. Monte-Carlo methods are detailed in [6, 23]. The method of cubic spline functions are detailed in [5, 8]. The method of quadratic approximation is given in [24], and the method of multiple points in [25, 27]. The classical method of analytical synthesis is treated in [2, 13, 26].

In symposium proceedings and in the Journal Mechanism and Machines Theory, there are always articles on the synthesis and optimization of mechanisms for generating trajectories.The above equations for the articulated quadrilateral become much more complicated for other more complex mechanisms. Therefore, the synthesis of mechanisms will still concern researchers for a long time.

- **An Example.**

The system solutions are sought so that the tracer point describes a straight-line segment [1]. The system given by Eq. 15.10 can also be written

$$f_i = (FG - HK)^2 + (QL - PF)^2 - (QG - KP)^2 = 0 \tag{15.15}$$

or depending on the sides:

$$f_i[x_0, y_0, x_2, y_2, a, b, e, d, \delta x(i), y(i)] = 0; \quad i = 1 \dots 9 \tag{15.16}$$

where the following notations were used:

$$\begin{aligned}
Q &= 2b(x - x_0)\\
K &= 2b(y - y_0)\\
F &= (x - x_0)^2 + (y - y_0)^2 - a^2 + b^2\\
P &= 2e[(x - x_2)\cos\delta + (y - y_2)\sin\delta]\\
G &= 2e[(y - y_2)\cos\delta - (x - x_2)\sin\delta]\\
H &= L = (x - x_2)^2 + (y - y_2)^2 - d^2 + e^2
\end{aligned} \tag{15.17}$$

The Newton–Raphson method is still used, provided that solutions as close as possible to the final ones are adopted as initial data. Considering the initial solutions denoted by the exponent zero, the system given by Eq. 15.16 can be written:

$$f_i\left[x_0^0 + \delta x_C, y_0^0 + \delta y_0 + \ldots + y(i)\right] = 0 \tag{15.18}$$

And after Taylor series development:

$$f_i\left[x_0^0, y_0^0, \ldots \delta^0\right] + \delta x_0 \left.\frac{\partial f_i}{\partial x_0}\right|^0 + \ldots + \delta\delta \left.\frac{\partial f_i}{\partial \delta}\right|^0 = z_i;\ \mathrm{i} = 1 \ldots 9 \tag{15.19}$$

The system can be written as a matrix:

$$\underbrace{\begin{vmatrix} \frac{\partial f_1}{\partial x_0} & \frac{\partial f_1}{\partial y_0} & \cdots & \frac{\partial f_1}{\partial \delta} \\ \frac{\partial f_2}{\partial x_0} & \frac{\partial f_2}{\partial y_0} & \cdots & \frac{\partial f_2}{\partial \delta} \\ \vdots & \vdots & & \vdots \\ \frac{\partial f_9}{\partial x_0} & \frac{\partial f_9}{\partial y_0} & \cdots & \frac{\partial f_9}{\partial \delta} \end{vmatrix}}_{(Z)} \underbrace{\begin{vmatrix} \partial x_0 \\ \partial y_0 \\ \vdots \\ \partial \delta \end{vmatrix}}_{(V)} = \underbrace{\begin{vmatrix} Z_1 - f_1(x_0^0, y_0^0, \ldots \delta^0) \\ Z_2 - f_2(x_0^0, y_0^0, \ldots \delta^0) \\ \vdots \\ Z_9 - f_9(x_0^0, y_0^0, \ldots \delta^0) \end{vmatrix}}_{(W)} \tag{15.20}$$

With the notations above: $(Z)(V) = (W)$.
Where from

$$(V) = (Z)^{-1}(W) \tag{15.21}$$

The free terms denoted by x (i) and y (i) were noted by Zi. After calculating the vector (V), the values of the corrections are known: $\delta x_0, \delta y_0 \ldots \delta\delta$.

So, replacements are made:

$$\begin{aligned} x_0^1 &= x_0^0 + \delta x_0 \\ y_0^1 &= y_0^0 + \delta y_0 \\ &\vdots \\ \delta^1 &= \delta^0 + \delta\delta \end{aligned} \tag{15.22}$$

after which the calculations are resumed with the new values for unknowns.

It is specified that in the Taylor series development the terms of the higher order derivatives were neglected, being considered as having small values, provided that the initial values are close to the system solutions.

In the case of the quadrilateral, the elements of the matrix (Z) are calculated from the Eq. 15.19, obtaining:

$$\frac{\partial f_i}{\partial x_0} = -2RG(x - x_0) + S[-2bL + 2P(x - x_0)] + 2bGU \tag{15.23}$$

$$\frac{\partial f_i}{\partial y_0} = R[-2G(y - y_0) + 2bH] + 2PS(y - y_0) - 2bPU \quad (15.24)$$

$$\frac{\partial f_i}{\partial x_2} = R[2eFsin\delta + 2K(x - x_2)] + S[-2Q(x - x_2) + 2eFcos\delta] - U[2eQsin\delta + 2eKcos\delta] \quad (15.25)$$

$$\frac{\partial f_i}{\partial y_2} = R[-2eFcos\delta + 2K(y - y_2)] + S[-2Q(y - y_2) + 2eFsin\delta] - U[-2eQcos\delta + 2eKsin\delta] \quad (15.26)$$

$$\frac{\partial f_i}{\partial a} = -2aGR + 2aPS \quad (15.27)$$

$$\frac{\partial f_i}{\partial b} = R[2bG - 2H(y - y_0)] + S[2L(x - x_0) - 2bP] - U[2G(x - x_0) - 2P(y - y_0)] \quad (15.28)$$

$$\frac{\partial f_i}{\partial e} = R\left(\frac{FG}{e} - 2eK\right) + S\left(2eQ - \frac{FP}{e}\right) - U\left(\frac{QG}{e} - \frac{KP}{e}\right) \quad (15.29)$$

$$\frac{\partial f_i}{\partial d} = 2dKR + 2dSQ \quad (15.30)$$

$$\frac{\partial f_i}{\partial \delta} = -RFP - SFG - U(-PQ - KG) \quad (15.31)$$

where the following notations were made:

$$R = 2(FG - HK)$$
$$S = 2(LQ - PF)$$
$$U = 2(QG - KP)$$
$$L = (x - x_2)^2 + (y - y_2)^2 - d^2 + e^2 \quad (15.31)$$

The 9 lines of the matrix (Z) are calculated with these relations, changing only the values of x (i) and y (i), i.e. the coordinates of the imposed points.

The elements of the vector (W) are calculated with the expression:

$$W(i) = -\left[(FG - HK)^2 + (QL - PF)^2 - (QG - KP)^2\right] \quad (15.32)$$

also differing in the values of x (i) and y (i).

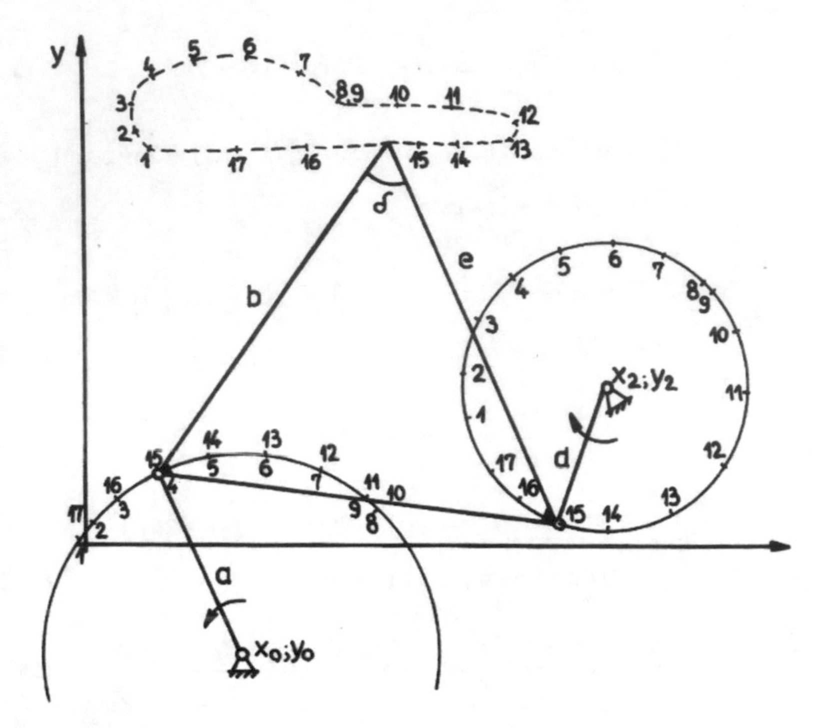

Fig. 15.2 Drawing a straight line segment [1]

If this curve is required to partially overlap with a line, the result is that a maximum of 6 points can be imposed on the connecting rod curve common to that of this line, because the line can intersect the connecting rod curve in a maximum number of points equal to the degree of the connecting rod curve.

Imposing 6 points on the connecting rod curve to draw a straight line, it results that the system (15.16) will have 6 equations with 6 unknowns. This leads to the convenient choice of three of the 9 unknowns and the determination by calculation of the other 6 unknowns, in this case the matrix equation becomes simpler, the vectors (V) and (W) having 6 lines each, and the matrix (Z) having 6 rows and 6 columns.

The calculation program can be easily adapted for different calculation variants, in the sense that any 6 of the 9 unknowns of the vector (V) in Eq. 15.20 can be chosen successively as unknowns.

To find mechanisms that draw the line parallel to the abscissa in Fig. 15.2, the following convenient values were adopted for three of the 9 unknowns: $x_2 = 100$; $y_2 = 30$; $\delta = 60$. For the unknowns to be determined, the following were adopted as initial values: $A = 40$; $B = 80$; $D = 50$; $E = 40$, $x_0 = 0$; $y_0 = 0$.

Table 15.1 indicates the results of the last iterations, and Fig. 15.2 represents the mechanism with the dimensions determined at iteration 22. It is found that indeed, a point of the connecting rod of this mechanism traces the imposed trajectory.

Table 15.1 Results of the last 12 iterations

Iteration	A	B	D	E	X_0	Y_0	Error
11	75.2	67.1	32.0	81.0	44.2	−41.5	0.5833419
12	62.9	69.1	31.8	80.3	39.6	−34.0	0.1731836
13	54.6	71.7	31.2	79.7	36.1	−29.7	0.0489891
14	48.7	73.7	30.4	79.2	33.4	−27.1	0.0133398
15	44.6	75.0	29.5	78.8	31.4	−25.3	0.0036058
16	41.9	75.6	28.8	78.5	30.2	−24.0	0.0009711
17	40.1	75.6	28.2	78.2	29.5	−23.0	0.0002520
18	39.0	75.5	27.8	78.1	29.1	22.4	0.0000581
19	38.6	75.3	27.6	78.0	29.0	−22.1	0.0000109
20	38.6	75.3	27.6	78.0	29.0	−22.0	0.0000012
21	38.6	75.3	27.6	78.0	29.0	−22.0	0.0000000
22	38.6	75.3	27.6	78.0	29.0	−22.0	0.0000000

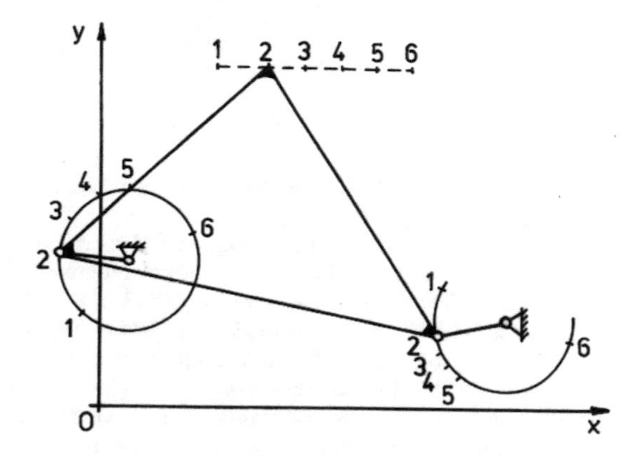

Fig. 15.3 Another mechanism that draws the same line [1]

The same trajectory is described by the mechanism of Fig. 15.3, result based on the calculations in Table 15.2, where the following initial data were adopted:

$$A = 70;\ B = 60;\ D = 30;\ E = 80;\ x_0 = 30;\ y_0 = -20;$$
$$x_2 = 90;\ y_2 = 20;\ \delta = 80.$$

Other mechanisms were obtained for tracing that straight line segment, with different accuracies. It is specified that, by using the Roberts-Chebyshev method, two more mechanisms can be constructed for each calculated mechanism, tracing the same trajectorys. The choice of initial values for unknowns is of great importance in terms of computing volume. It is recommended to choose dimensions of the order of magnitude of the coordinates of the points on the respective curve in order to improve the convergence of the system.

Table 15.2 Results of other 10 iterations

Iteration	A	B	D	E	X_0	Y_0	Error
2	29.7	36.2	32.0	71.8	21.0	18.0	0.3043016
3	9.0	54.2	26.9	55.8	10.2	32.3	0.0256598
4	16.4	75.6	13.6	68.8	−2.1	22.6	0.0103035
5	14.1	68.2	13.0	68.0	3.3	28.3	0.0002768
6	16.0	61.1	16.6	70.4	7.0	32.4	0.0001397
7	15.3	62.0	15.6	70.2	6.5	32.4	0.0001043
8	15.3	62.1	15.6	70.2	6.5	32.3	0.0000046
9	15.3	62.1	15.6	70.2	6.5	32.3	0.0000001
10	15.3	62.1	15.6	70.2	6.5	32.3	0.0000000

From Fig. 15.2 it is noted that the curve passing through the six imposed points is drawn, otherwise the curve having the shape imposed by the dimensions of that mechanism.

References

1. Popescu I, Mîlcomete DC (2006) Cercetări privind sinteza şi optimizarea mecanismelor. Editura Sitech
2. Popescu I (1977) Proiectarea mecanismelor plane. Ed. Scrisul Romanesc, Craiova
3. Popescu I Analytic methods for synthesis of fourbar linkage for realisation of some imposed curves. In: SYROM'77, vol 1, part 3, pp 543–552
4. Popescu I (1979) Analytic methods of synthesis of the plane mechanisms which outline imposed trajectories. In: "Proc. 5th world congress theoretical machinery and mechanics, Montreal, Canada, 1979, vol 2, The American Society of Mechanical Engineers, New York, pp 1062–1064
5. Popescu I, Predoi M (1979) Metoda de analiza cinematica a mecanismelor, pe baza diagramelor cinematice si a functiilor spline cubice. In: Buletinul stiintific, seria tehnica-matematica, institutul de invatamint superior Sibiu, vol. II, pp 197–205
6. Popescu I Metode iterative folosite la sinteza mecanismelor pentru traiectorii impuse. In: SYROM'81, vol I-1, pp 215–224
7. Popescu I, Predoi M Sinteza mecanismelor prin folosirea metodei Broyden. In: "Lucrarile simpozionului proiectarea asistata de calculator PRASIC'82, Mecanisme si organe de masini", vol I, pp 85–90, Brasov
8. Predoi M, Popescu I Folosirea functiilor spline cubice de doua variabile la analiza si sinteza mecanismelor In: "Lucrarile simpozionului Proiectarea asistata de calculator PRASIC'82, 12–14 noiembrie—Mecanisme si organe de masini", vol I, pp 109–114
9. Artobolevskii II, Levitskii NI, Cerkudinov SA (1959) Sintez ploskih mehanizmov. Gosu. Izd. Fiz. Mat. Lit, Moskva
10. Andreicenko GP (1983) Metod statisticeskih ispîtanii pri sinteze krivosipno—coromîslovogo mehanizma. Izvestia V.U.Z. Masinostroenie, nr. 12, Moskva, p 50
11. Bagchi TP (1993) Taguchi methods explained—practical steps to robust design. New Delhi
12. Gabriele GA, Ragsdell KM (1977) The generalized reduced gradient method. A reliable tool for optimal design. ASMEJ Engn. Indus. 99:394–400
13. Watanabe K (1992) Application of natural equations to the synthesis of curve genetating mechanisms. Mech. Mach. Theory 27:261–273

14. Cossalter V, Doria A, Pasini M, Scattolo C (1992) A simple numerical approach for optimum synthesis of a class of planar mechanisms. Mech. Mach. Theor. 27(3):357–366
15. Krishnamurty S, Turcic DA (1992) Optimal synthesis of mechanisms using nonlinear goal programing techniques. Mech. Mach. Theor. 27(5):599–612
16. Yu-Kuang L, Ji-Chang Z A new optimum synthesis method of planar mechanisms for the generation of paths and rigid—body position. SYROM'85, vol I-1, pp 215–222
17. Popescu I. Optimizarea mecanismelor pentru traiectorii impuse. In: "Lucrarile simpozionului Proiectarea asistata de calculator PRASIC'82, 12–14 noiembrie—Mecanisme si Organe de masini", vol I, pp 97–102
18. Popescu I (Aug 1983) The mechanisms optimization for trajectories and imposed associated positions. In: "Computer aided analysis and optimization of mechanical system dynamics". University Of Iowa City, S.U.A
19. Popescu I Probleme privind sinteza optimala a mecanismelor. In: The Fifth IFTOMM international symposium on linkages and computer aided design methods (Theory and Practice of Mechanisms), SYROM'89, Bucarest, Romania, July, 6–11, 1989, vol I 3, pp 721–728
20. Popescu I Rezultate obtinute la sinteza optimala a mecanismelor. In: SYROM'89, vol I 3, pp 713–720
21. Popescu Iulian. Accelerarea convergentei procesului de calcul la optimizarea mecanismelor . In MERO'91. Al X-lea simpozion national de roboti industriali, cu participare internationala, Bucuresti, 18–20 aprilie 1991, vol III, pp 793–802
22. Popescu I. Influenta aproximatiei initiale asupra convergentei, la optimizarea mecanismelor. In: MERO'91. Al X-lea simpozion national de roboti industriali, cu participare internationala, Bucuresti, 18–20 aprilie 1991, vol III, pp 813–822
23. Ermakov SM (1976) Metoda Monte Carlo şi probleme înrudite. Editura Tehnică, Bucureşti
24. Doronin VI, Konnova GV (1978) Primenenie lineinovo metoda cvadraticeskovo priblijeniia dlia sinteza mehanizma. In Izvestia VUZ Maşinostroeniia. 10
25. Jiag H, Jiang GK A new method of four—bar linkage synthesis with ten points using algebric curves. SYROM '89, Bucureşti, pp 433–438
26. Mohsen MO Sznthesis of cupler curve generating mechanism. SYROM'81, Bucureşti, vol I-1, pp 159–170
27. Hernandez A, Amezua E, Ajuria MB, Llorente JL (1994) Multiple points on the coupler curve of transitional four—hinge planar linkages. Mech. Mach. Theor. 29(7):1015–1032

Printed in the United States
by Baker & Taylor Publisher Services